Alkaloids: Chemical and Biological Perspectives

ALKALOIDS: CHEMICAL AND BIOLOGICAL PERSPECTIVES

Volume Three

Edited by

S. WILLIAM PELLETIER
Institute for Natural Products Research

and

The Department of Chemistry
University of Georgia, Athens

A Wiley-Interscience Publication ˙

JOHN WILEY & SONS

New York   Chichester   Brisbane   Toronto   Singapore

CHEMISTRY

7205-850X

Library of Congress Cataloging in Publication Data:
  (Revised for volume 3)
Main entry under title:

Alkaloids: Chemical and Biological Perspectives.

  "A Wiley-Interscience publication."
  Includes bibliographies and indexes.
  1. Alkaloids. I. Pelletier, S. W., 1924-
QD421.A56 1983   574.19′242   82-11071
ISBN 0-471 89302-1 (v.3)

Printed in the United States of America

10 9 8 7 6 5 4 3 2 1

*Dedicated to
my colleagues, friends, and students
throughout the world
who work with alkaloids*

Contributors

T. A. Blumenkopf, Department of Chemistry, University of California, Berkeley, California

B. Colasanti, West Virginia University Medical Center, Morgantown, West Virginia

M. A. El Sohly, Research Institute of Pharmaceutical Sciences, School of Pharmacy, University of Mississippi, University, Mississippi

G. B. Fodor, Department of Chemistry, West Virginia University, Morgantown, West Virginia

C. H. Heathcock, Department of Chemistry, University of California, Berkeley, California

R. B. Herbert, Department of Organic Chemistry, The University, Leeds, England

M. M. Joullie, Department of Chemistry, University of Pennsylvania, Philadelphia, Pennsylvania

R. F. Nutt, Medicinal Chemistry, Merck Sharp & Dohme Research Laboratories, West Point, Pennsylvania

L. E. Overman, Department of Chemistry, University of California, Irvine, California

M. Sworin, Department of Chemistry, University of Missouri, St. Louis, Missouri

P. G. Waterman, Phytochemistry Research Laboratory, University of Strathclyde, Glasgow, Scotland

Preface

Volume 3 of *Alkaloids: Chemical and Biological Perspectives* presents timely reviews of seven important alkaloid topics.

Chapter 1 reviews both the chemistry and pharmacology of the pyridine and piperidine alkaloids, compounds that are widespread in nature. To our knowledge this chapter is the first that treats both the chemistry and the pharmacology of major representatives of these two classes of alkaloids.

Chapter 2 treats the indolosesquiterpene alkaloids, a class of compounds that may formally be divided into indole and sesquiterpene subunits.

The cyclopeptide alkaloids, reviewed in Chapter 3, are a rapidly growing group of closely related polyamide bases of plant origin. Alkaloids of the class are sometimes called "ansapeptides" and "phencyclopeptines," the latter term representing the fundamental para-bridged 14-membered ring nucleus common in most of these macrocyclic alkaloids.

Cannabis sativa L., the plant from which marijuana, hashish, and hash oil are obtained, has been known in medicine for thousands of years. It has been recommended for treatment of rheumatism, beriberi, neuralgia, asthma, malaria, hysteria, insomnia, tetanus, epilepsy, and constipation, though in 1942 *Cannabis* was denied admission into the U.S. Pharmacopeia because of lack of acceptable medical use in the United States. Though over 400 chemical compounds are known to exist in *Cannabis,* only a few alkaloids have been isolated and characterized. Chapter 4 reviews the alkaloids, amino acids, amino sugars, proteins, and enzymes that occur in *Cannabis,* and summarizes the biological properties of the alkaloids.

Since the synthesis of alkaloids has become increasingly important during the past 20 years, we have included three chapters on alkaloid synthesis in this volume. Chapter 5 presents an excellent review of the biosynthesis and total synthesis of the *Lycopodium* (club moss) alkaloids. Chapter 6 reviews syntheses directed toward indolizidine and quinolizidine alkaloids found in *Tylophora, Cryptocarya, Ipomoea, Elaeocarpus,* and related species. A final chapter surveys recent work on the total synthesis of the important pentacyclic *Aspidosperma* alkaloids.

Each chapter in this volume has been reviewed by an authority in the field. Indexes for both subjects and organisms are provided.

The editor invites prospective contributors to write him about topics suitable for review in future volumes of this series.

S. WILLIAM PELLETIER

Athens, Georgia
October 1984

Contents

Chapter One

The Pyridine and Piperidine Alkaloids: Chemistry and Pharmacology

Gabor B. Fodor
Department of Chemistry
West Virginia University
Morgantown, West Virginia 26505

Brenda Colasanti
West Virginia University Medical Center
Basic Sciences
Morgantown, West Virginia 26506

CONTENTS

1. INTRODUCTION

The pyridine alkaloids and the hydrogenated species are widespread in nature. Among others, the first alkaloid to be synthesized was a piperidine, coniine. The chemistry of the unsaturated aromatic and the saturated group has usually been reviewed together. Starting in 1950 the Manske Alkaloids series gave three reviews, two by Leo Marion in 1950 and 1960 [1, 2], and one in 1969 by W. Ayer [3]. The latest review was written by D. Gross in 1970 in Fortschritte [4]. Since then there has been no extensive treatise of the chemistry of the piperidine alkaloids, except for some subgroups; several representatives of the class have been recognized and the biosynthesis of many compounds has been established. The stereochemistry including absolute configuration of many alkaloids is known. Nevertheless there is no review that treats both the chemistry and pharmacology of the major representatives of these two classes. Therefore it is of interest to update our picture of these important areas.

There are a few limitations. Because an excellent review has appeared recently in the present series on the biosynthesis of tobacco alkaloids [5] that treats a substantial number of the pyridines, like nicotine itself, and the piperidines such as anabasine, we have a priori excluded details of the biosynthesis of the tobacco alkaloids from this review. We have not followed the biogenetic pattern as it has been presented on other alkaloids in two recent monographs [6], [7], nor have we followed the plant source as a chief indicator for the sequence in which the individual alkaloids were treated. We feel that the pyridine alkaloids should be treated separately from the piperidine alkaloids. There is no close biochemical connection between the two groups except for some tetrahydropyridines that are derived in nature by the reduction of a pyridine ring, for example, anatabine. Therefore, we wish to present, in Table 1, the pyridine alkaloids and, in Table 2, the piperidines that we shall discuss. We are aware of the fact that every system is arbitrary. Starting with N-methylnicotinic acid betaine, it is easy to recognize that most of the pyridine alkaloids that we are dealing with are derived from nicotinic acid in one way or another. The family of nicotine alkaloids shall be presented in chemical sequence. Our major point of view is chemistry which is immediately followed, at the end of the description of the individual representative, by pharmacology. Concerning structure determination, we shall put emphasis on recently discovered alkaloids and on modern chemical and spectral methods. Syntheses will be reported, but only one in most cases. Preference will be given to the most recent method that, in addition to supporting a structure, may eventually lead to practical application.

2. PYRIDINE ALKALOIDS (TABLE 1)

We are proceeding from simple derivatives like the betaine of N-methylnicotinic acid to an ester with glucose, to arecoline, then to a nicotinonitrile that has been oxygenated to ricinine, to gentianine that is reminescent of the nicotinic acid

Table 1. Pyridine Alkaloids

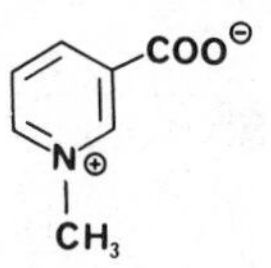

1 TRIGONELLINE

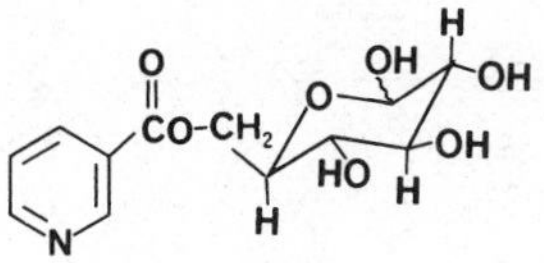

2a BUCHANANINE

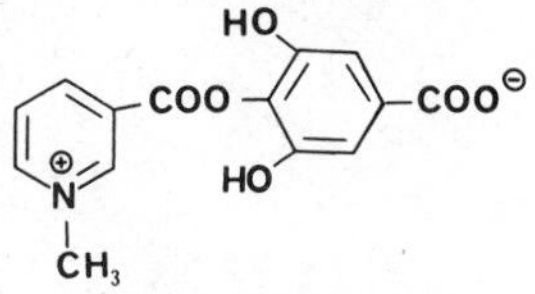

2b PRECATORINE

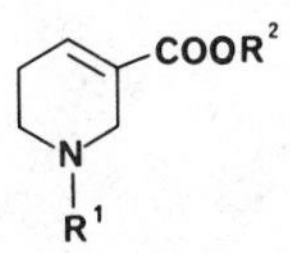

	R¹	R²	
3 a	H	H	GUVACINE
b	H	CH₃	GUVACOLINE
c	CH₃	H	ARECAIDINE
d	CH₃	CH₃	ARECOLINE

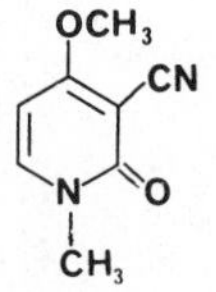

4 RICININE

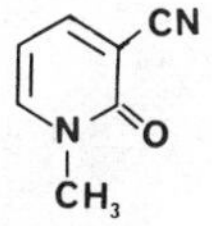

5 RICINIDINE

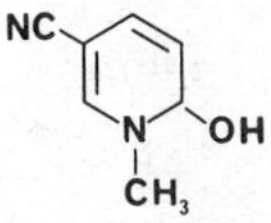

6 NUDIFLORINE

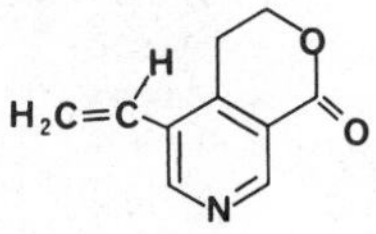

7 GENTIANINE*

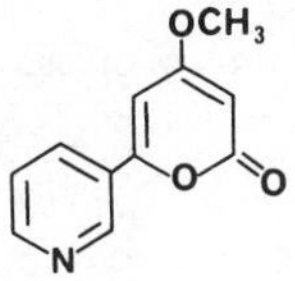

8 ANIBINE

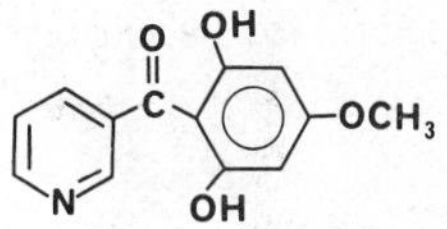

9 DUCKEIN

* = non-natural product

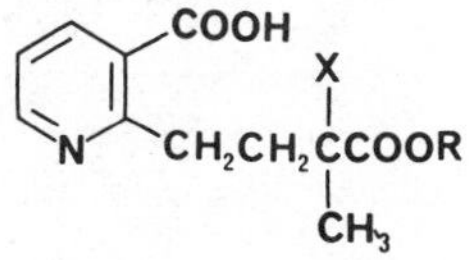

	X	
10 a	H	WILFORIC ACID
b	OH	WILFORDIC ACID

5

Table 1. (*Continued*)

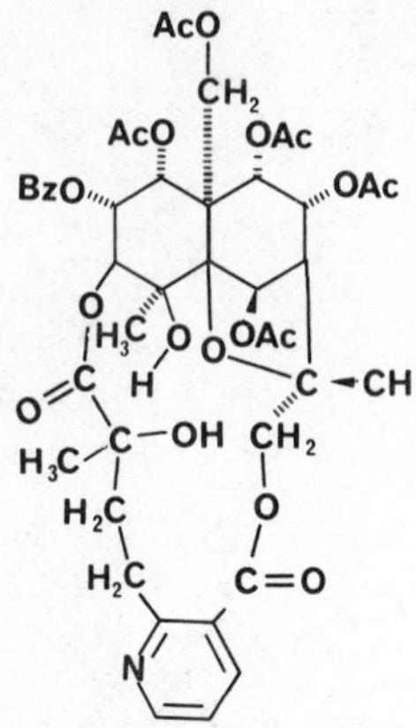

11 WILFORDINE

EVONINIC ACID

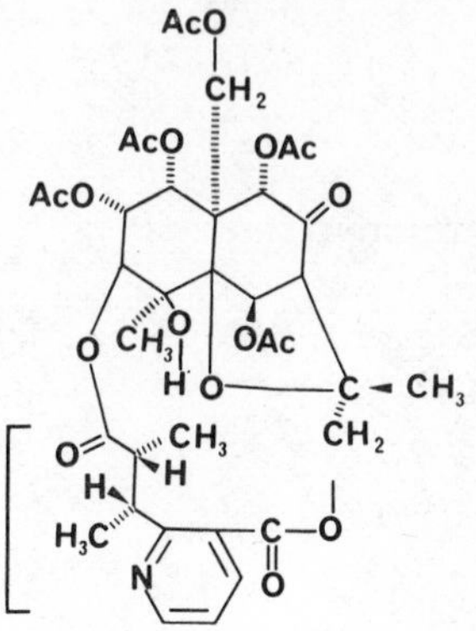

12 EVONINE

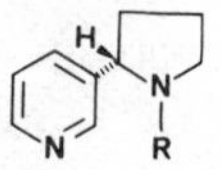

R
13 CH₃ S(-) NICOTINE
16 H S(-) NORNICOTINE

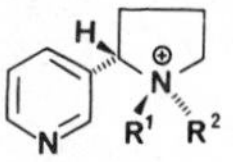

NICOTINE N-OXIDES
	R¹	R²
14	CH₃	O
15	O	CH₃

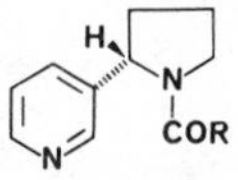

	R	NORNICOTINES
17	H	N-FORMYL
18	CH₃	N-ACETYL
19	n-C₅H₁₁	N-HEXANOYL
20	n-C₇H₁₅	N-OCTANOYL

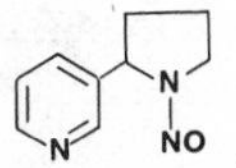

21 N'-NITROSO NORNICOTINE

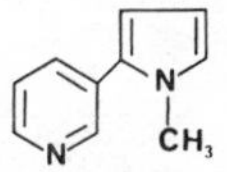

22 NICOTYRINE

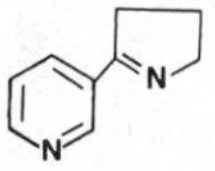

23 MYOSMINE

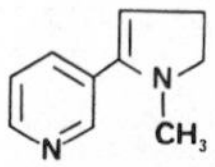

24 N'-METHYL-MYOSMINE

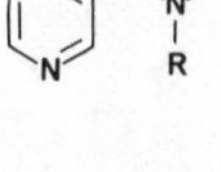

R
25 H S(-)ANABASINE
26 CH₃ N'-METHYL ANABASINE

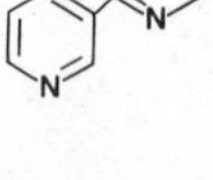

27 ANABASEINE

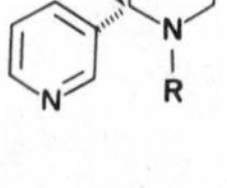

R
28 H ANATABINE
29 CH₃ N'-METHYL ANATABINE

6

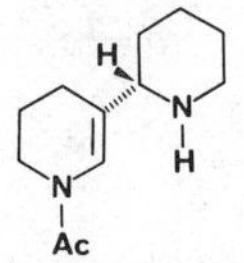

Table 1. (*Continued*)

30   CH₃CO   AMMODENDRINE
31   trans-cinnamyl   ORENSINE ≡ (+) ADENOCARPINE
32   cis-cinnamyl   ISOORENSINE

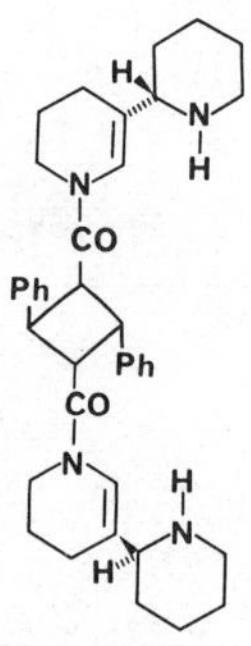

33 SANTIAGUINE

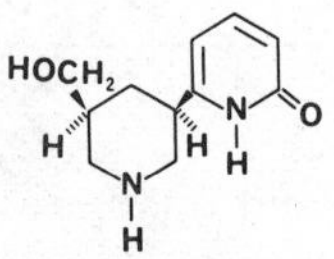

34 KURARAMINE

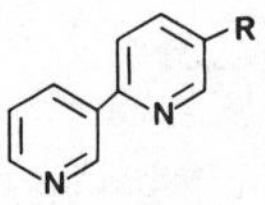

R
35   H   2,3'-DIPYRIDYL
36   CH₃   5-METHYL-2,3'-DIPYRIDYL

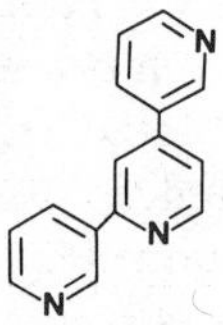

37 NICOTELLINE

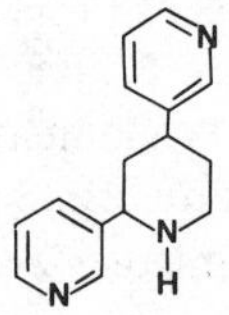

38 ANATALLINE

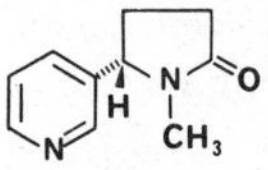

39 COTININE

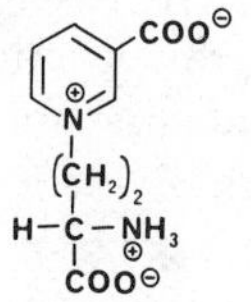

40 NICOTIANINE

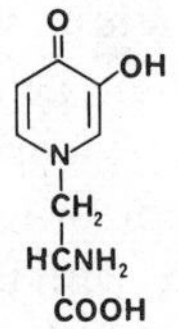

41 MIMOSINE and
LEUCAENINE

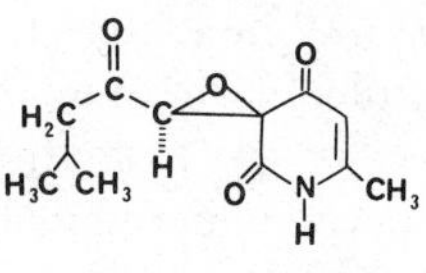

42 FLAVIPUCINE

7

Table 1. (*Continued*)

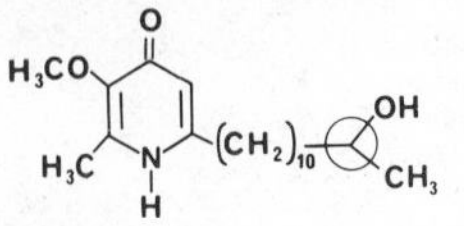

43 S (-) MELOCHININE

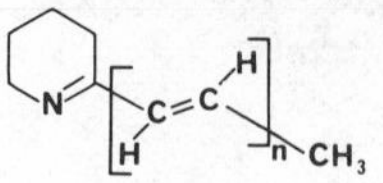

44 n=3 NIGRIFACTIN
45 n=2 ALKALOID of
STREPTOMYCES NA-337

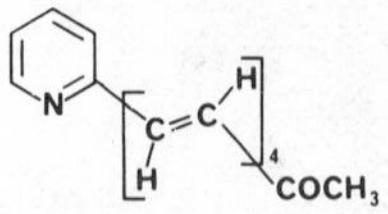

46 NAVENONE A

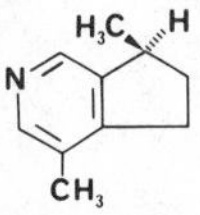

47 S(-) ACTINIDINE 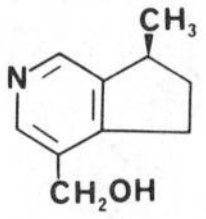   **48** S(-)TECOSTIDINE 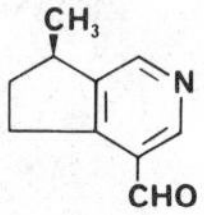   **49** R(-)BOSCHNIAKINE 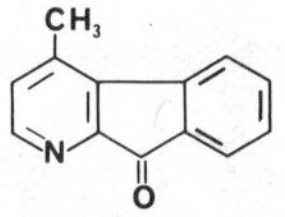   **50** ONYCHINE

skeleton, to the tobacco alkaloids, and finally to pyridines, dihydropyridines, and pyridines, some of which are structurally unrelated to nicotinic acid.

2.1. Trigonelline

2.1.1. Chemistry. Trigonelline (1) was discovered in the seeds of *Trigonella foenum,* but it is also present in the seeds of *Pisum sativum* and *Cannabis sativa* and in coffee, soybeans, and potatoes. It gives methylamine upon heating with barium hydroxide, and based on its elemental analysis, ($C_7H_7NO_2$), it was assumed to be the methyl betaine of nicotinic acid. This was indeed proven by comparison with authentic *N*-methylnicotinic acid that was obtained by quaternization of nicotinic acid with methyl iodide followed by neutralization [8].

2.1.2. Pharmacology. Although trigonelline is formed in the liver in both dogs and man after oral administration of nicotinic acid (niacin), trigonelline is not the major metabolite of this essential vitamin. The principal metabolic pathway of moderate doses of nicotinic acid as well as of nicotinamide, the chemical form of nicotinic acid after its incorporation into adenine nucleotides, involves the formation of *N*-methyl nicotinamide. This major metabolite, in turn, can undergo further degradation, with the formation of several pyridones. After administration of extremely high doses of nicotinic acid, little metabolism occurs, and the majority of the administered vitamin is excreted in the urine unchanged [9].

Nicotinic acid is widely recognized mainly as an essential dietary constituent.

A lack of nicotinic acid in the diet leads to the disease entity known as pellagra in man and to an equivalent condition termed black tongue in dogs. To perform its function as a vitamin, nicotinic acid is first converted in the body to both nicotinamide adenine dinucleotide (NAD) and nicotinamide adenine dinucleotide phosphate (NADP). These two compounds then serve as coenzymes for a number of enzymes that catalyze oxidation–reduction reactions essential for tissue respiration.

Nicotinic acid itself produces several important attendant pharmacological effects not seen after administration of nicotinamide. This vitamin rapidly reduces plasma levels of cholesterol and triglycerides by several diverse mechanisms [10]. Nicotinic acid also produces vasodilation. The blush areas are much more affected than the extremities, however, and neither skin nor muscle blood flow is consistently increased [11]. Nicotinic acid is used therapeutically for both hypolipidemic effect and, without basis, as a vasodilator. Its therapeutic use as prophylaxis and treatment of pellagra is, fortunately, rarely necessary in the United States today. It has been claimed that high doses of nicotinic acid possess therapeutic value in the treatment of schizophrenia. The assumption underlying the presumed effectiveness of this vitamin is that, by providing excess methyl acceptors, a postulated abnormal transmethylation of catecholamines is inhibited [12]. The actual usefulness of nicotinic acid after a host of clinical trials in schizophrenic patients, however, remains controversial.

Unlike nicotinic acid or nicotinamide, neither N-methyl nicotinamide nor trigonelline reverses the condition of black tongue in dogs. Because *Trigonella* seeds ("chilbe") have long been included in the diet of diabetic Yemenites in Israel in accordance with regional medical folklore, trigonelline was examined for, and reported to possess, hypoglycemic activity [13]. In subsequent studies trigonelline was found to exert some degree of an effect in both laboratory animals and diabetic patients [14]. In rabbits this alkaloid counteracted the hyperglycemic effect of cortisone when administered concomitantly with the latter steroid or 2 h earlier, but not when given 2 h after cortisone administration. After administration of trigonelline to diabetic patients, however, a transient hypoglycemic effect occurred in only 50% of the subjects. In a more comprehensive study several alkaloids of *Trigonella* seeds, as well as nicotinamide, were examined for hypoglycemic activity in both alloxan-diabetic and nondiabetic rats [15]. In this species trigonelline produced only a mild and transient hypoglycemia in the diabetic animals, while both nicotinic acid and nicotinamide exerted a profound effect of much longer duration in the same experimental group. Of these three compounds, only nicotinic acid exerted an effect in normal rats. On the basis of these results, the mild hypoglycemic activity of trigonelline was ascribed to its slowing down of the metabolism of nicotinic acid.

It is noteworthy that the acute toxicity of both trigonelline and nicotinic acid in the rat is quite low. The total minimal daily dietary requirement of nicotinic acid in man is in the range of 10–20 mg. In contrast, the lethal dose of either trigonelline or nicotinic acid in 50% of the rodents tested amounted to quantities ranging from 5 to 9 g/kg of body weight [15].

2.2. Buchananine

Buchananine (2) from *Cryptolepis buchanani* gave upon hydrolysis glucose and nicotinic acid. Since it consumed two molecules of periodic acid, the primary hydroxyl was involved in the esterification with nicotinic acid [16]. It has been synthesized [17] from nicotinoyl chloride and 1,2-isopropylidene glucofuranose, followed by acetal cleavage. There is nothing known about its biological or pharmacological activity. Precatorine (2b) from *Abrus precatorius* is 4-*N*-methyl nicotinoyl gallic acid betaine [18].

2.3. Chemistry of the Areca nut alkaloids

Guvacine (3a), guvacoline (3b), arecaidine (3c), and arecoline (3d) are the major alkaloids of *Areca catechu* nut. Arecaidine is the free carboxylic acid *N*-methyl-1,2,5,6-tetrahydronicotinic acid (3c) and arecoline is its methyl ester (3d) while guvacine and guvacoline are their *N*-nor derivatives. The structures of 3c and 3d are based on an unambiguous total synthesis [19] from 3-methyliminodi-propionaldehyde-tetraethylacetal that upon reaction with hydrochloric acid gave 1-methyl-Δ^3-tetrahydropyridine-3-aldehyde. The latter by oxidation was converted into the free carboxylic acid that proved identical with arecaidine. Action of methyl iodide on guvacine gave arecaidine methyl betaine, which proved the constitution of guvacine as 3a [20].

2.4. Pharmacology of the Areca Nut Alkaloids:
Arecoline, Arecaidine, Guvacine

Of the alkaloids occurring within the *Areca* nut, arecoline has by far received the most attention with regard to pharmacological properties. Arecoline is one of three major natural alkaloids known to possess cholinomimetic activity. While the other two alkaloids, that is, muscarine and pilocarpine, act predominantly at muscarinic receptor sites, arecoline acts at nicotinic receptor sites as well.

The muscarinic actions of arecoline encompass a variety of organ systems. Prominent effects are exerted on the cardiovascular system. Low doses produce vasodilation within the major vascular beds and thus a fall in both systolic and diastolic blood pressure. Pressor–receptor mechanisms cause reflex activation of sympathetic activity, with resultant tachycardia. At higher doses arecoline also exerts a direct depressant effect on the heart, and both blood pressure and heart rate fall.

Arecoline exerts stimulatory effects on the gastrointestinal tract, with an increase in both resting tone and force of contractions of the smooth muscle lining the tract. Bronchial smooth muscle is also stimulated; the resultant bronchoconstriction may precipitate asthmatic attacks. Arecoline similarly stimulates the salivary, lacrimal, and sweat glands. Diaphoresis is marked because the direct effect on the sweat gland is enhanced by cutaneous vasodilation.

Other prominent peripheral muscarinic actions of arecoline include pupillary constriction, urinary bladder contraction, and decreased bladder capacity. Centrally, arecoline induces a characteristic cortical EEG arousal response that is also muscarinic in nature, as the response is blocked by atropine.

The nicotinic effects of arecoline are seen after the administration of large doses. Nicotinic receptor stimulation results in the release of catecholamines from both postganglionic sympathetic nerve fibers and the adrenal medulla; the net effect is an increase in sympathetic activity. An additional nicotinic action consists of skeletal muscle stimulation.

Structure-activity relationships for the production of muscarinic effects by both naturally occurring and synthetic arecaidine derivatives have been examined [21, 22]. Arecoline (i.e., arecaidine methyl ester) is a tertiary amine that is active in the protonated form. The activity of the carboxylic acid, arecaidine, is markedly reduced. Tertiary esters with a side chain longer than an ethyl group likewise show considerably less activity. Quaternization considerably decreases the intrinsic activity of arecaidine methyl ester and confers antagonistic activity to the remaining esters. Hydrogenation of the double bond in the ring, as in the case of arecolidine, markedly lowers both affinity and intrinsic activity.

Although both arecaidine and guvacine possess only weak cholinomimetic properties, these two alkaloids are potent inhibitors of the uptake of gamma amino butyric acid (GABA) into brain slices *in vitro* [23]. After electrophoretic administration both amino acids also enhanced the inhibitory actions of GABA on the firing of spinal neurons [24]. In the latter study the uptake of GABA by cerebellar slices was likewise inhibited by arecaidine, as expected, and the effect of GABA on the firing of cerebellar Purkinje cells was correspondingly enhanced by electrophoretically administered arecaidine. After intravenous administration, however, arecaidine had no effect on synaptic inhibitions presumably mediated by GABA. On the basis of these results, it was concluded that the reduction of GABA inactivation by arecaidine and guvacine resulting from GABA uptake inhibition does not likely play a major role in the mediation of the behavioral effects of the *Areca* nut.

In spite of the large differences in the pharmacological profiles of arecaidine and arecoline, some of the behavioral effects exerted by these two alkaloids are quite similar. Both compounds decrease spontaneous motor activity and food intake in rodents [25, 26], and both are effective in reversing chlorpromazine sedation [25, 27]. The effects of these alkaloids on barbiturate sleeping time, however, were quite divergent, with arecoline shortening the duration [27] and arecaidine prolonging it [25].

Focus on the behavioral effects of the *Areca* nut alkaloids arose from the time-honored consumption of a masticatory mixture containing *Areca* nut by Oriental natives for its euphoric effects. The masticatory mixture, dubbed *betel quid,* consists of the leaves of *Piper betle* (a climbing species of pepper), *Areca* nut, catechu, and lime; some users also add tobacco. This mixture is commonly known as *paan* in India.

The physiological effects resulting from the chewing of betel quid are primarily referable to the prominent cholinomimetic activity of the major *Areca* nut alkaloid, arecoline. The neuropharmacological mechanisms involved in the production of the psychic changes, by contrast, remain obscure. Although arecaidine probably contributes to this effect because it is formed from arecoline after the chewing of betel, the inhibition of GABA uptake produced by arecaidine is thought not to be involved [24].

A relationship between betel quid chewing and oral cancer has been suggested since the early 1900s [28, 29]. Cellular changes such as leukoplakia have subsequently been produced by a number of investigators after application of betel quid to the buccal mucosa of a variety of species [30]. Tobacco did not have to be present in the betel mixture in order for carcinogenic properties to appear [31]. The *Areca* nut alkaloid arecoline was later shown to induce *in vitro* neoplastic transformation [32], and potentiation of its genotoxicity by other constituents of betel quid has recently been reported [33].

Although the major *Areca* nut alkaloids are quite valuable as pharmacological tools, current clinical use of any of these compounds is restricted to veterinary medicine, where arecoline is employed as an anthelmintic. The *Areca* nut itself, however, still retains an important position in the Indian way of life [30]. In addition to its use for chewing purposes, it still plays an important role in many of the religious ceremonies of India. The medicinal properties of the *Areca* nut, which have been extolled in older Indian literature for treatment of conditions ranging from bad breath to urinary tract disorders, are still utilized in several remote parts of the country.

2.5. Ricinine

2.5.1. Chemistry. Ricinine **(4)** has been known for over 100 years. It was extracted from the *Ricinus communis* plant, which contains 1.1%. The structure elucidation was done by conventional methods, such as distillation with zinc dust [1]. Coloration with ferric chloride was significant indicating an α-pyridone structure. Among several total syntheses of ricinine, the simplest is the condensation of two molecules of cyanoacetyl chloride to give 6-chloro-*O,N*-bisnorricinine. This upon methylation on *N*(1) and *O*(4) followed by reductive dechlorination with zinc and acid gave rise to ricinine [34, 35]. This synthesis proves its structure and it is certainly the simplest way to get to the molecule. There is no need for the preparation of ricinine by synthetic methods, however, because the *Ricinus* extract gives an abundant amount of the natural product.

The *biosynthesis* of ricinine is a fascinating subject. The methyl group originates from methionine [36], but the total pyridine skeleton arises from quinolinic acid, that is, pyridine-2,3-dicarboxylic acid, via the nicotinic acid mononucleotide (Scheme 1). The oxidation pathway is not perfectly elucidated;

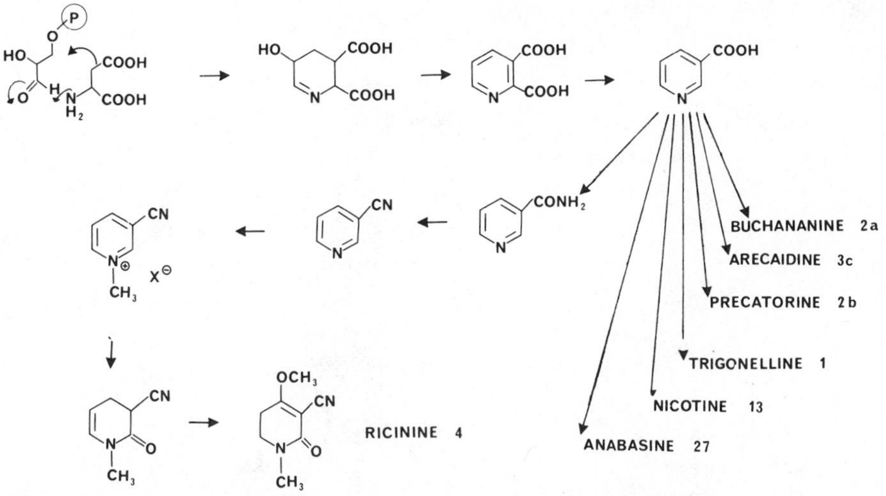

HO   COOH
N
H₂

HO   COOH
N   COOH

COOH
N   COOH

COOH
N

CONH₂
N

BUCHANANINE   2a

ARECAIDINE   3c

PRECATORINE   2b

TRIGONELLINE   1

NICOTINE   13

ANABASINE   27

CN
N

CN
N

CN
N⊕   X⊖
CH₃

CN
N   O
CH₃

OCH₃
CN
N   O
CH₃

RICININE   4

Scheme 1. Biosynthesis of nicotinic acid and of its derivatives. Adapted from figures 12 and 16, reference 429, with permission of Springer-Verlag, Heidelberg.

however, the origin of the carbon skeleton in nicotinic acid mononucleotide is certain [37–40].

2.5.2. Pharmacology. Ricinine has no pharmaceutical use. However, the triglyceride of ricinolic acid has been long applied as a cathartic [41]. Like the high molecular weight protein ricin that is extracted from the castor bean seed, the alkaloid ricinine, which is more abundant in the plant seedlings, is also highly toxic [42–44]. Ingestion of ricinine causes nausea, vomiting, hemorrhagic gastroenteritis, hepatic and renal damage, convulsions, coma, hypotension, respiratory depression, and ultimately death.

2.6. Ricinidine (5) and Nudiflorine (6)

Leaves of *Trevia nudiflora* Linn of the Euphorbiaceae family contain two 2-pyridone alkaloids of the molecular formula $C_7H_6N_2O(M^+134)$ that are closely related to ricinine. One is ricinidine [45] and the other was named [47, 48] nudiflorine. Ricinidine **(5)** was known since 1923 as a product of demethoxylation of ricinine and was obtained by synthesis [46]. Nudiforine was characterized by its IR spectrum, showing a conjugated amide carbonyl band at 1670 cm^{-1} and a band at 1610 cm^{-1} for C=CH bond. A sharp absorption band at 2200 cm^{-1} was indicative of a nitrile group. The UV maxima at 254 and 306 nm are typical of a 2-pyridone nucleus. The ^{1}H NMR spectrum shows a sharp singlet at δ 3.57 for the *N*-methyl group; the remaining three protons resonate in the aromatic region and give an eight-line spectrum. The coupling pattern of the latter is reminiscent of a 1,2,4-trisubstituted benzene system. Therefore, the

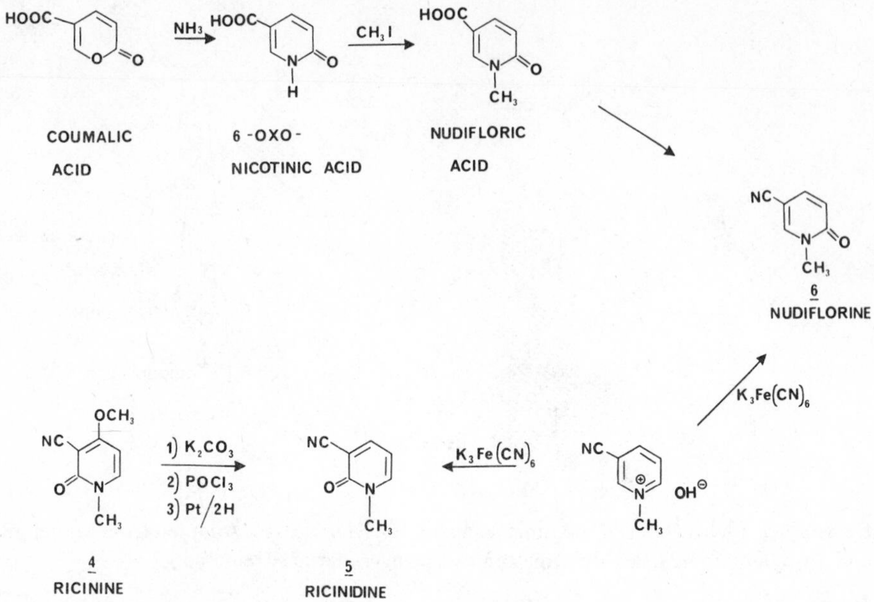

Scheme 2. Syntheses of ricinidine (**5**) and nudiflorine (**6**).

structure of 1-methyl-5-cyano-2-pyridone (**6**) was assigned to the new alkaloid [48]. This was confirmed by synthesis of nudifloric acid, the product of hydrolysis of nudiflorine, starting with malic acid. Heating gave coumalic acid and ammonolysis of the latter afforded 6-oxo-nicotinic acid. *N*-methylation led to *N*-methyl-2-pyridone-5-carboxylic acid, which proved to be identical with nudifloric acid. Furthermore, nudifloric acid was converted to the acid chloride and this, via the amide, into the nitrile **6,** identical with the natural alkaloid [48]. A second synthesis started with 1-methyl-3-cyano-pyridinium iodide, which was oxidized with potassium ferricyanide probably via the carbinolamine to a mixture of two regioisomeric 2-pyridones, 75% ricinidine and 25% nudiflorine (Scheme 2). This method of oxidation was previously applied to py-*N*-methylnicotinium iodide and to py-*N*-methyl anabasinium iodide (cf. Schemes 7 and 11). Since 5 years later ricinidine [47] proved to be a natural product [45], the ferricyanide oxidation also means that the total synthesis of *both* alkaloids has been achieved.

2.7. Gentianine

2.7.1. Chemistry. Extraction of different plants, for example, *Gentiana kirilowi* [49], *Dipsachus azureus* [50], *Anthocleista procera* [56], after addition of ammonia gave a base, gentiamine. Oxidation with permanganate led to pyridine-3,4,5-tricarboxylic acid. The constitution of a pyridine alkaloid was proposed by Konovalova [50]. The original structural concept proved erroneous in

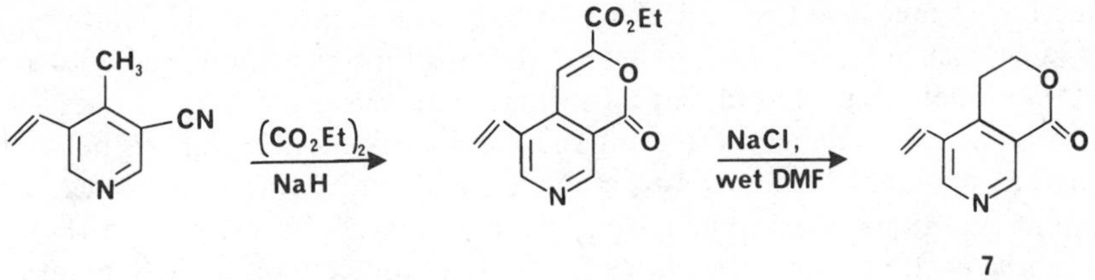

Scheme 3. Synthesis of gentianine.

the light of further investigations by Govindachari [51] and was corrected to **(7)**. The first synthesis of gentianine by the Indian authors [52] started with 4-methyl-5-vinylnicotinic acid and formaldehyde. A second variant of this synthesis was based on the condensation of 3-cyano-4-methyl-5-vinylpyridine with oxalic ester followed by treatment with acid [53]. The product was treated with sodium chloride in DMF and surprisingly gave the *saturated* lactone and not the expected dehydro derivative [54]. The use of ethyl formate led directly to the alkaloid [54] (Scheme 3).

2.7.2. Biosynthesis. Janot proved that gentianine is an artifact [55]; it forms only when ammonia is used before the extraction. Indeed, ^{15}N-labeled ammonia was incorporated into the gentianine molecule [56]. Probably pyrones, for example, gentiopicroside or gentioflavine, are the precursors for the pyridone alkaloid.

2.7.3. Pharmacology. Extracts prepared from plants of the family Gentianaceae have for centuries been used medicinally by inhabitants of the Middle East. Because gentianine is monoterpene derived, the purported medicinal activities of these extracts must be related mainly to other chemical constituents within the leaves and twigs of the plants. Experimental studies have substantiated this realization. The polyoxygenated xanthones, which comprise over 90% of the chemical characters isolated from the plant, have been found to contribute to a large number of the ascribed actions of the plant extracts [57, 58]. In addition, in light of the outstanding antimalarial activity of some plants within the Gentianaceae family, gentianine was early subjected to testing for this property. As it turned out, this alkaloid exerted no significant effect on the sexual activity of *Plasmodium gallinaceum* in mosquitoes [59], nor did it affect the activity of *Plasmodium berghei* in mice [60]. In the latter study antiamoebic activity was likewise shown to be lacking.

Gentianine does exert significant pharmacological activity within the central nervous system. This alkaloid was initially noted to produce apparent central stimulation at low doses but depression at higher doses [59]. After more intensive pharmacological screening in rodents, a wide array of central effects of gentianine was revealed [61]. A dose of 20 mg/kg intraperitoneally markedly reduced spontaneous motility and produced sedation, ptosis, and hypothermia. The same dose potentiated hexobarbital sleeping time but diminished both amphetamine stereotypy and toxicity. In behavioral tests gentianine inhibited the ability of mice to remain on a rotating rod. Footshock-induced fighting behavior as well as avoidance responses to a conditioned stimulus were inhibited. The alkaloid additionally exhibited significant antipsychotic activity. Although gentianine itself did not exert either analgesic or anticonvulsant activity at the 20-mg/kg dose level, this alkaloid did potentiate the analgesic effect of subthreshold doses of morphine and the anticonvulsant effect of subthreshold phenytoin doses. Higher doses of gentianine (50–100 mg/kg) produced hindlimb paralysis and catalepsy in addition to sedation. At these doses significant anticonvulsant activity was demonstrated in both the electroshock and chemshock seizure tests.

Gentianine has been examined for both acute and chronic toxicity in rodents [60, 61]. No overt adverse effects became evident during a 3-week period of administration of 20 mg/kg daily by the intraperitoneal route. The lethal dose is in the range of 300–400 mg/kg; lethality was preceded by clonic convulsions. The lactone ring of gentianine was considered to be responsible for the toxicity, as toxicity was substantially diminished after administration of 3-hydroxymethyl-4-hydroxyethyl-5-vinylpyridine, the lithium aluminum hydride reduction product of gentianine [60].

Gentianine does appear to be responsible in part for some of the effects attributed to Gentianaceae plants by ancient folklore, in particular those mediated centrally. A large number of the effects of these plants, however, appear to be due to other constituents, including the polyoxygenated xanthones, which comprise over 90% of the chemical compounds isolated from the plant [57, 58].

2.8.   Anibine and Duckein

These alkaloids isolated from *Aniba duckei* [62] are related Lauraceae alkaloids. After isolation anibine (8) was hydrolyzed to 1-(3'-pyridyl)butane-1,3-dione. It has been synthesized [63] by thermal condensation of 3-acetyl pyridine with benzylmalonic acid di-2,4-dichlorophenyl ester followed by methylation and debenzylation by aluminum chloride (Scheme 4).

Duckein proved to be 2-nicotinoyl-phloroglucinol-5-methyl ether (9) [64]. We are unaware of any pharmacological investigation of this alkaloid or of anibine.

2.9.   Wilforine and Wilfordine

2.9.1.   Chemistry.   Wilforine and wilfordine were isolated from the roots of *Tripterygium wilfordi* by countercurrent distribution [65]. Four ester alkaloids were subsequently separated. Although these alkaloids were isolated many years ago, the structure determination was very much delayed [66] by the complexity of the nonalkaloidal skeletal element, which is an esterifying group (R) of

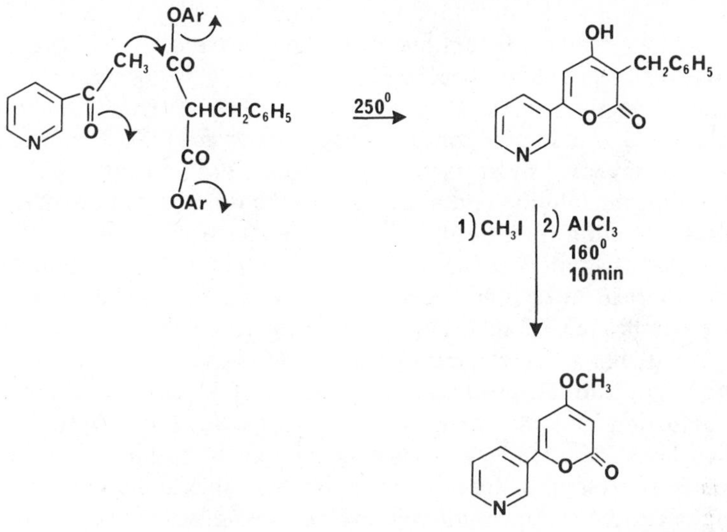

Scheme 4. Synthesis of anibine.

nicotinic acid 2-isovaleric acid [(wilfordic acid) **(10a)**] and hydroxywilfordic acid **(10b)**. Hydrolysis led to the dicarboxylic acids and to a ketopolyol (polyhydroxy compound) $C_{15}H_{26}O_{10}$. (See structural formula **11** for wilfordine.) The acids have been analyzed by UV and IR spectra, and later on the structure has been definitely elucidated. They were reviewed recently in a separate chapter [67] on Celastraceae alkaloids; therefore, we do not wish to duplicate that review. In addition to the two major components, wilforine and wilfordine, other Celastraceae plants gave a large number of other alkaloids, for example, evonine **(12)**. The esterifying acid in this case is evoninic acid.

2.9.2. Pharmacology. Although preparations derived from *T. wilfordii* Hook have been used by the Chinese both as a drug and an insecticide for hundreds of years, few studies have been undertaken to determine the activity of the alkaloids isolated from the plant. Shortly after the introduction of *Tripterygium* into the United States, interest in determining the fractions active as an insecticide arose, and an alcoholic extract was soon found to be effective [68]. The toxicity of the material relied upon its ingestion by the insect; poisoning did not occur after contact alone. Both wilfordine and wilforine were subsequently shown to be insecticidal [69, 70], and, after their discovery, wilforgine and wilfortrine were found to be active as well [71]. Recently both the alkaloid fraction and two diterpenoid triepoxides derived from *T. wilfordii* have been found to be cytotoxic both *in vivo* when tested against leukemias in the mouse and *in vitro* against cells obtained from human carcinoma of the nasopharynx [72].

Interest in utilizing *T. wilfordii* for the treatment of several medical conditions conventionally controlled by corticosteroids has just recently arisen in

China [73, 74]. In this utilization of the plant as a drug, extracts of the plant were, and apparently will continue to be, administered.

In the treatment of both rheumatoid arthritis and ankylosing spondylitis (i.e., arthritis of the spine), the effectiveness of extracts derived from *T. wilfordii* compared quite favorably with that of the corticosteroids [73]. The drug solution utilized was prepared by extraction of dry roots of the plant into alcohol; the resulting filtered solution contained 12% of the original plant material.

Patient selection for determination of the effectiveness of *T. wilfordii* extracts in the latter study resembled that utilized in the United States. Subjects included those in the early and intermediate stages of the disease process, some of whom had failed to respond to either conventional nonsteroidal or steroidal drug therapy. Patients with concurrent disease processes were excluded. The results indicated that pain relief was quite good in 75% of the subjects (133 total cases). Both reduction of joint swelling of some degree and improvement of joint function were evident in 88% of the patient population, and 80% of the 25 cases that were on prolonged steroid medication were able to discontinue the latter group of drugs. Although radiologic evidence demonstrated better calcification of osteoporotic bones, only symptomatic relief was possible, as is the case with other antirheumatic agents. Only 20% of the subjects experienced long-lasting relief after maintenance on the plant extract.

A variety of adverse effects, some rather severe, were observed during the course of treatment of arthritic patients with tinctures of *T. wilfordii* [73]. Frequent mucocutaneous reactions, including perioral blisters, ulcers of the mucous membrane of the mouth, conjunctivitis, and epidermal petechial hemorrhage, occurred. Anorexia and stomach discomfort also occurred quite frequently. Menstrual disturbances, most commonly amenorrhea, occurred in about 50% of the female subjects. Finally, pyogenic infections developed, but in a much smaller percentage of the total patient population.

The active principle of the *T. wilfordii* extract utilized in the latter study was considered to be more soluble in alchol than in water and readily destroyed by boiling. The dose required after those investigators' method of preparation amounted to only 10% of that utilized by another Chinese clinic utilizing the plant extracts for arthritis, at which clinic the drug was prepared by boiling.

Another disease process recently reported to be amenable to treatment with preparations derived from *T. wilfordii* is the clinical entity known as systemic lupus erythematosus [74]. The latter condition is an autoimmune disease that causes a variety of abnormalities in immune function. Among the multitude of symptoms and signs are joint pain, lumbago, Raynaud's phenomenon (i.e., swelling of the digits of the hand due to circulatory impairment), elevated body temperature, and malaise. Internally, hepatic, renal, cardiopulmonary, and hematologic functions are affected.

In the study of Wanzhang and co-workers [74] the 103 cases included in the report were selected by usual criteria, as described above. The *T. wilfordii* preparations utilized consisted of either syrup or tablets formulated from the roots and stems of the plant exclusive of the cortex. The results indicated that the majority of the cases (91%) were treated effectively, with marked improvement noted in 54% of the subjects. Of the patients concurrently treated with corticosteroids, 20% were able to discontinue these drugs, while 70% were able to reduce the dosage level.

Side effects of therapy of systemic lupus erythematosus with preparations derived from *T. wilfordii* were similar to those noted in the study of Juling and colleagues [73] and included predominantly gastrointestinal discomfort, nausea, mild abdominal pain, and amenorrhea. Mucocutaneous manifestations were probably not observable because the lupus condition itself is accompanied by a multitude of erythematous skin lesions.

In the study of Wanzhang's group [74] several experiments relating to the mechanism of action of *T. wilfordii* preparations were undertaken. Moderate anticoagulative activity and inhibition of cell division in sarcomas were demonstrated. During the course of these studies, the therapeutic index of the plant preparations was shown to be quite high, that is, a wide margin of safety between therapeutic and toxic doses existed. Cellular immunity as determined by T cell function and immunoglobin levels markedly improved. But because the *T. wilfordii* preparations exhibited corticosteriodlike action without steroid side effects, as well as immunosuppressive activity in the absence of the adverse effects of most immunosuppressive drugs, any conclusion with regard to mechanism of action from the point of view of immunity was not readily forthcoming.

2.10. Evonine

Evonine is an ester (12) of evoninic acid with a $C_{15}H_{24}O_{10}$ polyol. Evoninic acid is a nicotinic acid derivative bearing the 2,3-dimethyl-propionic acid chain which is attached to different complex nonalkaloidic polyhydroxy compounds as a cyclic diester. Its structure was elucidated by spectroscopic techniques [67].

2.11. Nicotine and Related Pyridine Alkaloids

The major alkaloid of tobacco (*Nicotiana tabacum*) is S(−)-nicotine. In addition, Späth and his associates isolated a number of related alkaloids in the thirties, for example, nicotyrine, 3,2-dipyridyl, (−)-nornicotine, anabasine and anatabine. Later, *N*-oxides of nicotine alkaloids, and dehydronicotine, myosmine, anabaseine, were discovered. Nicotine and the related pyridine alkaloids are quite widespread in Solanaceae (Chenopodiaceae) and in *Lycopodium*, that is, in vascular cryptogams.

2.11.1. Chemistry. Nicotine (13) was isolated in pure form [75] in 1828. The correct molecular formula $C_{10}H_{14}N_2$ was determined [76] as early as 1843. It is levorotatory. The structure was determined by oxidation with hydrogen peroxide to an amino acid derived of pyridine, named nicotinic acid by Weidel [77] in 1873. The $C_5H_{10}N$ residue of nicotine was recognized by Pinner [78] in

1892 as a *N*-methyl pyrrolidyl group, by bromination of nicotine followed by hydrolysis with baryta of "dibromoticonine" to malonic acid and methylamine, besides nicotinic acid. Another piece of evidence for the pyrrol moiety was obtained from distillation of the zinc chloride double salt of nicotine from calcium hydroxide that gave pyrrole, according to Laiblin [79]. The first synthesis of nicotine by Pictet and his associates [80, 81] started with 3-aminopyridine and mucic acid via *N*-3-pyridyl-pyrrole, which was thermally rearranged to 2-(3-pyridyl)-pyrrole **(22)** (nicotyrine) and this, in turn, was converted in several steps into (±)-nicotine. A modern synthesis of nicotine by Nakane and Hutchinson [82] (Scheme 5) starts with 3-acetylpyridine oxime methyl ether, which is converted by ketene ethylthioacetalmonoxide into the oxime ether of (3-pyridyl)-(3′-*bis*-ethylthio)-propylketone monoxide. Reduction of the oxime function by diborane to the amine followed by treatment with formic acid resulted (1) in removal of the thiol functions, (2) condensation to a pyrroline, and (3) reduction to a pyrrolidine and formylation of the secondary amine. Hydrolysis of the formamide gave (±)-nornicotine **(16)** while Eschweiler methylation afforded (±)-nicotine (Scheme 5).

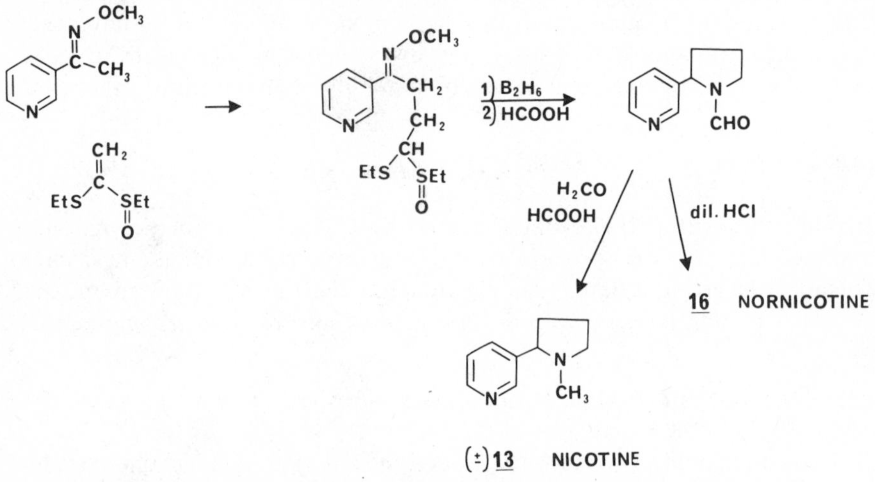

Scheme 5.   Synthesis of nicotine and nornicotine.

An elegant, biomimetic synthesis [83] by Leete (1972) involves 1,4-dihydropyridine that was prepared *in situ* from glutaraldehyde and ammonia. Treatment with *N*-methyl Δ^1-pyrrolinium acetate resulted in imminium salt addition to the C=C bond. In the presence of air (±)-nicotine formed. 1,4-Dihydronicotinic acid, the natural precursor of nicotine, should react in an analogous manner but with loss of carbon dioxide (Scheme 6).

2.11.2.   Stereochemistry.

Nicotine reacts with methyl iodide primarily on the pyridine nitrogen [84]. The oxidation of the pyridine-methiodide of nicotine with potassium ferricyanide and then with dichromate gave L-(−) *N*-methylpro-

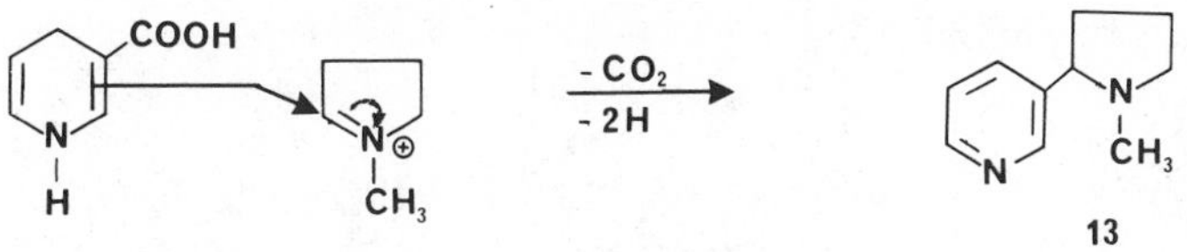

Scheme 6. Biomimetic synthesis of nicotine.

line [85] (Scheme 7). Thus, natural nicotine has the *S* absolute configuration
(13). The conformation of the pyrrolidine ring was determined as an envelope
form [86, 87]. The two rings are almost perpendicular to each other [86].

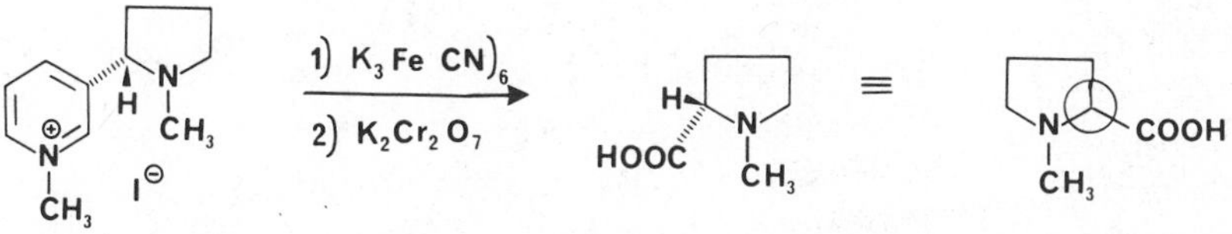

Scheme 7. Correlation of the configuration of (−)-nicotine with *S*(−)-hygrinic acid.

2.12. Nicotine Derivatives

Nicotine *N*-oxides (14 and 15) diastereomeric on the pyrrolidine nitrogen were
obtained from different *Nicotiana* species [88]. Their ^{1}H NMR spectra were
studied [89]. They form *in vivo* in mammals from nicotine [90]. Irradiation of
nicotine in the presence of benzophenone [91] gave nicotyrine (22).

2.13. Nornicotine

2.13.1 *S*(−)-Nornicotine.   The molecular formula $C_9H_{12}N_2$ suggested a difference of a methylene group between nicotine and this minor alkaloid that was
discovered in 1931 by Ehrenstein [92]. Studies by Späth [93] supported the view
that nornicotine is a natural product. The oxidative degradation of (−)-nicotine
usually resulted in partial racemization; optically pure (−)-nornicotine (16)
was finally obtained [94] when potassium permanganate or silver oxide were
used. The origin of nornicotine in the plant is most likely the dimethylation of
nicotine both in the plant and during the curing of tobacco [5].

2.13.2 R(+)-Nornicotine.   The antipode (+)-16 of (−)-nornicotine was isolated from *Duboisia hopwoodii* F. V. Muell in Australia [95]. (+)-Nornicotine
was isolated from tobacco [96]; thus both antipodes are synthesized by the
plant. (+)-Nornicotine is an intermediate in several nicotine syntheses [82, 97].

A new total synthesis of (+)-nornicotine, myosmine, and nicotine published
by Leete [97] used 3-formylpyridine as starting material (Scheme 8). A Strecker-
type addition of morpholine and potassium cyanide is followed by a Michael

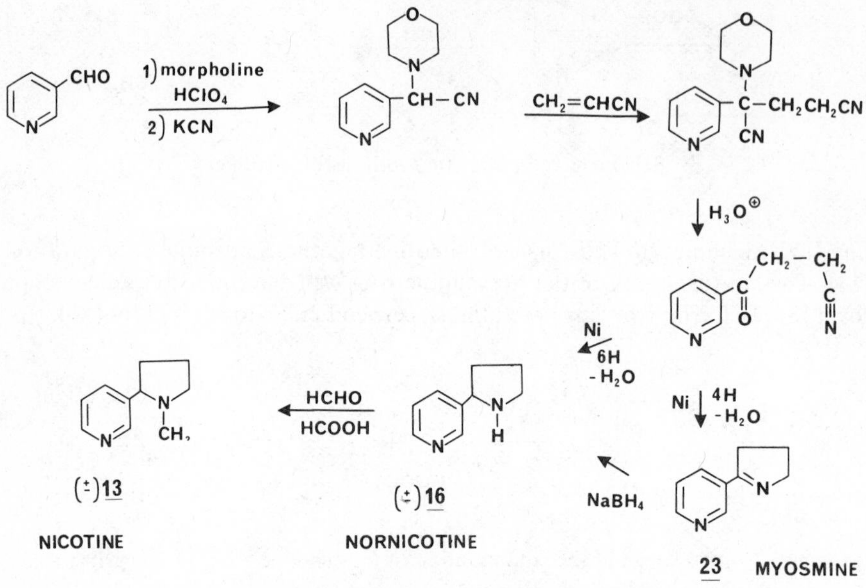

Scheme 8. Synthesis of myosmine, nornicotine and nicotine.

addition of the α-aminonitrile as an acyl anion equivalent to acrylonitrile. Recovery of the carbonyl function and reduction of the nitrile group to the aminomethyl group followed by intramolecular reductive amination gives rise to myosmine (23) and (±)-nornicotine (16) that are interconvertible; nornicotine is selectively methylated by formaldehyde and formic acid to (±)-nicotine (Scheme 8).

2.14.   Nornicotine Derivatives

Several *N*-acylnornicotines have been isolated: formyl (17) and acetyl (18) nornicotine from burley tobacco [98] and the *n*-hexanoyl (19) and *n*-octanoyl (20) amides from flue-cured aired tobacco [99]. *N*-nitrosonornicotine (21) is a tobacco smoke alkaloid; it is strongly carcinogenic [100]. It is also found in fresh tobacco leaves.

2.15.   Nicotyrine

The compound (22) was synthesized [80, 81] 30 years before it was discovered [101, 102] in the tobacco plant. Nicotyrine, 2-(3′-pyridyl)-*N*-methylpyrrole, is a product of dehydrogenation of nicotine with silver oxide [103] or with sulfur. Inversely, catalytic hydrogenation [102] of 22 gave (±)-nicotine (13). Photosynthesis of nicotyrine was achieved from 3-iodopyridine and *N*-methylpyrrole [104].

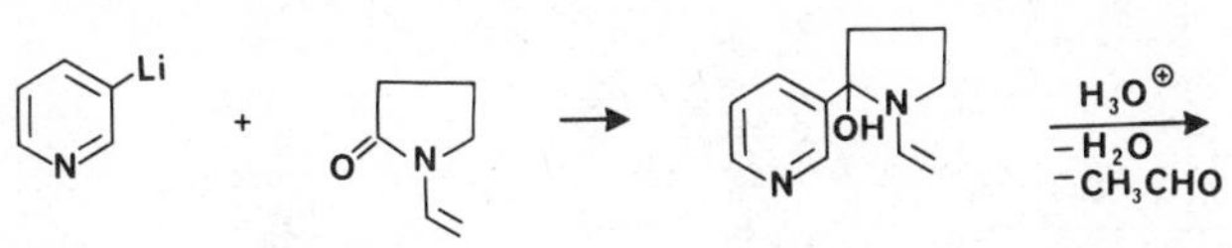

Scheme 9. Synthesis of myosmine from 3-bromopyridine via 3-lithiopyridine with *N*-vinyl-2-pyrrolidone.

2.16. Myosmine

This alkaloid was isolated first from tobacco smoke, and its structure was elucidated by the Späth school [105]. More recently it was identified in several *Nicotiana* species [105–108]; thus, myosmine (**23**) is not an artifact. The presence of the double bond in **23** was proven by its IR spectrum; there is no NH band [109] and strong band at 1626 cm^{-1} shows the presence of a C=N bond [110]. A simple synthesis [111] of **23** starts with 3-pyridyl-Li and *N*-vinyl-2-pyrrolidone, an intermediate of the polymer industry. The carbinol that forms first is dehydrated and devinylated by acid to **23** (Scheme 9). A somewhat similar approach starts with *N*-vinyl-2-pyrrolidone and ethyl nicotinoate via 3-nicotinoyl-1-vinyl-2-pyrrolidone. The hydrolytic step must involve ring opening, decarboxylation, devinylation, and cyclization (Scheme 10) [112].

2.17. *N'*-Methylmyosmine

This alkaloid was claimed to be a minor alkaloid of extracts of Turkish tobacco [113]. However, an authentic product (**24**) obtained by partial reduction of nicotyrine (**22**) proved to be nonidentical [114] with the alleged natural product.

N-Methylmyosmine is certainly a product of bacterial degradation of nicotine [115] or of the chemical oxidation of nicotine with oxygen [116].

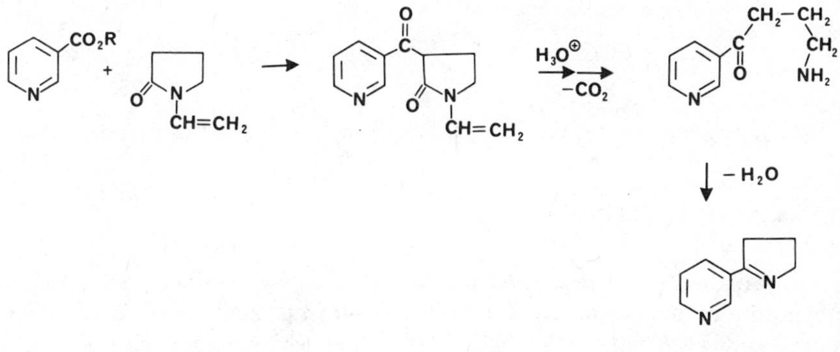

Scheme 10. Synthesis of myosmine from nicotinic acid and *N*-vinyl-2-pyrrolidone.

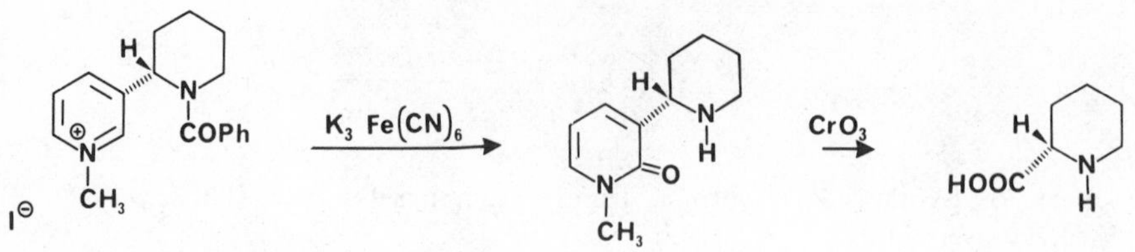

Scheme 11. Configurational correlation of (−)-anabasine with S(−)-pipecolic acid.

2.18. (−)-Anabasine

This *Nicotiana* alkaloid [92, 96] is an isomer of nicotine. It proved to be identical [117] with the "nicotimine" of Pictet [118] and with (−)-anabasine [119], the main alkaloid of *Anabasis aphylla*, (Chenopodiaceae) isolated by Orechoff [120]. It was recently isolated from *Solanum carolinense* by Evans [121]. Oxidation with permanganate gave nicotinic acid [123]. Oxidation with ferricyanide of anabasine py-methiodide to methyl anabasone followed by action of chromic acid led to pipecolic acid [123] (Scheme 11). Catalytic dehydrogenation produced 3′,2-dipyridyl [96] [122]. Thus, the structure of 3′-2-pyridyl piperidine **(25)** was assigned to anabasine. A modern synthesis of anabasine by Scully [124] is given on Scheme 12. Nucleophilic addition of 3-pyridyl lithium to Δ^1-piperideine (that was generated *in situ* from *N*-chloropiperidine and potassium peroxide in crown ether) gave (±)-anabasine with 58% yield (Scheme 12). The absolute configuration of (−)-anabasine was deduced [125] from its conversion into S(−)-pipecolic acid (Scheme 11). The piperidine ring of anabasine is derived from L-lysine and the pyridine ring is derived from nicotinic acid in the biosynthesis [5] of anabasine.

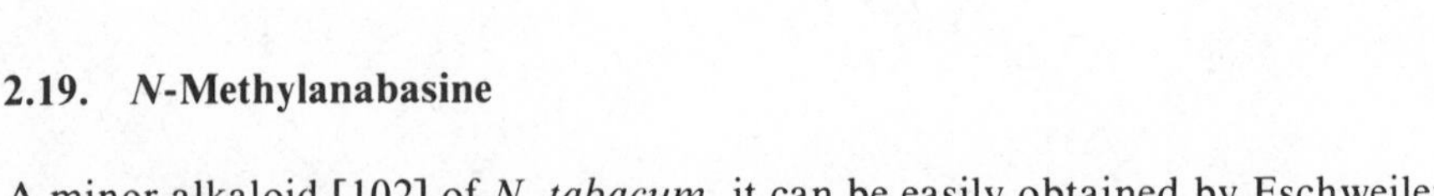

$(\pm)\ \underline{\mathbf{25}}$

Scheme 12. Synthesis of (±)-anabasine from α-piperideine.

2.19. *N*-Methylanabasine

A minor alkaloid [102] of *N. tabacum*, it can be easily obtained by Eschweiler methylation of anabasine by formaldehyde–formic acid or by methylmagnesium iodide [126]. Methylanabasine **(26)** reacts with benzoyl chloride to give 3-(5′-methyl-5′-benzoyl-2′-pentenyl)pyridine [127].

2.20. Anabaseine

1′,2′-Dehydroanabasine is a minor *Nicotiana* alkaloid. It has also been found in animals, for example in ants of the genus *Aphaenogaster* [128]. Feeding of *N. tabacum* plant with (±)-anabasine produced **27**. Anabaseine is very unstable.

2.21. S(−)-Anatabine

This minor *Nicotiana* alkaloid has been isolated from crude nicotine [92]. It is optically active. The molecular formula is $C_{10}H_{12}N_2$, corresponding to de-hydroanabasine [96]. One of the nitrogen atoms was benzoylated, so it is a secondary amine. Dehydrogenation by Pd-asbestos led to 3′-2-dipyridyl. Hence, anatabine is a (3′-pyridyl)-2-tetrahydropyridine wherein the double bond cannot be in either the 1,2 or 5,6 position since it is optically active. Decision in favor of the $\Delta^{3,4}$ versus the $\Delta^{4,5}$ tetrahydropyridine ring was achieved by mild oxidation of *N*-benzoyl anatabine to nicotinic and hippuric acids, respectively [96]. The total synthesis of (±)-anatabine **(28)** was achieved [129] from the *bis*-urethane of 3-formylpyridine and butadiene by heating in the presence of boron trifluoride. The intermediate Schiff base underwent a hetero Diels–Alder synthesis with butadiene. Hydrolysis of the resulting *N*-ethoxycarbonyl (±)-anatabine led to (±)-**28** (Scheme 13).

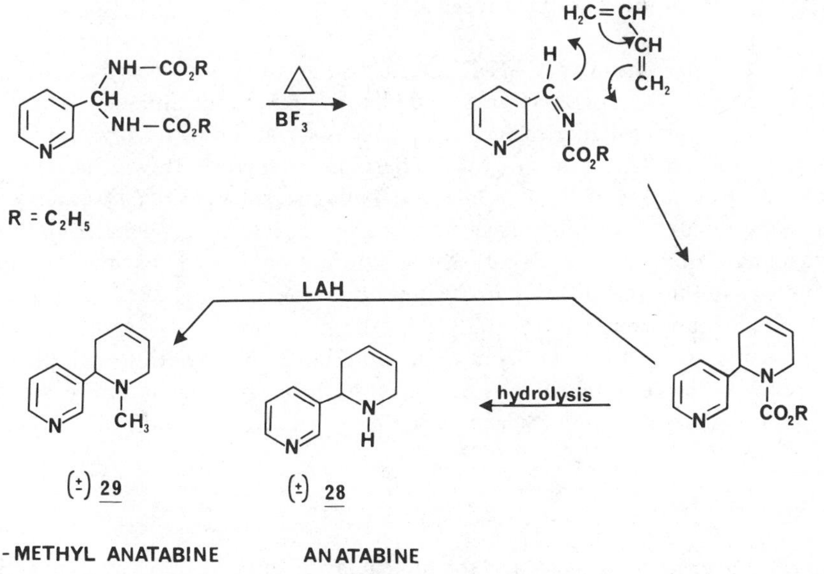

Scheme 13. Synthesis of anatabine.

2.22. $S(-)$-N'-Methylanatabine

The strongly basic fraction of crude nicotine contains a base $C_{11}H_{14}N_2$ that was identified with the product of N-methylation of anatabine with formaldehyde and formic acid [102]. Therefore, structure 29 has been assigned to this base. The new synthesis [129] of anatabine offered a simple approach to 29 by reduction of the intermediate N-ethoxycarbonyl anatabine (Scheme 13) with lithium aluminum hydride.

2.23. Ammodendrine

Ammodendron conollyii, of the Leguminosae family, contains two major alkaloids: ammodendrine (30), related to anabasine and sparteine, a quinolizidine alkaloid [130]. The molecular formula of ammodendrine is $C_{12}H_{20}NO_2$. It is optically active. With methyl iodide one of the nitrogen atoms can be methylated by substitution, while the other gives a methiodide. Catalytic hydrogenation of ammodendrine results in the uptake of 1 mol of hydrogen and in the formation of ($\pm$)-hexahydroanabasine. Therefore, ammodendrine is a monounsaturated 3'-2-dipiperidyl. The location of the double bond was apparent from the synthesis of ($\pm$)-ammodendrine by Schöpf [131, 132] from the trimer of Δ^1-piperideine (Scheme 14). Based on these facts, the structure of 1,2,5,6-tetrahydroanabasine (30) has been assigned to this alkaloid.

2.24. ($\pm$)-Orensine and ($\pm$)-Isoorensine

Their molecular formulas ($C_{19}H_{24}N_2O$) shows that these *Adenocarpus* alkaloids are isomers. In an analogous manner to the synthesis of ammodendrine the trimer, "isotripiperideine," was cleaved after N-acylation with *trans*-cinnamoyl chloride to give ($\pm$)-orensine [132]. Therefore, structure 31 was ascribed to orensine. Resolution of 31 gave (+)-and (−)-adenocarpine [133]. *Isoorensine* (32) gave upon hydrolysis cinnamic acid and a base that was hydrogenated to hexahydroanabasine [134]. The synthesis of isoorensine proved that the "double" bond was in the same position in the piperidine ring as in 31; therefore, the two compounds are stereoisomers [135] in the cinnamic acid residue. Indeed, isoorensine was synthesized from "isotripiperidine" by acylation with phenylpropiolic chloride followed by cleavage and partial hydrogenation to the *cis*-cinnamoyl derivative (32) (Scheme 14).

2.25. Santiaguine

An *Adenocarpus* alkaloid, it has the molecular formula $C_{38}H_{48}N_2O_4$ [136]. Hydrolysis gave α-truxillic acid and a base $C_{10}H_{18}N_2$. The latter upon catalytic hydrogenation took up 1 mol of hydrogen leading to 2,3'-dipiperidyl.

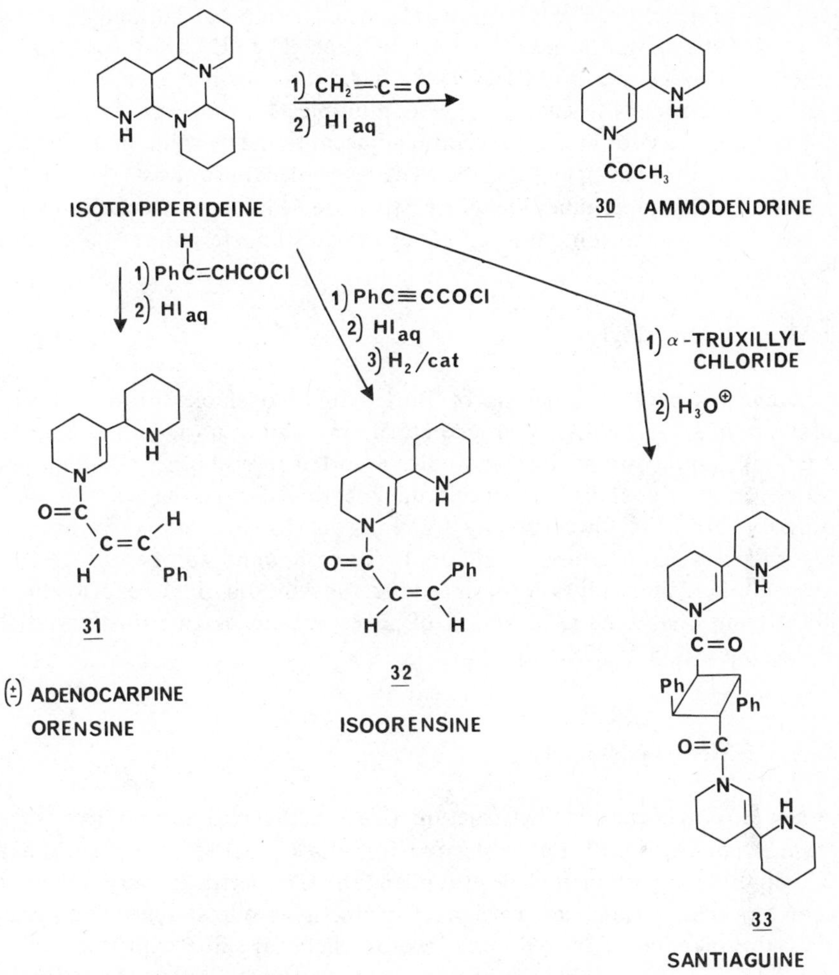

Scheme 14. Synthesis of ammodendrine, orensine (adenocarpine), isoorensine and santiaguine, from "isotripiperideine."

N-Methylation of santiaguine followed by acid hydrolysis led to tetrahydro-*N'*-methylanabasine. The synthesis followed the same pattern as outlined in Scheme 14; by using α-truxillyl chloride in the *N*-acylation of "isotripiperideine" (±)-santiaguine was formed [137]. The natural product is dextrorotatory.

2.26. (+)-Kuraramine

An alkaloid of *Sophora flavescens*, it has two chiral centers. The constitution and the relative configuration **(34)** have been determined by Murakoshi et al. [138]. The IR spectrum showed bands at 1645, 1610, and 1550 cm^{-1}, consistent with a

2-pyridone structure. The UV spectrum showed maxima at 226 and 304 nm, also characteristic of the same structural element. The ^{1}H NMR spectrum was interpreted as follows: δ 13.03 (NH), δ 7.0–4.5 three olefinic protons: δ 3.49 and 3.56 CH$_2$OH; δ 2.34 (N-methyl); δ 3.06 (multiplet) for three equatorial protons; δ 1.27 one axial proton; δ 2.90, proton adjacent to the pyridine ring. The chemical shifts of the high-field protons closely resembled those of cis-1,3-dimethyl-5-hydroxymethyl piperidine. Therefore, structure 34 has been ascribed to kuraramine. The absolute configuration so far remains undetermined.

2.27. 2,3'-Dipyridyl

Gas chromatographic analysis of Burley tobacco smoke proved [139] the presence of 2,3'-dipyridyl (35) together with myosmine, nornicotine, anabasine, anatabine, and cotinine. Earlier studies reported several other alkaloids [140]. However, fractional distillation of crude nicotine allowed the isolation of 2,3'-dipyridyl from its lower boiling fractions [141]. Thus, dipyridyl has to be regarded as a genuine tobacco alkaloid. The compound was identified with the authentic [142] dipyridyl known since 1892; the same dipyridyl was also obtained [122] from anabasine (25): action of silver acetate or zinc dust resulted in dehydrogenation of 25 to 35.

2.28. 5-Methyl-2,3'-dipyridyl

A new variant of the nicotine structure, this 5-methyl pyridine has been isolated from *N. tabacum* [98]. The molecular formula C$_{11}$H$_{10}$N$_2$ differs from that of 2,3'-dipyridyl by one methylene group, and the UV spectra are very similar. The ^{1}H NMR spectrum indicated a signal for a methyl group attached to a pyridine ring. Comparison of the aromatic proton chemical shifts with those of 2,3'-dipyridyl showed the absence of one proton in the 5-position. Therefore, the synthesis of 5-methyl-2,3'-dipyridyl was carried out starting from diazotized 3-aminopyridine by coupling with 3-picoline (Scheme 15). The major product (36) was identical with the alkaloid [98].

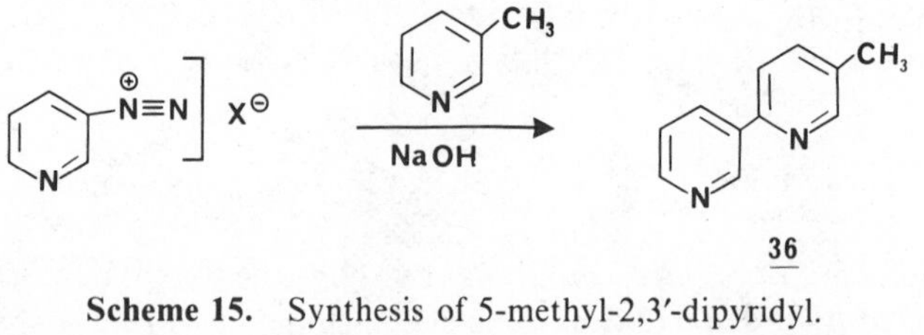

Scheme 15. Synthesis of 5-methyl-2,3'-dipyridyl.

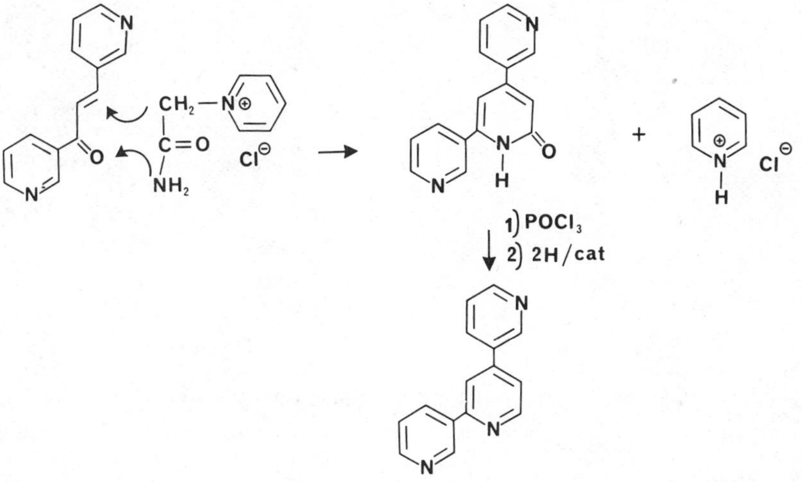

37   NICOTELLINE

Scheme 16. Synthesis of nicotelline.

2.29. Nicotelline

This crystalline base of *Nicotiana* sp., known [1] since 1901, has the molecular formula $C_{15}H_{11}N_3$ [143]. Oxidation gave nicotinic acid and pyridine-2,4-dicarboxylic acid; therefore, the constitution of 3,2':4',3-terpyridyl (37) was assigned to the base. The first synthesis [144] started with 3-nicotinoylvinyl pyridine and *N*-(carbamylmethyl) pyridinium chloride, which gave 4,6-*bis*-3-pyridyl-2-pyridone. Treatment with phosphoryl chloride followed by reductive dehalogenation led to nicotelline (Scheme 16).

2.30. Anatalline

One of the hexahydronicotellines was discovered in the roots of *N. tabacum* and was named anatalline [145]. The molecular formula is $C_{15}H_{17}N_3$. In accordance, the mass spectrum indicated m/z 239 as the parent radical ion. The UV spectrum was indicative of two pyridine rings. This, combined with elemental analysis, showed that a third, saturated heterocyclic ring must be present. The IR spectrum showed an NH absorption at $3230\,\text{cm}^{-1}$. The mass spectrum resembled that of anabasine quite closely. Interpretation of the ^{1}H spectrum and double irradiation confirmed the view of a 2,4-disubstituted piperidine ring. Oxidation consistently gave only nicotinic acid; therefore, the two pyridine rings were 3-substituted and not attached to one another. Finally, dehydrogenation with Pd gave nicotelline. Therefore, the constitution 38 of 2,4-*bis*-3'-pyridyl-piperidine was assigned [145] to the alkaloid.

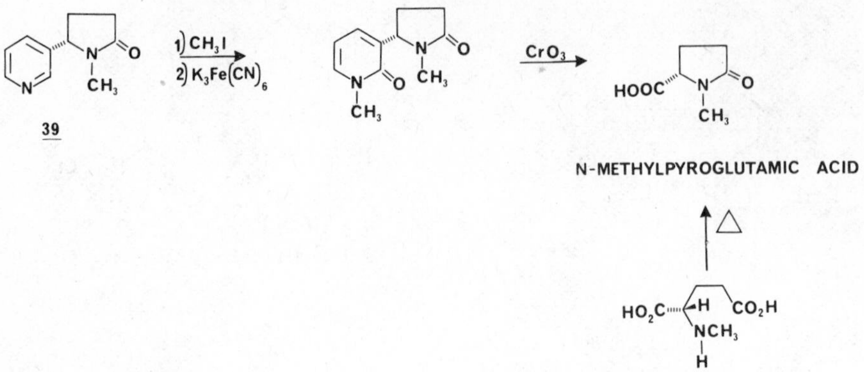

Scheme 17. Configurational correlation of cotinine with N-methylpyroglutamic acid.

2.31. Cotinine

Cotinine ($C_{10}H_{12}N_2O$) was first described by Pinner as its dibromo derivative [78] that formed during the bromination of nicotine in acetic acid. The correct lactam structure 39 for cotinine was based by Pinner on the elemental analysis and on the fact that it forms a monopicrate and monohydrochloride [146]. This conclusion has been confirmed by IR, UV, and chemical correlation studies [147] on cotinine, which was isolated later from cured tobacco. The lactam carbonyl is apparent from an IR band at 1700 cm^{-1}. The UV spectrum in aqueous acid showed λ_{max} 262 nm characteristic of the pyridine ring. Furthermore, following the principle of Karrer's degradation [85] of nicotine to $S(-)$-N-methylproline, cotinine methiodide was degraded to N-methyl-5-pyrrolidone-2-carboxylic acid (Scheme 17). The latter proved to be identical with N-methyl-pyroglutamic acid, prepared thermally from N-methylglutamic acid.

Cotinine is known as a tobacco smoke alkaloid [139], but it has also been discovered in fresh leaves of *N. glutinosa* [148]. Circular dichroism curve of cotinine showed the same absolute configuration [149] as that of nicotine.

2.32. Nicotianine

Different fractions of methanolic tobacco plant extracts showed positive ninhydrin test, but their R_f values were not consistent with any known α-amino acid. One of them was isolated and named nicotianine [150]. The molecular formula was $C_{10}H_{12}N_2O_4$. The UV spectrum (λ_{max} 265 nm) was reminiscent of the absorption curve of nicotinic acid. The IR spectrum showed no C—O-stretching vibration corresponding to the nonionized carboxyl group(s) but a broad absorption near 1630 cm^{-1} that can be ascribed to a zwitterionic carboxylate group. The monohydrochloride, however, has an IR band at 1700 cm^{-1} for a non-dissociated carboxyl group. The ^{1}H NMR spectrum (D$_2$O) indicates five ali-

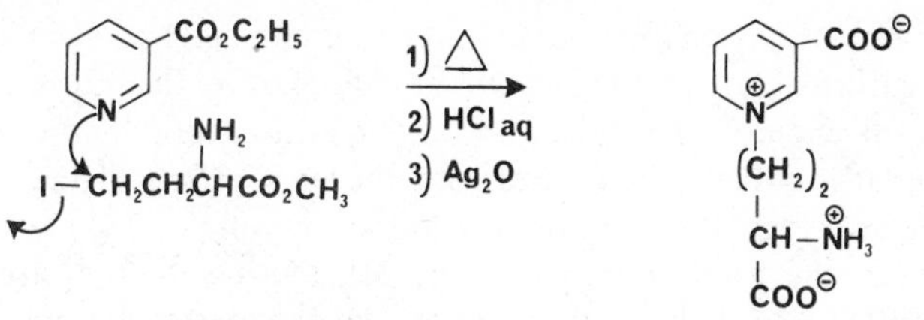

Scheme 18. Synthesis of nicotianine.

phatic and four aromatic protons. Thus structure **40** of L(+)-*N*-(3'-amino-3'-carboxypropyl)-3-carboxypyridinium betaine was assigned to nicotianine. This structure was corroborated by total synthesis. Methyl-2-amino-4-iodobutyrate was prepared from bromoserine [151] and then heated with ethyl nicotinate to obtain the quaternary pyridinium amino acid ester. Hydrolysis gave **40** identical with nicotianine (Scheme 18). Nicotianine appears to be the first naturally occurring pyridinium amino acid. However, its biogenetic role is so far undetermined [5].

2.33. Pharmacology of the Tobacco Alkaloids

The pharmacological effects of nicotine **(13)**, the predominant pyridine alkaloid of common tobacco (*N. tabacum*), are often complex and unpredictable. Contributory factors include the inherence of both stimulatory and depressant phases of action as well as the multiplicity of neuroeffector and chemosensitive sites acted upon in the body. The end response of a given organ or tissue to administered nicotine is often the sum total of a variety of frequently opposing effects.

Certain paradoxical novelties in the actions of nicotine at early-recognized cholinergic receptors ultimately led to the classification of the involved subpopulation of recognition sites as nicotinic cholinergic receptors. It was early recognized that nicotine causes an initial transient stimulation of all autonomic ganglia and a subsequent more persistent ganglionic blockade. The adrenal medulla is affected in a similar fashion. Nicotine initially elicits a release of medullary catecholamines; afterwards, and more prominently with larger doses, catecholamine release in response to nerve stimulation is blocked.

The actions of nicotine in a variety of isolated organs may be explained by the release of catecholamines, both from the adrenal medulla and from peripheral adrenergic nerve terminals. In these tissues classical sympathomimetic responses are produced.

The cardiovascular responses to nicotine are complex. This alkaloid may accelerate heart rate either by excitation of sympathetic or depression of parasympathetic cardiac ganglia. On the other hand, heart rate may be slowed either by stimulation of parasympathetic or by depression of sympathetic cardiac ganglia. The actions of adrenal medullary catecholamines released by nicotine tend to favor elevation of both heart rate and blood pressure. In addition, activation by nicotine of chemoreceptors of the aortic and carotid bodies reflexly augments a sympathomimetic response, with the production of tachycardia, vasoconstriction, and resultant elevation of blood pressure.

In contrast with the predominant actions of nicotine on the circulatory system, the net effects of this alkaloid on the gastrointestinal tract are due primarily to parasympathetic stimulation. The end result is increased tone and motor activity of the smooth muscle lining the gastrointestinal tract.

As in most other peripheral tissues, nicotine produces a biphasic effect on the neuromuscular junction. In this system the initial stimulatory phase is quite brief and is rapidly superceded by neuromuscular blockade involving depolarization and subsequent receptor desensitization.

Nicotine stimulates a variety of sensory receptor systems. The mechanoreceptors responsive to stretch or pressure of the skin, mesentery, stomach, tongue, and lung are affected. Thermal receptors of the skin and tongue are also stimulated by nicotine, as are chemoreceptors of the carotid body and pain receptors. While the ganglionic blocker hexamethonium prevents this effect of nicotine, this blocking agent exerts no effect on the activation of the sensory receptors by physiological stimuli.

The effects of nicotine on the central nervous system are predominantly stimulatory. Moderate doses produce tremor, while somewhat higher doses lead to convulsions. The respiratory center is also stimulated, as are the nearby chemoreceptor trigger zone and the emetic center. Action at the latter two sites results in nausea and vomiting.

Acute nicotine toxicity rarely occurs from smoking tobacco. Cigarettes usually contain only 1–2% nicotine, resulting in the delivery of 0.05–2.5 mg, although nicotine contents as high as 5–6 mg have been documented. On the other hand, the acute fatal dose of nicotine is approximately 60 mg in the adult. Nicotine poisoning does occasionally occur, however, usually as a result of accidental ingestion of insecticide sprays containing this alkaloid predominantly or as a result of ingestion of tobacco products by children. Clinical signs of toxicity are a direct extension of the pharmacological effects. As a terminal event, central nervous system stimulation is followed by depression, and death occurs via respiratory paralysis of both central and peripheral origin.

At the doses of nicotine inhaled from cigarette smoking, effects on the major organ systems are predominantly stimulatory. Nicotine absorbed by smoking mothers has been found in breast milk in amounts sufficient to affect nursing infants. Although smoking increases the incidence of abortion, significantly reduces the weight of the newborn child, and increases the probability of both perinatal mortality and sudden infant death, the relation of nicotine to these processes is unclear.

Of the host of diseases related to the chronic smoking of tobacco by mankind, the relative role of nicotine as a causative factor still remains unclear. This alkaloid is only one of about 4000 compounds emitted after tobacco has been burned. Nicotine per se has been implicated to some extent in the etiology of some cardiovascular diseases and in chronic obstructive lung disease. On the other hand, nicotine does not appear to be involved in the propagation of certain neoplastic diseases or in the related facilitatory ciliotoxic action of tobacco smoke within the pulmonary system.

The pharmacological actions of a host of tobacco alkaloids structurally related to nicotine has been thoroughly reviewed by several groups [152, 153]. Of the total of about 20 compounds, few exceeded the potency of nicotine in any of the organ systems tested.

Nornicotine (16), one of the tobacco alkaloids of the highest concentrations in smoke in comparison with the remainder of the trace alkaloids, produced pharmacological effects resembling those of nicotine in a variety of organ preparations. This alkaloid was about one-half as potent as nicotine in several systems, including adrenal medullary release of catecholamines, contraction of frog rectus muscle, and inhibition of both cat knee jerk and cat flexor reflex. On the other hand, nornicotine was about twice as potent as nicotine in the rat phrenic nerve–diaphragm preparation. Toxicity of nornicotine was likewise about 2 or 3 times greater than that of nicotine in several animal species, including the rabbit and the rat.

Of the remaining tobacco alkaloids studied, anabasine exhibited the greatest potency with regard to the production overall of pharmacological effects. This alkaloid was about two-thirds as potent as nicotine in evoking release of catecholamines from cat adrenal medulla. Its potency in the rat phrenic nerve–diaphragm preparation, and that of its pressor action in pithed rat, equaled those of nornicotine, but in other systems, such as contraction of frog rectus muscle and inhibition of cat spinal reflexes, anabasine was only about one-half as potent. In contrast, its potency in inducing contraction of guinea pig ileum was about 4 times greater than that of nornicotine; for comparison, the latter alkaloid was 20-fold less potent than nicotine itself. Although anabasine was only slightly less toxic than nicotine in the limited number of animal studies conducted, the pattern of toxicity revealed less stimulatory effects and more depression.

Of the remaining tobacco alkaloids exhibiting any significant pharmacological effects, only metanicotine, myosmine, and nicotyrine approached the range of potency of nicotine in several of the organ systems studied. Metanicotine evoked adrenal medullary catecholamine release at doses 5 times greater in magnitude than nicotine and induced contraction of guinea pig ileum and pressor action in pithed rat at doses about 25- to 30-fold greater in magnitude. At the latter doses the remainder of the organ systems discussed above were not affected. Myosmine was relatively active in the rat phrenic nerve–diaphragm preparation at doses only three-fold greater than those for nicotine, but exhibited pressor action in the pithed rat, frog rectus muscle contraction, and inhibition of cat spinal reflexes at doses 20–30 times greater than those for nicotine. Toxic doses of myosmine were correspondingly much higher, at least in the rat, and convulsions during toxicity were somewhat less severe. On the other hand, nicotyrine exhibited potencies only 5–10 times less than those for nicotine in inhibition of cat spinal reflexes, but pressor action required doses 40 times greater than those for nicotine.

Although nicotine has proven to be teratogenic in laboratory rodents, with the production of congenital defects of the skeletal system [154], no deformities were induced by this alkaloid in farm animals under study, including pigs, sheep, and cows [155, 156]. In a recent study grazing livestock were exposed under controlled conditions to *N. glauca*, a plant unique among the naturally occurring tobaccoes in containing anabasine as its principal or sole alkaloid [157, 158], and congenital deformities of the skeletal system did not arise [159]. Anabasine is one of the few alkaloids in tobacco containing a piperidine ring. The presumed teratogenic activity of anabasine in light of this structural feature adds further support to the earlier proposed hypothesis that, among naturally occurring plant pyridines and piperidines, both degree of unsaturation of the ring structure (i.e., pyridine in the case of anabasine) and side chain length greater than propyl (i.e., piperidine in anabasine) are critical for the production of teratogenicity [160].

2.34.   (−)-Mimosine and (±)-Leucaenine

2.34.1.   Chemistry. (±)-Leucaenine from *Leucaena* and the levorotatory form, mimosine, were isolated [161] from *Mimosa* species in 1937. The alanine side chain was proven by the ninhydrin test and the basic skeleton by zinc distillation, which gave pyridine. Degradation upon exhaustive methylation gave, among other products, *N*-methyl-3-hydroxy-4-pyridone. Thus, the structure of leucaenine has been established [161–165] as *N*-(3-alanyl)-3-hydroxy-4-pyridone **(41)**. It has a zwitterionic structure [166] as expected from a fairly strong base having a carboxylate group. The *biosynthesis* of mimosine is from lysine. However, aspartic acid and α-aminoadipic acid are also incorporated into its skeleton [167].

2.34.2.   Pharmacology. The pharmacological actions of leucaenine, which appears in the biological literature almost exclusively under its alternative nomenclature of mimosine, have been thoroughly investigated. Although this alkaloid confers toxicity to the Leucaena plant, there still remains much interest in the utilization of *Leucaena* as an animal feed [168, 169]. This plant flourishes in poor soils and is remarkably drought resistant. Its high protein and mineral contents, coupled with low fiber concentrations, are quite valuable properties. Attempts are currently being made to produce cultivars of *Leucaena* that will be considerably lower in mimosine content [170].

The toxicity of *Leucaena* plants was recognized as early as 1896, at which time Morris [171] alluded to one of the most obvious toxic effects in affected animals, that is, loss of hair. This effect has been observed in many species, including humans (women) who consumed unripe seeds obtained from the shrub [172, 173]. Ruminants are somewhat more resistant than other species to this phenomenon, as well as to other toxic effects of *Leucaena* in general, because of extensive degradation of the active principle, mimosine, in the rumen [174].

After intravenous or abomasal administration of mimosine to sheep, depilation had a much earlier onset than after maintenance of the animals exclusively on finely ground *Leucaena glauca* [174]. In all instances, however, the involved animals developed loss of appetite and weight loss and subsequently died. Histological examination of skin samples revealed a marked decrease in mitotic activity. Mimosine has subsequently been demonstrated to inhibit DNA, RNA, and protein synthesis in a variety of biological preparations [175–177]. This alkaloid likewise proved effective in inhibiting tumor growth in rats, and its utilization as a chemotherapeutic agent has subsequently been suggested [178]. The possibility of utilization of mimosine as a tool for obtaining synchronously dividing clones has also recently arisen [177].

Decreased fertility has been observed in animals having *Leucaena* included in the diet [179, 180]. Egg production by poultry is also reduced [181]; see Jones [169]. The reduction in fertility in particular was shown to be due to mimosine in a study in the rat [182]. When conception did occur in rats fed 15% *Leucaena*, fetal death and resorption rates were significantly increased [183], and addition of mimosine itself to the diet of rats resulted in teratogenesis [184].

Because of the eye's visibility, the cytotoxic effects of mimosine on ocular structures were readily detected in experimental animals. The development of cataracts in rats fed diets containing *Leucaena* was noted by both Yoshida [185] and Matsumoto and colleagues [186]. In a comprehensive study undertaken by von Sallmann and co-workers [187] the existence of a variety of changes in ocular function as a result of addition of mimosine to the diet of rats was documented. Early changes in lens structure indicative of incipient cataract formation were accompanied by extensive inflammation of the cornea and iris; the magnitude of the inflammatory response led to corneal neovascularization. The cataract induced by mimosine resembled other experimental cataracts histopathologically only inasmuch as mitotic activity in the lens epithelium was depressed. Cataract formation apparently did not rely on a mimosine-induced vitamin deficiency, as cataractogenesis was not aborted by administration of high levels of either pyridoxine or niacin.

One of the signs and symptoms of *Leucaena* toxicity early attributed to mimosine, that is, enlarged thyroid gland and goitrous offspring [188–190], has been recently found to be due entirely to the formation of the major mimosine metabolite, 3-hydroxy-4-(1*H*)-pyridone [191]. This potent antithyroid compound, like thiouracil, inhibits the uptake of iodine by the thyroid at the iodine binding step. The remaining overall toxicity of 3-hydroxy-4-(1*H*)-pyridone was quite low inasmuch as its chronic administration resulted in only goiter and reduced weight gain as overt symptoms. Cytotoxicity *in vitro* was likewise low [176].

Mimosine produces several noteworthy biochemical effects that were earlier thought to be involved in the production of toxicity (see Thompson et al. [192]

and Hegarty and Peterson [193]). This alkaloid complexes with pyridoxal phosphate and may thus interfere with enzymes requiring the latter compound. Activity of several other enzymes is also affected. Mimosine may chelate metal ions and in this manner inhibit metal-containing enzymes. The enzymes cystathionase and cystathionine synthetase are more directly inhibited. Finally, mimosine may interfere with tyrosine or phenylalanine metabolism by affecting several enzymes. The mimosine metabolite, 3-hydroxy-4(1H)-pyridone, independently inhibits an additional enzyme within the same system, catechol-*O*-methyl-transferase [191].

2.35. Flavipucine

Flavipucine is a metabolite of *Aspergillus flavipes* [194]. The structure (42) has been confirmed by X-ray crystallography [195] and by synthesis [196] (Scheme 19). 3-Methylpropylglyoxal reacted with 4-hydroxy-6-methyl-2-pyridone and acetic anhydride to give the diacetate. Elimination of acetic acid by K-*t*-butoxide and epoxidation with peracetic acid or *t*-butyl-hydroperoxide led to the antibiotic-alkaloid. The intermediate olefin was trapped as a Diels-Alder adduct. The relative configuration, as *Z*, was determined by a chemical procedure [197], and is in complete agreement with X-ray crystallographic data. It is a somewhat unusual pyridine-2,4-dione, the C(3) of which is a bridgehead of an epoxide ring that is attached to a C$_4$ side chain with a carbonyl group. The compound

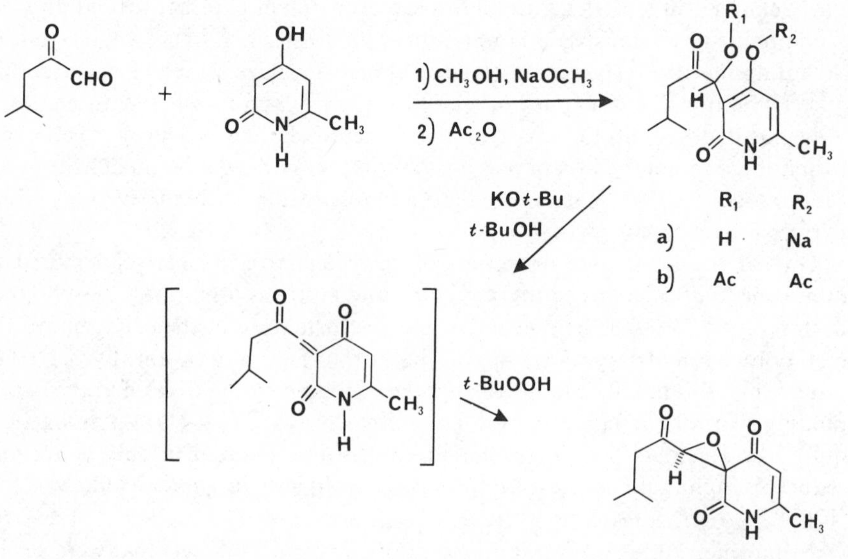

Scheme 19.   Synthesis of flavopucine.

has antibiotic activity [198] but it is extremely labile and rearranges into "isoflavopucine."

2.36. S(−)-Melochinine

Melochinine (43), an alkaloid from the plant *Melochia pyramidata*, is a 4-pyridone derivative that contains a methoxyl group in a 3-position, a methyl group at C(2), and the 11-hydroxydodecyl side chain in the 6-position [199]. It is closely related to the long side chain piperidine alkaloids, like cassine (87), and it has been correlated [200] with it via 1-(2-pyridyl)-11-dodecanone. It occurs as the corresponding ketone melochininone and also as the D-glucoside. Melochinine has been synthesized [201] (Scheme 20) quite recently. 2,6-Dimethyl-3-hydroxy-4-pyrone was converted into the methylene-Li-dianion which underwent aldol condensation with *R*- and *S*-9-tetrahydropyranyloxydecanal to a carbinol. Methylation of the enolic C(3) hydroxyl followed by dehydration of the carbinol and hydrogenation led to the 4-pyrone isologue of melochinine. Ammonia replaced the ring oxygen to afford R(+)- and S(−)-melochinine, depending on which enantiomer of the 9-hydroxydecanal was used at the beginning [201]. The natural product is levorotatory (Scheme 20).

2.37. Nigrifactin

The very unstable alkaloid has been isolated from different *Streptomyces* [202]. The structure is of a 2-substituted Δ^1-piperideine (44) and was proven by two independent total syntheses [203, 204]. The first one starts with 1-chloro-5-oxo-*E,E,E*-6,8,10-dodecatriene and sodium azide, leading to the 1-azido compound, and this on reaction with triphenylphosphine, probably via a phosphinimine,

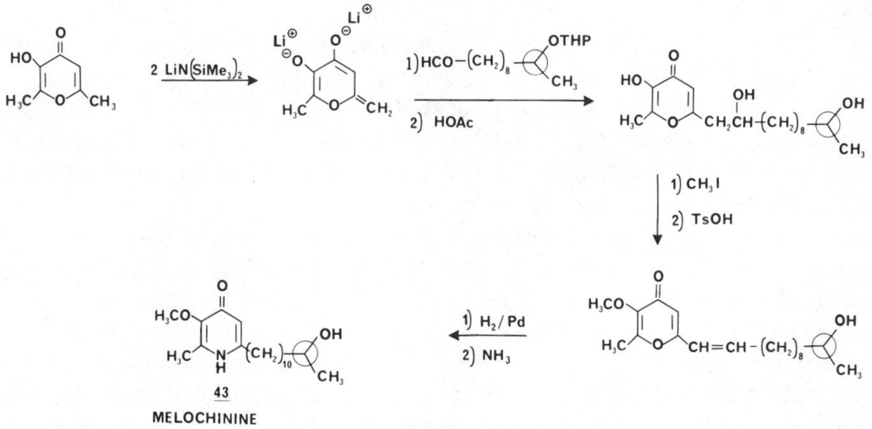

Scheme 20. Synthesis of melochinine.

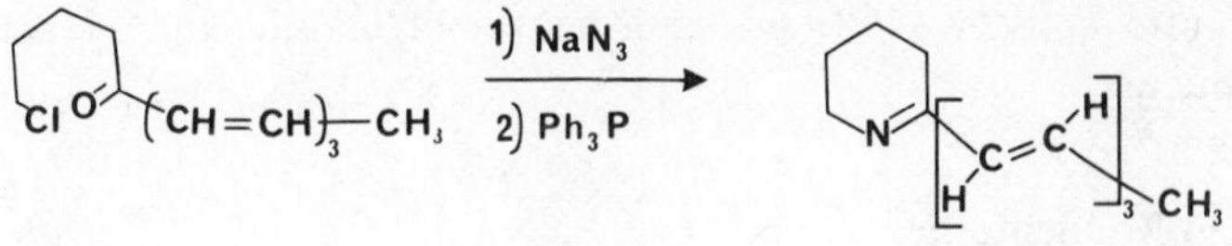

44

Scheme 21. Synthesis of nigrifactin.

cyclizes to nigrifictin [203] (Scheme 21). The second synthesis makes use of the acidity of methyl protons in 2-methyl-Δ^1-piperideine and adds E,E-2,4-hexadienal to the carbanion. Dehydration of the hydroxy-heptadienol side chain [204] leads then to the triene.

2.38. Alkaloid of Streptomyces NA-337

This compound was isolated from *Streptomyces* species, so it most likely has some kind of antibiotic activity. The structure of this compound as E,E-pentadienyl-Δ^1-piperideine **(45)** has been determined [205] by simple chemical operations, such as converting into pentylpiperidine by hydrogenation, and also by spectral data, particularly NMR spectroscopy.

2.39. Navenone A

Isolated from the sea slug *Navanox inermis* [206], it is an alarm pheromone. The structure as 2-pyridine-(E,E,E,E)-deca-3,5,7,9-tetraene-2-one **(45)** was established by ^{1}H NMR spectroscopy.

2.40. *S*(−)-Actinidine

The isoprenoid alkaloid actinidine **(47)** has been isolated from *Actinidia polygama* [207], the roots of *Valeriana officinalis* [208], as well as from the defense secretion of the Australian cocktail ant, *Iridomyrmex nitidiceps* [209]. It has interesting biological activity; **47** is a supposed cat attractant—therefore, its chemistry and pharmacology became significant. Among its many syntheses [210–212], we wish to present a most recent one [211] that involves an intramolecular cycloaddition of an acetylenic side chain across a pyrimidine skeleton, giving rise to hydroxyactinidine, which in turn was deoxygenated via the chloro compound to (±)-**47** (Scheme 22).

The absolute configuration of actinidine was proven by two partly independent correlations [207, 213]. *R*(+)-pulegone dibromide by Favorskii rearrangement [214] gave pulegenic acid ester which on ozonolysis was converted into the key compound, *R*(+)-2-methoxycarbonyl-3-methylcyclopentanone. Condensation with methyl cyanoacetate followed by hydrolysis gave 5-methyl-2,6-

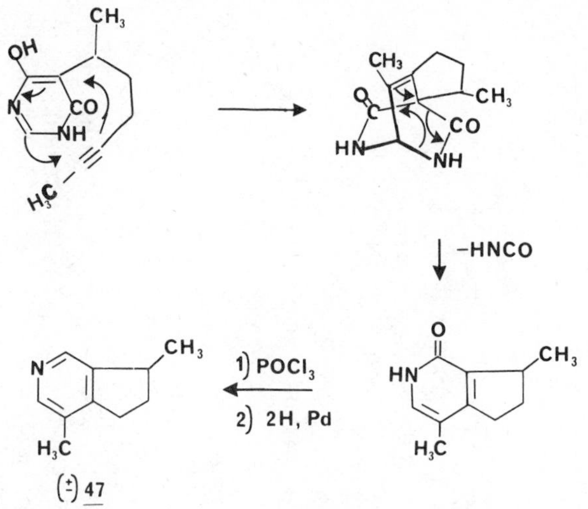

Scheme 22.   Synthesis of actinidine by rearrangement.

dihydroxy-3,4-methylcyclopentanopyridine [215]. Conventional deoxygenation via the dichloropyridine afforded $R(+)$-actinidine, which is the (+)-enantiomer of the natural product (Scheme 23). Another approach [213] starts with (+)-iridodial. This is a product of partial oxidation of (+)-citronellal followed by intramolecular Michael addition [216]. The cyclic tautomer of iridodial undergoes exchange of oxygen by nitrogen and dehydrogenation to give $R(+)$-actinidine (Scheme 24).

Loganin was assumed as a precursor; this assumption was not supported by tracer experiments [217]. The related terpene–piperidine alkaloid skytanthine is derived from geraniol [218].

2.41.   S(−)-Tecostidine

Tecoma stans roots contain several terpenoid pyridine and piperidine alkaloids, among others tecostidine. Tecostidine (**48**) is a 11-hydroxy derivative of actinidine. Chemical correlation by Cavill [213] proved that $R(+)$-tecostidine is related to (+)-pulegone and this tecostidine is the antipode of the natural product (Scheme 23).

2.42.   R(−)-Boschniakine

Boschniakine (**49**) was isolated from *Boschiniakia rossica* [219] and was converted into actinidine (**47**). Its absolute configuration has been determined [220] by O.R.D. and by chemical correlation. It is a formyl pyridindane. Japanese authors [219] followed essentially the same line of synthesis as that for actinidine

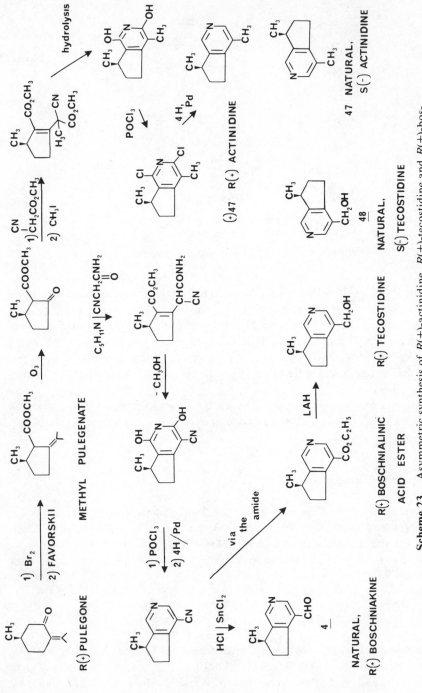

Scheme 23. Asymmetric synthesis of *R*(+)-actinidine, *R*(+)-tecostidine and *R*(+)-boschniakine from *R*(+)-pulegone.

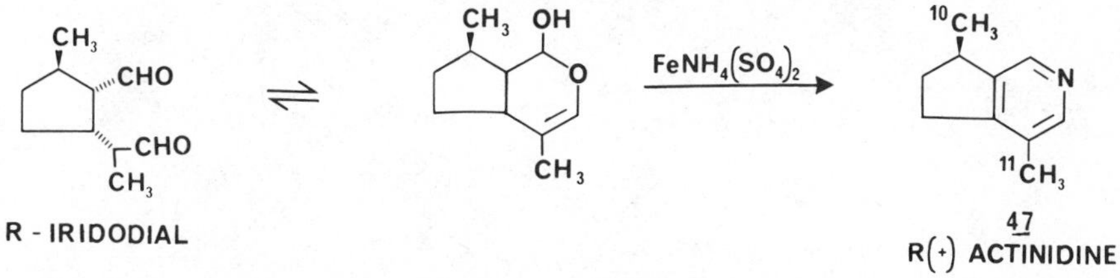

Scheme 24. Conversion of *S*-iridodial into *S*(−)actinidine.

[207] (Scheme 23). Starting with *R*(+)-pulegone, *R*(+)-boschniakine was obtained and proved identical with the natural alkaloid. There were two more related alkaloids isolated by Russian authors from different plants: plantagonine [221], a carboxypyridindane which has not been identified with either boschnialinic acid, nor with its antipode, and an isomer of boschniakine, indicaine [222], the configuration of which is so far unsettled. At the present time the constitution of both compounds has been corrected and ascertained by Torssell [223].

2.43. Onychine

An alkaloid of *Onychopetalum amazonicum*, the structure (50) of onychine has been established by synthesis [224].

3. PIPERIDINE ALKALOIDS

Piperidine itself occurs in many plants together with its methyl homologues: 2,6-*trans*-dimethyl and 2,3,6-trimethyl piperidine. Those are, however, of limited importance.

3.1. *N*-acyl Derivatives of Piperidine (51–60)

Table 2A–C. In addition to piperine, that is, *N*-piperyl piperidine (51), the main alkaloid of this series that was discovered in 1838, several closely related piperidides have been isolated mostly in the 1960s and 1970s (Table 2a,b). The circle of the acylating groups has considerably widened more recently including open-chain polyolefinic and acetylenic carboxylic acids (60a–d) (Table 2C). Variants of the piperidine ring include α-piperidone (57), 5,6-dehydro-α-piperidone (58), and 2,3-dehydropiperidine (59). The plants that produce these piperidides are primarily *Piper* and *Achillea*. We shall discuss some representatives of this class in more detail.

3.1.1. Piperine and Chavicine. Piperine occurs in *Piper nigrum* (black pepper) up to 11% and many other *Piper* spp. Isolated by Regnault [225] in 1838, its hydrolysis to piperidine and an unknown carboxylic acid, named piperic acid,

Table 2. Piperidine Alkaloids

(A)

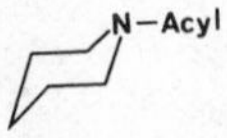

	Acyl		Name	Geometry of double bonds

General Formula

Structure	No.	Name	Geometry
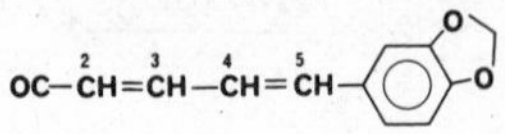 $OC-\overset{2}{C}H=\overset{3}{C}H-\overset{4}{C}H=\overset{5}{C}H$—(methylenedioxyphenyl)	51	PIPERINE	(2E,4E)
	52	CHAVICINE	(2Z,4Z)
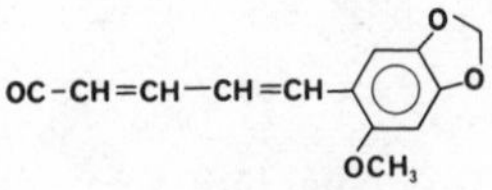 $OC-CH=CH-CH=CH$—(methylenedioxyphenyl, OCH$_3$)	53	WISANINE	(2E,4E)
	54	ISO-WISANINE	(2E,4Z)
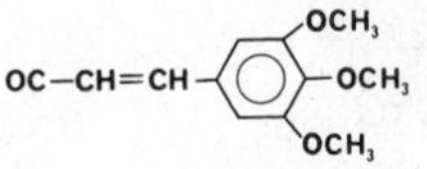 $OC-CH=CH$—(trimethoxyphenyl)	55	PIPLARTINE	(E)
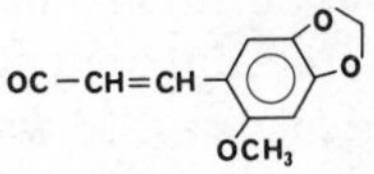 $OC-CH=CH$—(methylenedioxyphenyl, OCH$_3$)	56a	2'-METHOXY-4',5'-METHYLENDIOXYCINNAMYL PIPERIDINE	(Z)
$OC-CH=CH$—(dimethoxyphenyl, dimethylallyl)	56b	2',4'-DIMETHOXY-3'-γ,γ-DIMETHYLALLYLCINNAMYL PIPERIDINE	(E)

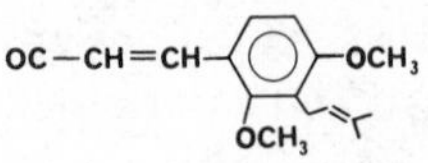

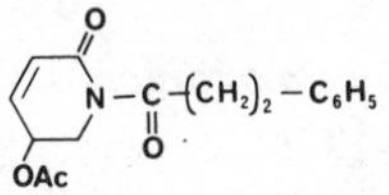

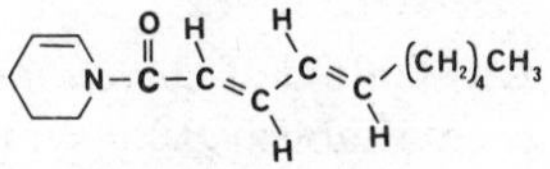

(B)

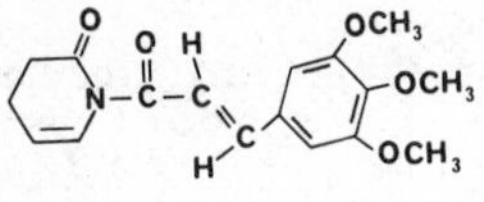

57	58	59
PIPERLONGUMINE	PIPERMETHYSTINE	E,E-DECA-2,4-DIENOYL-Δ^2-DEHYDRO PIPERIDINE

(C)

General Formula

	Acyl Group	Name and Geometry of double bonds
60 a	$OC-CH=CH(CH_2)_6CO(CH_2)_3CH_3$	12-OXO-2(E)-OCTADECENOYL
b	$OC-\overset{2}{C}H\overset{E}{=}\overset{}{C}H-\overset{4}{C}H\overset{E}{=}\overset{5}{C}H(CH_2)_2\overset{6,7}{C}\equiv\overset{8}{C}-\overset{9}{C}\overset{10}{H}\overset{Z}{=}\overset{11}{C}H-\overset{12,13}{(CH_2)_2}\overset{14}{C}H_3$	TETRADECA-2,4(E) 10(Z)-DIENE-8-YNEOYL
c	$OC-\overset{2}{C}H=CH-\overset{4}{C}H=CH(CH_2)_2\overset{8}{C}\equiv C-\overset{10}{C}\equiv C-\overset{12}{C}\equiv CCH_3$	TETRADECA-2,4(E)-DIENE-8,10,12-TRIYNEOYL
d	$OC-CH\overset{E}{=}CH-CH\overset{E}{=}CH(CH_2)_2C\equiv C-C\equiv C-CH\overset{E}{=}CHCH_3$	TETRADECA-2,4,12(E)-TRIENE-8,10-DIYNEOYL
e	$OC-CH\overset{E}{=}CH-CH=CH-CH=CH-C\equiv C-C\equiv C-CH\overset{E}{=}CHCH_3$	TETRADECA-2,4,6,12(E)-TETRAENE-8,10-DIYNEOYL

42

Table 2. (Continued)

(D)

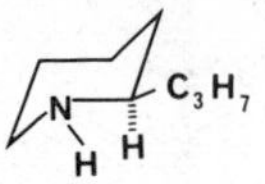

61

R (-) CONIINE

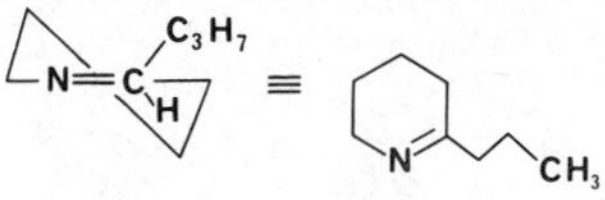

62

γ - CONICEINE

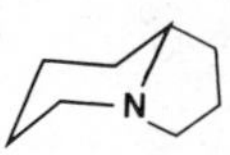

63

δ - CONICEINE*

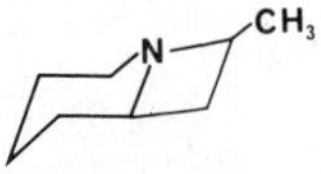

64

ε - CONICEINE*

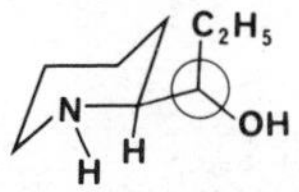

65

2S,1'R(+) CONHYDRINE

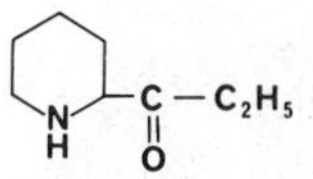

66

S(+) CONHYDRINONE

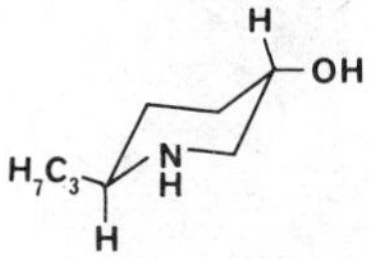

67

PSEUDO - CONHYDRINE

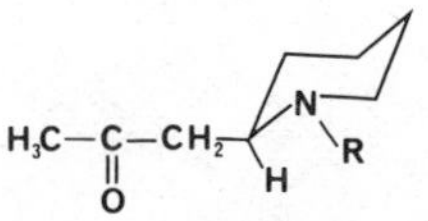

68

a PELLETIERINE
b N-METHYLPELLETIERINE

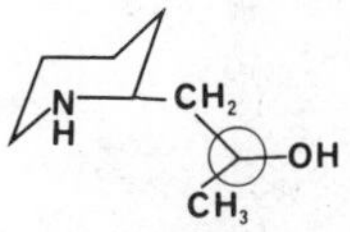

69

2S,2'S(+) SEDRIDINE

(E)

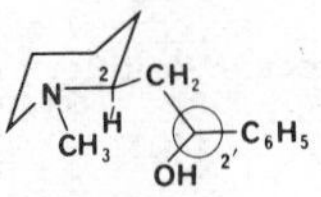

70

2S,2'S (-) SEDAMINE

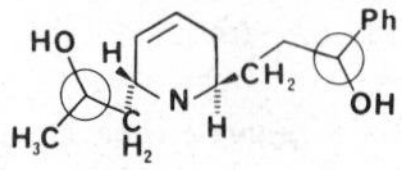

71

SEDININE

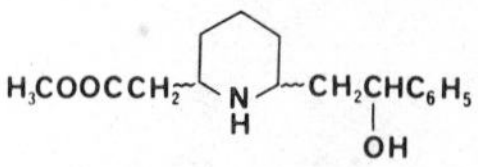

72

SEDERINE

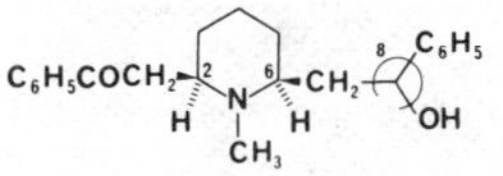

73

2R,6S,8S LOBELINE

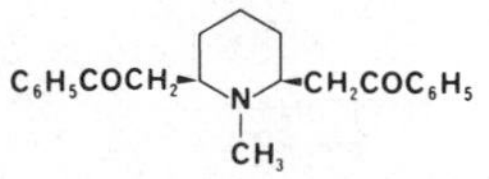

74

(meso) LOBELANINE

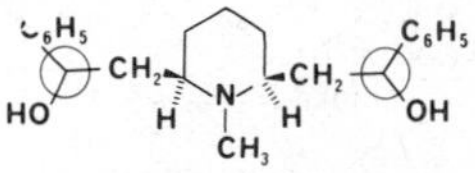

75

(meso) LOBELANIDINE

(F)

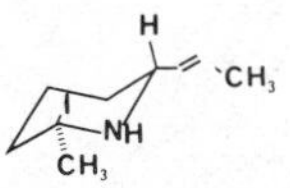

76

2R,6R(-)PINIDINE

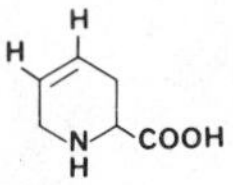

77

BAIKIAIN

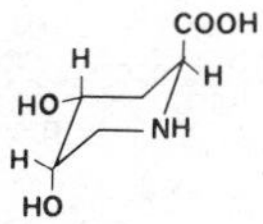

78

(2S,4R,5S)4,5 - DIHYDROXY
L-PIPECOLIC ACID

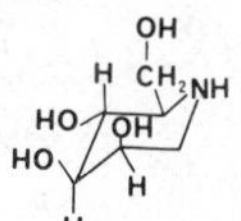

79

1,5 - DIDESOXY - AZA-
D - MANNOPYRANOSE

43

Table 2. (*Continued*)

(F) (*Continued*)

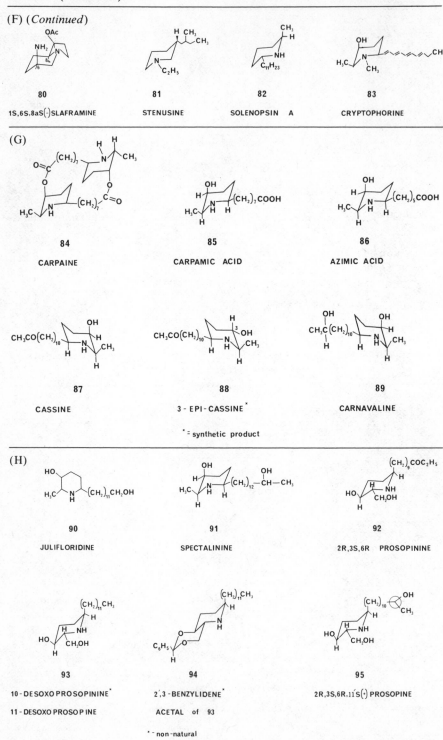

80

1S,6S,8aS(-)SLAFRAMINE

81

STENUSINE

82

SOLENOPSIN A

83

CRYPTOPHORINE

(G)

84

CARPAINE

85

CARPAMIC ACID

86

AZIMIC ACID

87

CASSINE

88

3 - EPI - CASSINE *

89

CARNAVALINE

* = synthetic product

(H)

90

JULIFLORIDINE

91

SPECTALININE

92

2R,3S,6R PROSOPININE

93

10 - DESOXO PROSOPININE *

11 - DESOXO PROSOPINE

94

2´,3 - BENZYLIDENE *

ACETAL of 93

95

2R,3S,6R,11´S(·) PROSOPINE

* - non-natural

44

Table 2. *(Continued)*

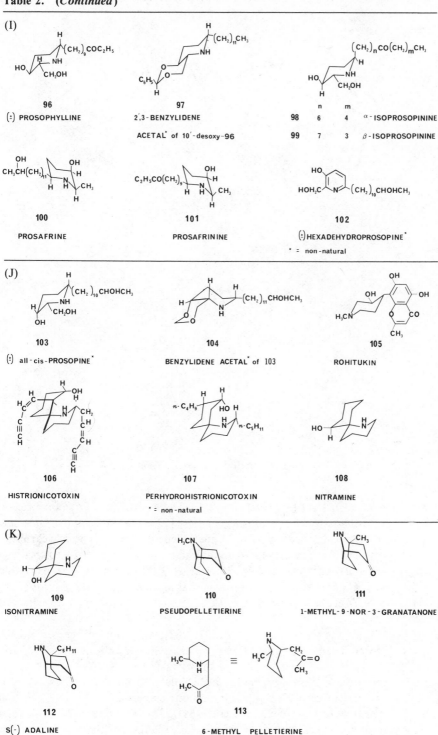

(I)

96
(±) PROSOPHYLLINE

97
2′,3-BENZYLIDENE
ACETAL* of 10′-desoxy-96

	n	m	
98	6	4	α - ISOPROSOPININE
99	7	3	β - ISOPROSOPININE

100
PROSAFRINE

101
PROSAFRININE

102
(±) HEXADEHYDROPROSOPINE*
* = non-natural

(J)

103
(±) all-cis-PROSOPINE*

104
BENZYLIDENE ACETAL* of 103

105
ROHITUKIN

106
HISTRIONICOTOXIN

107
PERHYDROHISTRIONICOTOXIN
* = non-natural

108
NITRAMINE

(K)

109
ISONITRAMINE

110
PSEUDOPELLETIERINE

111
1-METHYL-9-NOR-3-GRANATANONE

112
S(-) ADALINE

113
6-METHYL PELLETIERINE

45

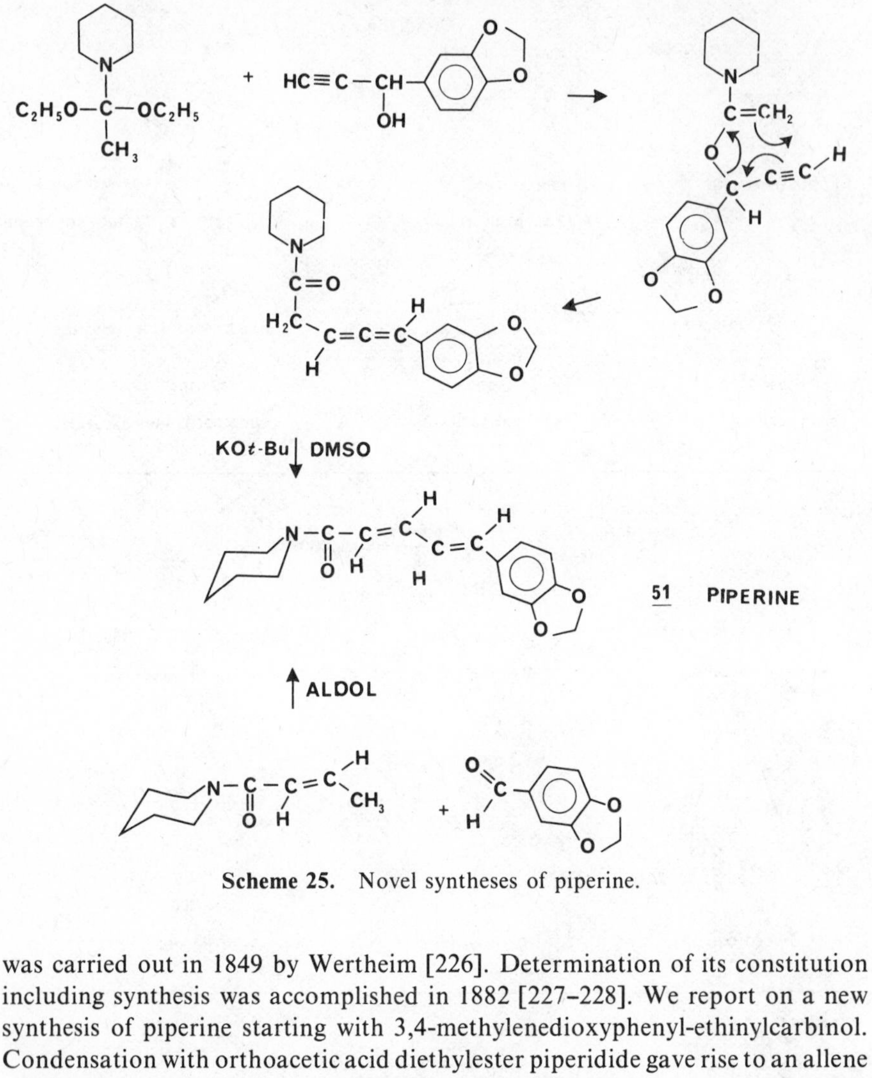

Scheme 25. Novel syntheses of piperine.

was carried out in 1849 by Wertheim [226]. Determination of its constitution including synthesis was accomplished in 1882 [227–228]. We report on a new synthesis of piperine starting with 3,4-methylenedioxyphenyl-ethinylcarbinol. Condensation with orthoacetic acid diethylester piperidide gave rise to an allene which then rearranges into the 63:35 mixture *E,E*-piperine and isochavicine (*E,Z*) piperyl-piperidine [229]. Another new approach [230] involves aldol condensation of *N*-crotonylpiperidine with piperonal by phase transfer catalysis (Scheme 25).

Chavicine is the *Z,Z*-isomer **(52)** [231, 232]. 4,5-Dihydropiperine is also known [233] and originates from *Piper guineense*. It has antimicrobial activity.

3.1.2. Wisanine.   Wisanine **(53)** has been isolated from *P. guineense*. It is the 2′-methoxy derivative of *E,E* piperylpiperidine (piperine) [234]. IR, UV, and NMR spectral studies confirmed structure **(53)** [235]. Its "mixed," 2*Z*,4*E*-isomer **(54)** has been isolated recently from *P. guineense* [236]. 4,5-Dihydrowisanine has been synthesized by the Wittig reaction [237].

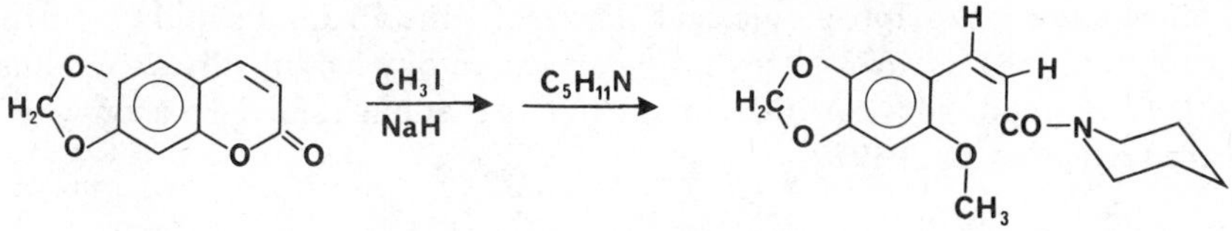

56

Scheme 26. Synthesis of *cis*-2-methoxy-4,5-methylendioxycinnamyl piperidine from 6,7-methylenedioxy coumarine.

3.1.3. Piplartine (55). Stems of *P. longum* contain two alkaloids; upon hydrolysis each gave 3,4,5-trimethoxycinnamic acid. The major alkaloid was piperlongumine, while the minor one was named piplartine [238]. However, another research group claimed [239] that the two alkaloids are identical.

3.1.4. *cis*-2′-Methoxy-4′-5′-methylendioxycinnamyl Piperidine. An alkaloid of *Z* configuration, 2′-methoxy-4′,5′-methylenedioxy-*cis*-cinnamoyl piperidine **(56a)**, was isolated from *P. peepuloides*. This compound has been synthesized by an elegant method [240] from 6,7-methylenedioxycoumarin by sodium hydride and methyl iodide, followed by conversion of the free acid or its ester into the piperidide (Scheme 26).

The *biosynthesis* of all the natural piperidides can be deduced from lysine.

3.1.5. 2,4-Dimethoxy-3-γ,γ-dimethylallyl-*trans*-cinnamic Piperidine. The Euphorbiacea plant, *Excoecharia agallocha L.* contains a piperidine alkaloid. The constitution **(56b)** of the acyl group was determined by NMR spectra and confirmed by synthesis [241]. Osthol, a coumarine derivative, was the starting material which can be converted by a procedure analogous to the one described in Scheme 26 into 2,4-dimethoxy 3-γ,γ-dimethylallyl-*cis*-cinnamic acid. This was isomerized to the *trans*-acid, and the latter via its acid chloride gave with piperidine the alkaloid **56b**.

3.1.6. Piperlongumine. Piperlongumine **(57)**, the major alkaloid of *P. longum* [238, 242], upon hydrolysis gave rise to 3,4,5-trimethoxycinnamic acid and 3,4-dihydro-2-pyridone. Chemically it is a diacylamine that would provide high reactivity in acylation reactions. Little is known, however, about the biological role of **57**.

3.1.7. Pipermethystine. Leaves of *P. methysticum* contain 0.17% of a new alkaloid, $C_{16}H_{17}NO_4$. Mass spectrum indicated the presence of an acetoxy group. The UV spectrum showed the presence of a conjugated system. IR data were characteristic of an amide, ester carbonyl groups, and an olefinic double bond. The PMR spectrum was helpful in finding signals for an α-pyridone ring among others. Hydrogenation over Pd/C resulted in the uptake of 1 mol of hydrogen. Finally, mild hydrolysis gave after separation on TLC 5-acetoxy-5,6-dihydro-2-

pyridone and 3-phenylpropionic acid. Therefore, the alkaloid should have the constitution (58) of N-3-phenylpropanoyl-5-acetoxy-5,6-dihydro-2-pyridone [243]. The compound is easily decomposed and that might be the reason why it escaped isolation until 1979.

3.1.8. E,E-2,4-Decadienoyl-$\Delta^{2(3)}$-dehydropiperidine.

In addition to the large number of polyene and polyacetylene C_{14}- and C_{18}-carboxylic acid piperidides, the Bohlmann group has discovered a 2(3) dehydropiperidide (59) of a simpler 2,4-decadieneoic acid [244]. Hydrolysis should lead to $\Delta^{2(3)}$- or $\Delta^{1(2)}$-piperideine besides the decadienoic acid; however, that would predictably undergo secondary reactions. Therefore, the spectral characteristics, particularly ^{1}H and ^{13}C NMR gave the clue to the structure (59).

3.1.9. Piperidides of Alkenoic and Alkeneynoic Acids (60a–e).

Thirteen *Achillea* species and *Leucocycles firmosus* allowed isolation of a large number of amide-type alkaloids [244, 245]. Twelve of these have been separated by TLC on silica gel–$AgNO_3$. Based on IR, ^{1}H NMR spectra, coupling constants of olefinic protons, double irradiation experiments, and mass spectra, five of the new compounds were identified as piperidides of alkadienoic, alkatrienoic, or alkynoic acids with 1, 2, or 3 acetylenic bonds. The entire structure determination is based on a combination of spectral methods; no chemical degradation, that is, hydrolysis and analysis of the products of hydrolysis, was attempted [245].

3.1.10. Pharmacology of the Piper Alkaloids.

Extracts prepared from *P. longum* have long entertained use in the Indian Ayurvedic system of medicine as an effective drug for the treatment of asthma and chronic bronchitis. The pepper originating from this climbing plant and from other *Piper* species continues to be utilized in India, and worldwide as well, as a spice.

In common with a great number of the piperidine alkaloids, piperine has been found to possess antimicrobial activity. A detailed study has been undertaken on the effects of piperine on the reproductive activity as well as the morphological and cultural properties of *Bacillus anthracis* in comparison with those of capsaicin, the pungent substance found in red pepper, or paprika [246a]. The effects of these two pepper constituents on the biochemical properties and the bound amino acids of *B. anthracis* were subsequently reported [246b].

P. longum, black pepper, and ginger are all used in the formulation of a large number of drug prescriptions in the Ayurvedic system of medicine in India. In some instances these acrids apparently increase the bioavailability of the drug with which they are compounded. *P. longum* was shown to increase the blood levels of the test drug vasicine by nearly 233%, while piperine similarly increased sparteine blood levels more than 100%. There are several possible mechanisms for production by piperine of an increase in bioavailability. The two most obvious include promotion of rapid absorption from the gastrointestinal tract and protection of the primary drug from metabolism or oxidation in its first passage through the liver after absorption.

Within the last decade piperine has been shown to exert central nervous system activity (see Pei [247]). In contrast with most of the piperidine alkaloids, piperine, and some synthetic derivatives as well, exerted anticonvulsant activity in rodents. Protection was afforded against a variety of seizures, including those induced by maximal electroshock, pentylenetretrazol, picrotoxin, strychnine, tubocurarine, and sodium glutamate. Piperine was likewise an effective antiepileptic agent in man.

In general, piperine resembles classical central nervous system depressants in its spectrum of pharmacological activity, although high doses of this alkaloid do not produce anesthesia. The central actions of other depressants are intensified by piperine. The mechanism of action of effective doses of piperine and its derivatives apparently involves central serotonergic pathways. Toxicity stems from central depression, and convulsions are not seen after large doses. One synthetic derivative in particular, antiepilepsirine, has been utilized clinically in China since 1975. This slightly modified piperine molecule appears to have been more effective than piperine in the treatment of a variety of epilepsies over a time span of observation greater than 5 years (see Pei [247]).

The physiological effects of inclusion of either black pepper or piperine in the daily diet of rats have recently been examined [248]. Black pepper at levels of 2 and 5% reduced food intake, presumably because of the pronounced aroma, while 0.05% piperine increased food consumption. At these concentrations, liver weight was increased due to higher total and neutral lipid content. Hematologic values were not adversely affected. Several histological changes, however, were apparent in animals fed either the black pepper or piperine, including loss of taste buds in the tongue, keratinization of the tongue, esophagus, and stomach epithelium, erosion of the gastrointestinal mucosa, and hepatic cellular infiltration. On the other hand, no adverse effects were seen in rats maintained on diets containing either 0.02 or 0.15% black pepper, that is, levels approximating, or 10 times higher than, those consumed daily by humans.

Among the pepper alkaloids other than piperine, wisanine has attracted some pharmacological attention, while 4,5-dihydrowisanine has been investigated for antibacterial activity. A preliminary study [249] showed that wisanine has tranquilizing and sedative effects in mice. In addition, the drug appeared to have some anticonvulsant properties. 4,5-Dihydrowisanine showed a marked growth inhibition of *Streptococcus pyogenes* in a dosage of 128 μg/ml [237]. A similar action was observed with wisanine on gram-positive microorganisms [250].

3.2. Alkyl Piperidines

3.2.1. Coniine.

$R(-)$-Coniine (61) (Table 2D) is an alkaloid of ancient vintage and is a member of the hemlock (*Conium maculatum*) alkaloids. Its structure determination occurred in the conventional way [1]: oxidation to butyric acid, dehyrdogenation with silver acetate to 2-propyl-pyridine, and oxidation of the latter to picolinic acid, that is, pyridine-2-carboxylic acid. The first synthesis, of

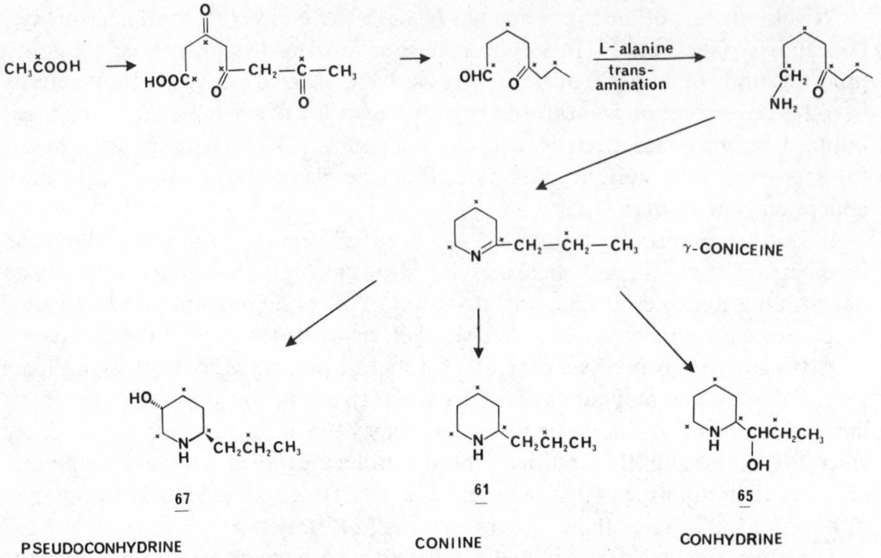

Scheme 27. Biosynthesis of the *Conium* alkaloids β-coniceine, coniine, conhydrine and pseudoconhydrine. Reproduced from reference 431, figure 6, with permission of Springer-Verlag, Heidelberg.

coniine was achieved by Ladenburg in 1886 and was the first total synthesis [251] of an alkaloid. Condensation of α-picoline (2-methylpyridine) with acetaldehyde followed by reduction with sodium in ethanol gave recemic 2-propyl-piperidine which has been resolved to coniine. Nature produces both the (+)- and (−)-modification. The *S* configuration was determined for dextrorotatory coniine 46 years later [252] by oxidation to *R*(−)-pipecolic acid. The *biosynthesis* of coniine occurs via the acetate route [253, 254]. The radiocarbon of 1-^{14}C-labeled acetic acid is incorporated in alternate carbons. The immediate precursor is most likely a triketone derivative that reacts with L-alanine through transamination [255, 256] to give rise to γ-coniceine **(62)** and this by reduction furnishes coniine (Scheme 27).

3.2.2. Coniceines. Coniine is always accompanied by γ-*coniceine* [1], which contains two hydrogens less and can be easily saturated *in vitro* to racemic coniine. It is a pro-chiral compound because on reduction it gives (±)-coniine. γ-Coniceine **(62)** and coniine **(61)** seem to form a balance in the biosynthesis during the different growth periods of the plant [265]. There are two additional derivatives of coniine and coniceine that do not occur in the plant and have no functional group other than the basic nitrogen. One was named δ-*coniceine*, the structure of which was determined by classical means [1] to be 1,2-pyrrolidino-1,2-piperidine **(63)**. Diels–Alder reaction of pyridine with 2 moles of methylacetylene dicarboxylate gave a tetracarbomethoxyquinolizidine and this upon ring contraction, decarboxylation, and hydrogenation led to γ-coniceine [257].

Cyanogen bromide degradation followed by dehalogenation gave (±)-coniine. The other isomer is ε-*coniceine* **(64)** which is formed upon the action of base on 2-(2'-bromopropyl)- [1] or (2'-mesyloxypropyl)-piperidine [258].

3.2.3. Pharmacology of Coniine and γ-Coniceine.

In light of the poisonous nature of *Conium maculatum* (hemlock), both the plant and the isolated alkaloids have commanded considerable attention over the centuries. The juice from the hemlock plant was a primary constituent of the deadly concoctions used by the ancient Greeks in executing criminals, and it was presumably an extract of hemlock that was used to cause the death of Socrates around 400 B.C. The major alkaloid of hemlock is coniine.

The well-known neuromuscular blocking effects of coniine were first described by Kolliker [259], who utilized frog preparations. Similar observations in frogs were subsequently made by Cushny [260] and by Moore and Row [261]. Although Langley and Dickenson [262] were unable to detect any effect of coniine on ganglionic transmission, the classical initial stimulation and subsequent blockade of autonomic ganglia produced by coniine were demonstrated by Cushny in 1899 [263].

Interestingly enough, Conium, or poison hemlock (U.S.P. 1831–1863, 1873–1916; N.F. 1916–1936), appeared in either the *U.S. Pharmacopoeia* or the *National Formulary* as an official drug for over a century. This plant preparation was assigned a variety of medicinal uses, including use as a sedative, a narcotic, an anodyne, an antispasmodic, and as an anaphrodisiac [264]. Its usage was also recommended for more serious conditions, including whooping cough, melancholia, neuralgia, delirium tremens, tetanus, asthma, and epilepsy. Around 1936, Conium was no longer officially recognized as a drug and rapidly fell into disuse. A major reason for its discontinuance was apparently the great variation in the potency of different preparations. Subsequently the alkaloid content and composition of *C. maculatum* was demonstrated to differ greatly, depending on the climatic conditions and even on the time of day of collection of the plants [265].

Recently coniine was found to be present in a volatile fraction of *Sarracenia flava* (pitcher plant), a plant known to be insectivorous [266]. After administration of the isolated coniine to fire ants in nanogram quantities, this alkaloid was shown to be an insect paralyzing component of the pitcher plant.

Since the pioneering work of Cushny [263] peripheral actions of coniine on smooth muscle have been shown to be due to initial ganglionic stimulation and secondary blockade in a variety of species, including mice, rats, guinea pigs, and cats [267–269]. Neuromuscular blockade produced by coniine was likewise shown to be preceded by initial stimulation [268].

The central effects of coniine still remain unclear to this day. This alkaloid definitively produces blockade of spinal reflexes by an action in the spinal cord [268–270], and increased membrane permeability to potassium ions has been suggested as the mechanism [271]. Strychnine-like excitatory effects on the cord have additionally been observed by some groups [267, 270] but not by others [268]. Because both coniine and strychnine blocked postsynaptic inhibition and depressed neuromuscular and ganglionic transmission, as does nicotine, it has been suggested that these two alkaloids act by a common mechanism involving a cholinergic link [270]. Experimental evidence, however, does not support this viewpoint; cholinergic antagonists failed to reduce direct postsynaptic inhibition in the spinal cord, and the postsynaptic inhibitory transmitter glycine, which is antagonized by strychnine, was not affected by coniine (see Curtis and Johnston [272]).

The actions of several hemlock alkaloids other than coniine have been examined to some extent. The alkaloid of the smallest content in the plant, that is, conhydrine, is the least potent with regard to toxicity [273]. γ-Coniceine is about 6 to 8 times more potent than coniine, while N-methylconiine is about one-half as potent by most routes of administration [268]. The toxic symptoms of these alkaloids are quite similar, and only slight differences are apparent with regard to pharmacological activity. While γ-coniceine has more pronounced stimulatory action on autonomic ganglia than coniine, N-methylconiine had a more predominant blocking effect. At the same time N-methylconiine was more effective than coniine in stimulating skeletal muscle in mammals, but considerably less effective in stimulating autonomic ganglia.

Human poisoning from accidental ingestion of hemlock still occurs occasionally in the United States [44]. The plant was originally introduced from Europe and currently grows as a weed in marshy areas and roadside ditches in the eastern part of the country, the Rocky Mountains, the Pacific coast, and southern Canada. Poison victims have usually mistaken either the leaves for parsley or the seeds for anise. Fortunately, the taste of both plants parts is highly unpleasant, and fatal amounts are rarely consumed. Signs of toxicity consist initially of vomiting, diarrhea, muscular weakness, nervousness, trembling, pupillary dilation, low pulse rate, and coldness of the extremities of the entire body. If the syndrome progresses, clonic and tonic contractions of separate limbs as well as convulsions of the entire body ensue; death occurs via respiratory paralysis.

The description of the demise of Socrates provided by Plato (see Church [274]) encapsulates a quite accurate version of the toxic effects of hemlock. Several inconsistencies, however, including failure to lose the ability to talk early on and the absence of convulsions, prompted a pharmacological analysis of the effects of coniine and opium in combination in rodents [267]. The conclusion reached by this study was that the lethal potion imbibed by Socrates actually consisted of a mixture of hemlock and opium.

The toxic syndrome occasionally occurring in livestock happening to graze on wild hemlock closely resembles that observed in man [43]. The plant is usually quite distasteful to animals, probably not so much on account of the taste but because of the characteristic odor, which resembles that of mouse urine. The hemlock alkaloid conferring this property to the plant is coniceine [275].

Probably more commonly observed than overt toxicity in livestock grazing on

C. *maculatum* is the production of teratogenic effects [276, 277]. Both the plant and coniine itself produced congenital defects in calves after maternal ingestion [277]. Maternal inhalation of either the plant or coniine has been shown to be insufficient for defects to occur. Structure-activity relationship studies undertaken at the same time indicated that both chain length and degree of unsaturation in the ring are critical for the production of teratogenicity. Overall, the results encouragingly suggested that, of the host of naturally occurring pyridines and piperidines, those possessing either a fully unsaturated ring or a chain length shorter than propyl are quite unlikely to pose a teratogenic threat to grazing livestock.

3.3. Piperidine Alcohols and Ketones

3.3.1. Conhydrine (65). After early degradation experiments [278] the two-dimensional structure of conhydrine has been determined by Hofmann degradation followed by ring opening of the epoxide and saturation to an octane-3,4-diol [279]. This octanediol proved to have an *erythro*-configuration, and since its formation involved two inversions, the original compound conhydrine must also have the *erythro*-relative configuration [280]. The synthesis of conhydrine has been carried out from 2-pyridyl ethylketone by two-step reduction: hydrogenation of the aromatic ring and reduction of the carbonyl function [281]. By a further, more stereoselective method 2-pyridyl ethylketone has been converted into 2-pyridyl-ethylcarbinol that has been resolved and the individual antipodes hydrogenated. Due to a significant asymmetric induction the two diastereo-isomers conhydrine and epiconhydrine are formed in unequal amounts [282]. Concerning the absolute configuration, the correlation of the piperidine carbon atom attached to the side chain was determined by Willstätter as early as 1901 by oxidation to $S(+)$-pipecolic acid [278]. Since the *erythro*-relative configuration of conhydrine was demonstrated before [280], there was indirect evidence for the $2S, 1'R$ configuration. A more direct approach, conversion of optically active conhydrine to $R(+)$-3-octanol, has been carried out more recently [282] that confirmed the *erythro*-configuration as well as the absolute configuration of the alkaloid. The biosynthesis of this compound [254] proceeded through a similar pattern as that for coniine, which led from 1-^{14}C-labeled acetic acid to alternately labeled conhydrine. The oxygen of **65** does not originate from the carboxyl group of acetic acid. This can be rationalized in two different ways. Either reduction of γ-coniceine to coniine took place, followed by oxidation at $C(1')$ of the side chain, or, alternatively shift of the 1,2 double bond to the $(2,1')$dehydroconiine occurred with the latter subsequently being hydrated to conhydrine.

3.3.2. (±)-Conhydrinone. This compound **(66)** was known for a long time as a product of oxidation of conhydrine, but it was isolated [283] from hemlock only recently. The total synthesis [283] involved catalytic reduction of the ring in

2-pyridyl-ethyl-ketone ethylenecycloketal, over Adams-Pt catalyst, followed by acid hydrolysis of the ketal.

3.3.3. (2S,5S) (+)-Pseudoconhydrine.

This hemlock alkaloid (67) was deoxygenated to S(+)-coniine with hydriodic acid and phosphorus. On the other hand, Hofmann methylation results in an opoxide which upon ring opening gives an optically active 1,2-octanediol [284].

The unbranched carbon skeleton was thus proven. The absolute stereochemistry of the C(2) atom was proven by its conversion to R(−)-pipecolic acid, and by oxidation to (+)-norleucine [285]. Destruction of the skeleton except for chirality at C(5), led to (+)-dimethylamino-2-octanol. Hofmann elimination of the latter gave an epoxide which on reduction with lithium aluminum hydride gave S(+)-2-octanol [286]. This result provided evidence for both the relative (*trans*)-configuration of the chiral centers and the absolute configuration as 2S, 5S (67). Among several syntheses of pseudoconhydrine, we refer to the most recent one: intramolecular amidomercuration [287] of an olefin. The intermediates are 2-acetoxymercurimethyl and 2-iodomethylpyrrolidine. Ring enlargement takes place via the 1,2-pyrrolido-1,2-aziridine. Hydrolysis leads to mixture of two racemates of 5-hydroxy-2-propylpiperidine (Scheme 28). The biosynthesis of pseudoconhydrine follows the acetate route via γ-coniceine, which is characteristic of the *Conium* alkaloids [255], but the oxygen at C(5) must be the result of a secondary transformation (cf. Scheme 27).

3.3.4. Pelletierine.

The roots of *Punica granatum L.* (pomegranate) contains several alkaloids, four of which (pelletierine, isopelletierine, methylisopelletierine, and pseudopelletierine) were isolated at an early time by Tanret [288].

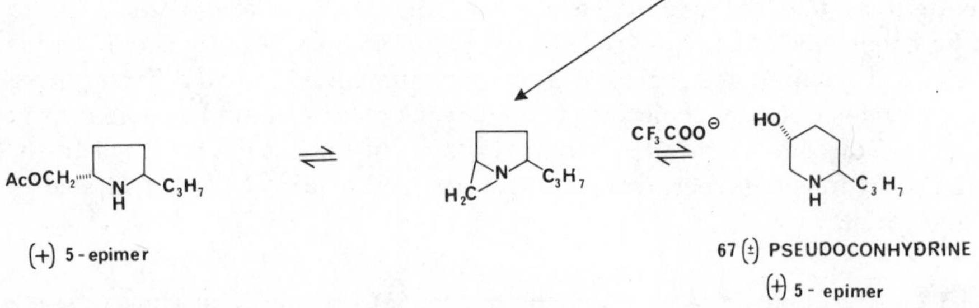

Scheme 28. Novel Synthesis of pseudoconhydrine via intramolecular amidomercuration.

Pelletierine sulfate was crystalline and optically active ($[\alpha]_D - 30°$), and methylisopelletierine was active also, while isopelletierine was inactive. Pseudo-pelletierine belongs to a different class of compounds; it is mesoid and will be discussed separately. Hess and Eichel [289, 291] found only optically inactive bases in *Pomegranate;* one of those was named "isopelletierine". By resolving the isopelletierine, Hess and Eichel [290] found much lower optical rotation for the (−)-form (5.33°) than Tanret did for pelletierine and first reported racemization thereof. The German authors believed that their pelletierine was 3-(2-piperidyl-propanal) while Tanret's active compound was isopelletierine. Unfortunately, seven attempts to synthesize that aldehyde have all failed. Finally, in 1961 Marion [292] shed light on this problem by a polarimetric, NMR, and IR spectral investigation of Tanret's original pelletierine sample. It proved to be optically active $R(-)$-(2-piperidyl)-2-propanone (**68**, R = H). "Methylisopelletierine" was correlated by the von Braun cyanogen bromide reaction with pelletierine, and therefore it has structure **68b**. Hess' isopelletierine was obviously racemic pelletierine, and the assumed 3(2-piperidyl)-propanal does not exist. In consequence, Marion [292] proposed that the name "isopelletierine" be eliminated from the literature. A new, biomimetic synthesis [293] of pelletierine (Scheme 29) starts with a Michael addition of ethyl sodioacetoacetate upon Δ^1-piperideine, followed by hydrolysis and decarboxylation. Labeled pelletierine forms in the *pomegranate* plant from labeled L-lysine via Δ^1-piperideine and acetoacetate [294]. The same route applies to *N*-methyl pelletierine (**31**; R = CH₃) [295–297]. Pelletierine is believed to be the biogenetic precursor of the complex *Lycopodium* alkaloids [298]. Pseudopelletierine as a bicyclic piperidine will be dealt with later in this chapter.

3.3.5. Pharmacology of the Pelletierines. The pelletierine alkaloid complex, as well as the parent plant, is one of the oldest of drugs employed in therapeutics. Medicinal usage of pomegranate roots for the treatment of roundworm was included in the famous Ebers papyrus, a historical document detailing variegated remedies considered useful to the inhabitants of Egypt around 1550 B.C. [299]. The pomegranate plant itself is thought by some to be the tree of life which flourished in Eden, and there are many biblical references to the botanical entity [300].

Pomegranate fruit, or *Granatum* (U.S.P. 1820–1842), was incorporated in the *U.S. Pharmacopoeia* as an officially recognized drug in 1820, the year of origin of this important publication. The pulp of the pomegranate fruit was used

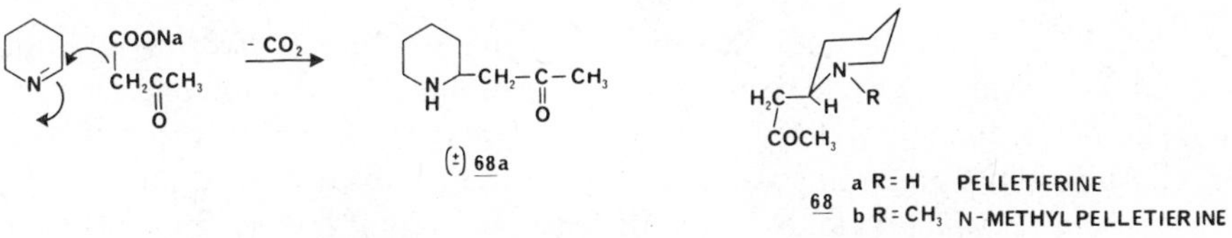

Scheme 29.   Biomimetic synthesis of pelletierine.

medicinally at that time as an acidulous refrigerant [301]. Several preparations of the stem and root barks were included in the *Pharmacopoeia* shortly thereafter (1831; 1842). The therapeutic indication for these parts of the plant resided in their historically recognized anthelmintic properties. Pelletierine tannate (U.S.P. 1905–1947; N.F. 1947–1950) was added to the Pharmacopoeia in 1905, only 20 years after the initial discovery of this mixture of alkaloids. Pelletierine, which is also known as punicine, received its official nomenclature in honor of Pierre Joseph Pelletier, the French chemist and pharmacist renowned for his isolation of a variety of plant alkaloids, among the most important of which are colchicine, strychnine, quinine, and caffeine.

Pelletierine was regarded by its discoverer, Tanret, as the anthelmintic constituent of pomegranate [288]. The accuracy of this initial assessment was demonstrated early on. The officially recognized drug, that is, pelletierine tannate, is actually a mixture, in varying proportions, of the tannates of four alkaloids: (−)-pelletierine, isopelletierine [(+)-pelletierine], methylpelletierine, and pseudopelletierine. The alkaloid content was equivalent to not less than 20% as the hydrochloride. The administered dose was formerly quite high, ranging from 500 mg to 1.6 g [264]. It was subsequently recognized that doses around 200–300 mg were just as effective.

Pelletierine is one of a subgroup of anthelmintics that acts locally within the gastrointestinal tract to eradicate worms. The drug essentially causes a marked increase in contraction of the smooth muscle of the parasite, and the result is relaxation of the invader's hold on the tract wall. Pelletierine is particularly effective against tapeworms, although *Ascaris* has been shown to be affected as well. In the actual treatment of tapeworms with pelletierine [302], the gastrointestinal tract was initially cleared by restriction of the diet to liquids and by the administration of cathartics. Pelletierine was then administered; 1–2 h later, a cathartic was again given. The tapeworm was delivered in a tub of tepid water, and finally, it was ascertained that the head of the parasite was intact in order to rule out the possibility of renewed tapeworm growth in the gastrointestinal tract.

If systemic absorption of pelletierine from the gastrointestinal tract occurs, toxicity ensues. The presence of residual fatty food in the tract enhances absorption. Some absorption apparently does occur even after administration of the 300-mg dose, as symptoms such as dizziness, twilight vision, tingling or numbness, and muscular twitchings were not uncommon [302]. After further absorption of pelletierine, muscle cramps become severe, and convulsions, although not generalized, may appear. As more effective and less toxic taeniacides are currently available, pelletierine is rarely used as an anthelmintic today.

3.3.6.   2S,2'S(+)-Sedridine.   *Sedum acre* contains the 2-(2-hydroxyalkyl)piperidines: sedridine **(69)**, sedamine **(70)**, and sedinine **(71)** as major, and sederine **(72)** as minor, alkaloids [304] (Table 2E). We treat the first two representatives as monosubstituted piperidines separately from the 2,6-disubstituted piperidines, that is, the *Lobelia* alkaloids. From the chemical standpoint sedinine **(71)** and sederine **(72)** belong to the latter group.

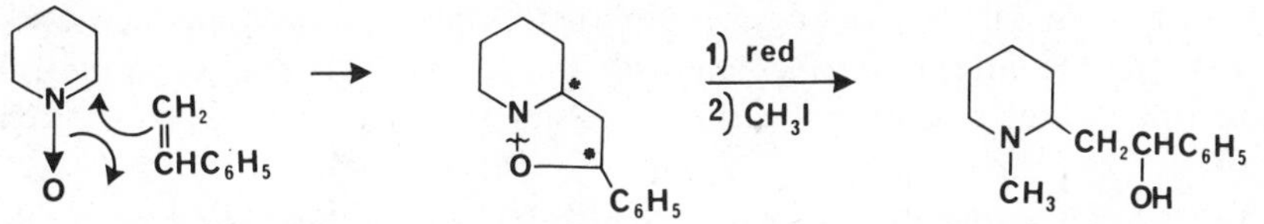

Scheme 30. Nitrone route to sedamine from Δ^1-piperideine N-oxide

(+)-Sedridine is (2S:2'S) 2-(2'-hydroxypropyl)piperidine **(69)**. Among three known syntheses, one is highly stereoselective: 2-picolyl-methylcarbinol was resolved and the (+)-antipode hydrogenated to (+)-sedridine [258]. The absolute configuration of (+)-sedridine was proven by correlations: on the one hand, oxidation to S(+)pipecolic acid [305], and on the other, by von Braun reaction of N,O-dibenzoyl sedridine to S(+)-2-octanol [306]. (−)-Allosedridine was converted into a p-nitrophenyl-1,3-oxazine, and the absolute configuration of the latter was determined by X-ray crystallography [307] independently of a ORD-CD study that led to the same conclusions [308].

3.3.7. (−)-Sedamine. This alkaloid is 2S:2'S (−)-1-methyl-2-(2'-hydroxy-2'-phenylethyl)-piperidine **(70)** [309]. The racemate, together with the 2-allo-diastereomer, is formed in a 1:3.5 ratio in a recent synthesis [310] from 2-piperideine-N-oxide and styrene via the isoxazolidine, followed by cleavage of the N—O bond by reduction and N-methylation (Scheme 30). Separation of the racemates, followed by their resolution, gave the four optically active stereoisomers. (−)-Allosedamine was correlated with (−)-N-methylpipecolic acid; thus, (−)-sedamine has the 2S configuration [311]. The 2'-hydroxyl configuration was assigned by correlation with (+)-2-hydroxy-2-phenylpropionic acid [312].

3.4. Di-, Tri-, and Tetra-substituted Piperidines

3.4.1. Sedinine. This alkaloid from *Sedum* species proved to be, by X-ray crystallography [312], $\Delta^{3(4)}$-*trans*-2-(2-hydroxypropyl)-6-(2-hydroxy-2-phenyl-ethyl)-piperidine **(71)**. The correct absolute configuration of all four chiral centers was also determined by X-ray crystallography.

3.4.2. Sederine. Sederine, a minor base of *S. acre*, is a 2-methoxycar-bonylmethyl-6-(2'hydroxy-2'-phenyl-ethyl)-piperdine **(72)** of yet uncertain stereochemistry [313]. The biosynthesis of all four Sedum alkaloids is from L-lysine. The side-chain of sedamine proved to be derived from phenylalanine [314].

3.4.3. Pharmacology of the Sedum Alkaloids. Extracts from *S. acre* were utilized in Europe in the past as one of the popular herbal medicines of the times. Ascribed effects include lowering of blood pressure, gastric stimulation, antiscorbutic activity, activity against diphtheria, and local anesthetic activity. In early studies the latter two effects were attributed to the alkaloid content of the plant, whereas the antiscorbutic effect was due to other chemical constituents [315].

A variety of *Sedum* species have been utilized worldwide as salad vegetables. *S. acre* appears not to have been so employed, presumably because of its strong bitter taste [316].

Because of the close chemical similarity between sedridine and isopelletierine, sedridine and a great number of synthetic derivatives have been extensively examined for anthelmintic activity [317]. The synthetic compounds were designed to contain either a piperidine or pyridine nucleus. Other structural changes included methylation of the ring nitrogen, replacement of the hydroxy hydrogen in the side chain by alkyl groups ranging from one to five carbons in length, and addition of alkyl groups of the same lengths to the 2-position of the side chain. The majority of the derivatives found effective against tapeworms in the mouse were only weakly active. These compounds contained either a piperidine or pyridine nucleus with either a hydrogen or methyl group on both the ring nitrogen and the ether function. Substituents on the 2-position of the side chain of these derivatives could be either a propyl or a pentyl group. One derivative, that is, 2-(β-methoxyethyl)-pyridine, did exert a strong vermicidal effect in both the mouse and the rat, but this additional finding did not permit formulation of any solid structure-activity relationships.

In a comparative study two of the effects attributed to extracts of *S. acre*, that is, gastric stimulation and lowering of blood pressure, could not be specifically ascribed to the alkaloid content of the plant with regard to sedridine, sedamine, and sedinine [318]. While sedinine stimulated atropine-reversible contraction of the isolated rabbit intestine, both sedamine and sedridine depressed spontaneous motility. Of the three alkaloids sedridine did on occasion evoke a rapid short-lived fall in blood pressure accompanied by transient suppression of respiration; this effect, however, was not always reproducible.

While one fraction of the *S. acre* extract utilized in the latter study [318] produced sedation in mice, all three of the pure *S. acre* alkaloids produced central nervous system stimulation. The effect of sedamine was weak and short-lived. Both sedridine (69) and sedinine (71) were considerably more active, and toxicity associated with these two alkaloids was manifested as convulsions.

Several other pharmacological properties of the major *Sedum* alkaloids were discerned in the above study [318]. Both sedamine and sedridine, like the veratrum alkaloids, increased the sensitivity of the isolated frog rectus muscle to potassium ions. In contrast, sedinine counteracted potassium-induced contractions. All three alkaloids exerted a negative inotropic effect in the isolated frog heart. In spite of the potassium sensitization observed in the rectus muscle preparation, neither sedamine nor sedridine demonstrated digitalis-like activity. Both sedamine and sedinine produced neuromuscular blockade in the isolated rat phrenic nerve–diaphragm preparation, while sedridine was without effect. In contrast, sedridine, and sedamine as well, stimulated respiration in intact animals by a central action.

In more recent studies the pharmacological effects of a novel group of synthetic derivatives of sedridine have been examined [319]. These derivatives contained either a piperidine or pyridine nucleus, and the hydroxy group at position 2 of the aminopropyl side chain was replaced by aminoalkyl substituents of varying lengths. As with the earlier synthesized sedridine derivatives [317], anthelmintic and antimicrobial activity was poor. Some of the newer derivatives, however, did possess an interesting array of central nervous system activity. Effects resembling those of amphetamine, including excitation, shortening of barbiturate sleeping time, reversal of the sympatholytic effects of tetrabenazine, and anorexia, were readily apparent. In contrast with amphetamine, stimulation of locomotor activity was only slight. Antidepressant activity was suggested inasmuch as the pressor and myocardial effects of catecholamines were potentiated. Because body temperature was elevated, an effect also produced by hallucinogens, psychotomimetic activity was likewise inferred. Peripheral adrenergic effects produced by some of these derivatives of sedridine included mild bronchodilation, elevation of blood pressure, and cutaneous vasoconstriction.

Structure-activity relationships for the above compounds were readily discernible. Derivatives with a pyridine nucleus and a small alkyl substituent on the amino group at position 2 of the side chain possessed the greatest degree of activity. While increasing the size of the aminoalkyl group progressively reduced the degree of stimulatory effects, substitution of cyclohexyl or benzyl groups abolished activity. Methylpyridinium compounds likewise lacked central stimulatory activity, but, with small amino alkyl groups, weak effects on circulation were still apparent. In general, derivatives containing a piperidine nucleus and an aminopropyl side chain exhibited neither central stimulation nor peripheral autonomic effects regardless of the size of the aminoalkyl group on position 2 of the side chain. On the other hand, introduction of an aminobenzyl substituent on position 2 of an aminoethyl side chain of a piperidine nucleus did confer mild central stimulatory activity.

3.4.4. Lobelia Alkaloids: Lobeline, Lobelanine, and Lobelanidine. *Lobelia inflata* was shown by H. Wieland and his associates to contain several important alkaloids [320]. The major alkaloid is lobeline (73). In addition, lobelanine (74), and lobelanidine (75) were isolated at an early date. Lobeline ($C_{22}H_{27}NO_2$) is an optically active tertiary base that forms a monobenzoyl derivative. Therefore, one of the two oxygen atoms is present in a hydroxyl group [320]. Heating to 125° C with dilute hydrochloric acid gave rise to acetophenone and methylamine.

Lobeline, upon reduction with sodium amalgam, takes up 1 mol of hydrogen. Lobelanine, a related alkaloid ($C_{22}H_{25}NO_2$), under the same conditions consumes 2 mol of hydrogens and in both cases the product is lobelanidine ($C_{22}H_{29}NO_2$). This proved that all three alkaloids are very closely related. Never-

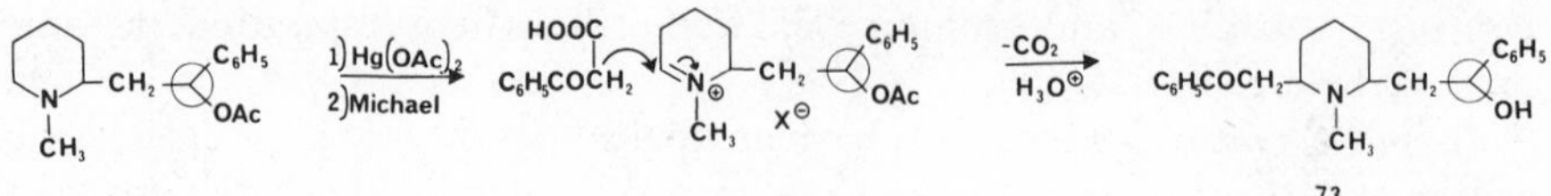

70

(-) SEDAMINE
ACETATE

73

(-) LOBELINE

Scheme 31.   Stereospecific conversion of sedamine into lobeline.

theless, Wieland ascribed [321] a wrong structure to lobeline. Later both lobeline and lobelanidine were converted by chromic oxide to lobelanine.

Lobelanine is a diketone, which on Hofmann elimination gave a diene, and this on hydrogenation led to 1,7-dibenzoyl heptane. The latter was oxidized to benzoic and benzoylheptanoic acids [322]. These facts were indicative of a 2,6-*bis*-phenacylpiperidine structure (74) for lobelanine and of its dihydro derivative (73) for lobeline [322]. Hence, lobelanidine must be the tetrahydro derivative (75) of lobelanine.

Lobelanine was synthesized in 90% yield by Robinson-type biomimetic synthesis by Schöpf [323], using methylamine, glutaraldehyde, and benzoylacetic acid at pH 4.

The stereochemical relationship of (−)-lobeline to lobelanine was demonstrated by the fact that the diketone was an optically inactive, mesoid compound. Therefore, the two phenacyl groups must be *cis* in 74, and accordingly the phenacyl and the dihydrophenacyl groups must also be *cis* in lobeline. Lobelanidine, although it is a *bis*-secondary alcohol, cannot be resolved. Given the *cis*-position of the two-dihydrophenacyl groups, the two chiral centers should be antipodal, that is, lobelanidine is a mesoid modification.

The absolute configuration of (−)-lobeline was established by Schöpf [324] by its synthesis from 2S,2S-(−)-sedamine (70) (Scheme 31). Lobeline undergoes mutarotation. This can be explained in terms of enolization leading to epimerization at the chiral ring C(6), close to the carbonyl group [325].

Biosynthesis: The label of 2-^{14}C-lysine was incorporated in both C(2) and C(6) of both lobelanine and lobeline [326, 327], indicating a symmetrical precursor.

3.4.5.   Pharmacology of Lobeline.   Indian tobacco, the common term for *Lobelia inflata*, derived its name as a result of observations made by American settlers of its use by native Indians. This source for lobeline was apparently employed by the Indians as a substitute for authentic tobacco only when required by necessity.

Lobelia, or Indian tobacco (U.S.P. 1820–1936; N.F. 1936–1953), was introduced into medicine in 1807. One primary early use of extracts of the plant was to induce emesis, and in the nineteenth century this was one of the most important of the available emetics [303]. By 1917, however, it was recognized that the dose required to induce emesis was too often associated with mortality, and this usage was contraindicated [264]. Remaining recommended uses of *Lobelia* during that era included as a remedy for spasmodic asthma, catarrh, bronchial spasms, and whooping cough after oral administration, for strangulated hernia and constipation as an enema, and for poison ivy externally.

In the early 1920s other therapeutic indications for *Lobelia* extracts originated in recognition of the ability of the active components to stimulate respiration. A variety of clinical conditions accompanied primarily by respiratory impairment were purported to be improved, including narcotic poisoning, coal gas asphyxia, pneumonia, and respiratory failure secondary to anesthesia [328, 329]. It soon became recognized, however, that doses significantly improving respiration were accompanied by a diversity of adverse effects [330, 331], and utilization of *Lobelia* as an analeptic was promptly discontinued.

Early after the isolation of lobeline from *Lobelia* plants, the similarity between the pharmacological actions of this alkaloid and those of nicotine was recognized [332]. Other alkaloids of *Lobelia*, including lobelanine, lobelanidine, and lobelidine, exerted similar but weaker effects [333].

Both lobeline and nicotine are classified pharmacologically as ganglionic stimulants. These alkaloids produce a wide array of pharmacological effects via initial stimulation of ganglionic nicotinic receptors and subsequent ganglionic blockade. Therapeutically, this class of drugs has little use. These agents retain importance, however, as research tools for both basic pharmacological investigations at the receptor level and more specific toxicologic studies on adverse effects of cigarette smoking.

Ganglionic stimulation by lobeline, and by nicotine, results from combination of the alkaloids with classical nicotinic cholinergic receptors localized on the postsynaptic membranes. Membrane depolarization and generation of an excitatory postsynaptic action potential that activates the effector organ are the end result. Classical nicotinic ganglionic blockade is comprised of two phases. The block originates from persistent depolarization of the ganglion cell, and during this period there is initially no response to any administered ganglionic stimulant and subsequently activity only in response to noncholinergic stimulants. In the second phase, after repolarization, the actions of nicotinic agonists remain blocked, but responses to muscarinic agents return. In the case of lobeline the initial depolarizing block is only minimal with regard to both degree and duration [334]; the actions of this alkaloid otherwise resemble those of nicotine.

With regard to the cardiovascular system, the effects of nicotine arise from catecholamine release from both the adrenal medulla (predominantly epinephrine) and adrenergic nerve terminals (predominantly norepinephrine). In the case of lobeline the majority of released catecholamines originates from the nerve endings, with very little being contributed by the adrenal medulla [334]. The end result after administration of either alkaloid, however, is the same: positive inotropic and chronotropic effects on the myocardium and increases in both cardiac output and systolic as well as diastolic blood pressure.

The respiratory effects of nicotine and lobeline are qualitatively similar. The stimulatory effects observed after low doses are due to activation of chemore-ceptors in the aortic arch and carotid bodies. Higher doses stimulate the central respiratory center directly. Toxic doses produce respiratory paralysis by both direct depression of the respiratory center and peripheral action at the neuromuscular junction of the respiratory muscles.

Aside from those related to respiration, several pharmacological properties common to both lobeline and nicotine arise from action within the central nervous system. The nausea and emesis noted in response to *Lobelia* in its early history were readily found to be medullary in origin [332]. After the administration of large doses of lobeline, central stimulatory effects result in the production of tremors and convulsions [335]. In this regard, it has been noted that another *Lobelia* alkaloid, lobelanine, exhibited a greater propensity toward the development of convulsions at relatively small doses [333].

With regard to the neuromuscular junction, it has been observed that lobeline does not significantly influence postsynaptic membrane resting potentials [336]. These results are consistent with later observations regarding the effects of lobeline on ganglionic transmission, wherein a depolarizing phase, if present, was only minimal [334]. Lobeline apparently decreases only quantal size at the end-plate region of the neuromuscular junction, whereas nicotine decreases both quantal size and quantal content [336].

Lobeline is currently included in the *U.S. Adopted Names* (USAN), a compendium listing the names of all available drugs, whether they are time honored or have just recently been developed. Lobeline's sole therapeutic indication today resides in its promotion as a substitute for nicotine in the cessation of cigarette smoking. There are early reports that lobeline appeared to reduce nicotine craving and should accordingly be quite useful in the elimination of smoking behavior [337]. In attempts to reproduce these findings, however, Wright and Littauer [338] noted that the doses stated as necessary for this effect produced too many gastrointestinal symptoms, that is, the early-recognized nausea and emesis, and lower doses in their experience did not prove effective. Although the latter observation was the primary conclusion of a host of experimental studies to be undertaken in the next three to four decades (see Bradshaw [339]), lobeline sulfate still remains available as an over-the-counter drug in several preparations combining antacids with this alkaloid to reduce the unpleasant gastrointestinal effects (i.e., Bantron®, Lobidan®).

In the past, cases of human poisoning associated with *Lobelia* arose almost entirely from overdoses of homemade medicinal preparations [44]. Signs of toxicity are an extension of the pharmacological effects of lobeline and include nausea, vomiting, anorexia, exhaustion and weakness, stupor, and tremors. After progression of the syndrome, convulsions, coma, respiratory paralysis, and death ensue.

Reports on cases of animal posioning as a result of *Lobelia* ingestion are rare, although a relatively recent communication did implicate *Lobelia berlandieri* as the primary cause for local poisoning of cattle and goats. In laboratory studies undertaken by the investigators of the latter study, toxicity was induced experimentally after feeding of levels of the plant equivalent to 0.5% of the animal's weight over a period of 3 days. Symptoms accompanying toxicity resembled those associated with an overdose of the alkaloid lobeline.

3.4.6. (−)-Pinidine (76).

(−)-Pinidine is an alkaloid of *Pinus sabiniana*. Dehydrogenation to 2-methyl-6-*n*-propylpyridine and oxidation to pyridine 2,6-dicarboxylic acid gave an idea of its carbon skeleton. Ozonolysis gave 6-methylpipecolic acid and acetaldehyde, suggesting a 2-methyl-6-propenyl piperidine (76) (Table 2F) structure [340]. The synthesis of pinidine by Leete [341] started with 2,6-dimethylpyridine, via the (6-methyl-2-picolyl)methylcarbinols, by hydrogenation, dehydration, and resolution with (−)-6,6′-dinitrodiphenic acid. The 2,6-*cis* relative configuration was concluded by Hill [342] from the conversion of the above mentioned carboxylic acid into *cis*-(meso) 2,6-dimethylpiperidine. The absolute configuration (2R) follows from catalytic hydrogenolysis of *N*-methyl-pinidinium iodide to (+)-2-dimethylaminononane, the antipode of the known (−)-2-dimethylaminononane. In view of the *cis*-configuration at C(2) and C(6) the chiral carbon C(6) must be *R*, too [342].

Biosynthesis. The label of 1-^{14}C-acetate is incorporated [343] into C(2), C(4), C(6), and C(8) of pinidine (Scheme 32). Using labeled malonate, C(9) contains more radiocarbon than C(6). The intermediates are not quite certain [344].

3.4.7. Baikain (77) and 4,5-Dihydroxy-L-pipecolic Acid (78).

4,5-Dihydroxy-L-pipecolic acid (78), a simple alkaloid, was isolated from *Calliandra haematocephala*. Baikain (77) was converted by osmium tetroxide [345] into this *cis*-diol (78).

3.4.8. 2-Hydroxymethyl-3,4,5-trihydroxypiperidine.

This trihydroxypiperidine has the configuration of 1,5-didesoxy-aza-mannopyranose (79). It was erroneously named "1,5-dideoxy-1,5-azamannitol" because mannitol is acyclic.

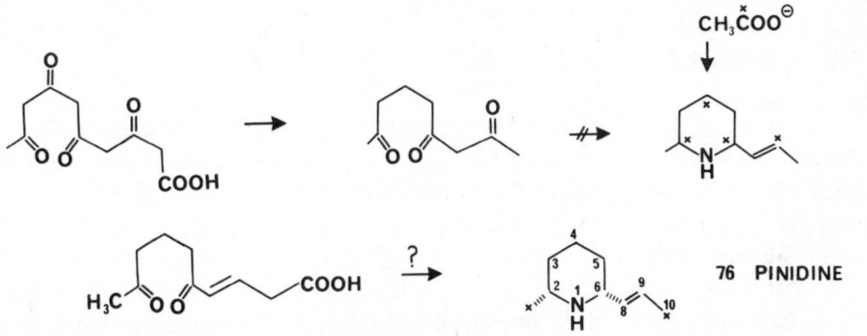

Scheme 32. Biosynthesis of pinidine. Reproduced from reference 431, figure 7, with permission of Springer-Verlag, Heidelberg.

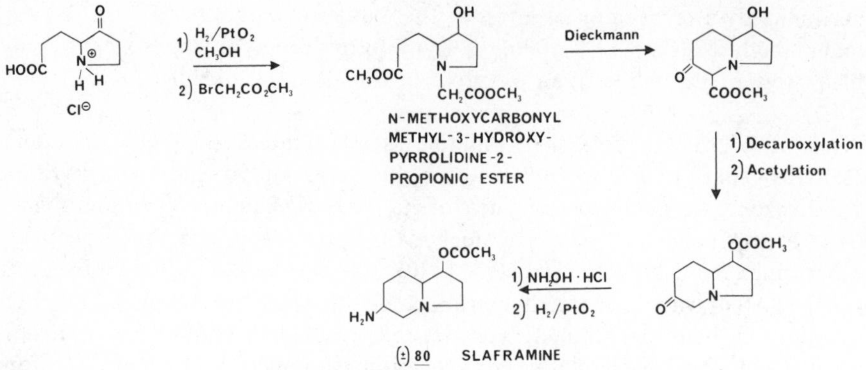

Scheme 33.   Synthesis of slaframine.

Isolated from seeds of *Lonchocarpus costaricensis* [346], its relative configuration has been determined by ^{1}H and ^{13}C NMR spectroscopy and its absolute configuration by the benzoate chirality method. The most likely biogenetic precursor is 5-desoxy-5-amino-D-mannose [346].

3.4.9. Slaframine.   The alkaloid has been isolated from *Rhizoctonia leguminicola* [347]. It stimulates salivation in livestock foraging on fungus-infected red clover. Degradation and spectral data suggested first an indolizine skeleton [348], but that was revoked by the same authors [349]. The correct absolute configuration was deduced by using Horeau's method [349]. Slaframine is an *O*-acetylated amino alcohol **(80)** that has the skeleton reminiscent of δ-coniceine. Synthesis of racemic slaframine proceeded by way of Dieckmann condensation of *N*-methoxycarbonylmethyl-3-hydroxypyrrolidyl-2-propionic ester [350] (Scheme 33). The biosynthesis starts from L-lysine via L-pipecolic acid and, ultimately, incorporation of malonate [351].

3.5. Piperidine Alkaloids with a Long Aliphatic Side Chain

3.5.1. Animal Alkaloids

Stenusine.   An alkaloid of the pygidial defense gland of the animal, *Stennus comma* enables it to move rapidly on water surface [352]. The structure *N*-ethyl-3-(2-methylbutyl)-piperidine **(81)**, is based on high-resolution ^{1}H NMR and IR spectroscopy, particularly on the mass spectrum of stenusine. The conclusive proof was its synthesis [353] from 2-methylbutyl malonic ester. Details of this synthetic scheme were reported in Vol. 1, Chapter Two, of this monograph series by T. H. Jones and M. S. Blum. The configuration of the tertiary carbon in the side chain was correlated with (−)-2-methyl-butan-1-ol, while the configuration of the C(3) still awaits determination.

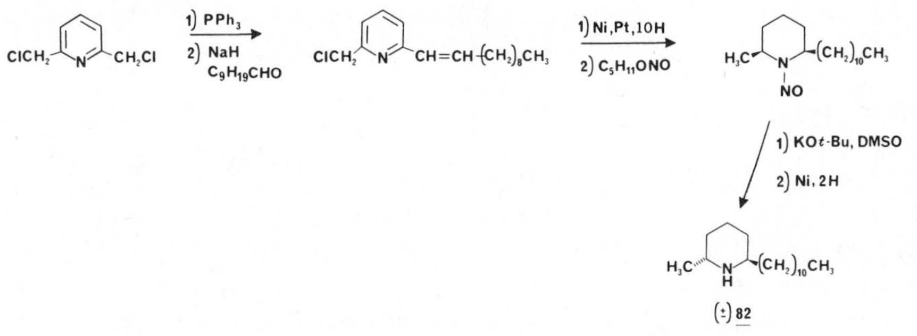

Scheme 34. Synthesis of solenopsin A.

Solenopsin A. This alkaloid was isolated [354] from the venom of fire ants along with a number of other representatives. The structure determination involved conventional chemical and spectral methods and led to the assignment of 2-methyl-6-*n*-undecylpiperidine **(82)** to this alkaloid. The most recent total synthesis start with 2,6-*bis*-chloromethylpyridine via a Wittig reaction on one of the two functionalities. The phosphonium salt was combined with *n*-dodecanal, followed by hydrogenation of the olefinic bond and of the ring. Hydrogenolysis of the chloromethyl group to methyl completed the synthesis of all-*cis* solenopsin [354] (Scheme 34). The secondary amine was isolated as *N*-nitrosamine, epimerized and hydrogenolyzed over Ni to afford (±)-solenopsin A **(82)**.

Pharmacology of the Fire Ant Venoms. Fire ant (*Solenopsis*) venoms have been recently shown to contain 2,6-disubstituted piperidines with long alkyl or alkenyl chains at position 6 [355, 356]. Some of these derivatives have more recently been shown to inhibit Na^+-K^+-ATPase activity in brain synaptosomal preparations [357]. As neither a long-chain aliphatic hydrocarbon, a simple amine, or simple piperidines and dipiperidines were effective, the combination of a piperidine moiety and a long alkyl side chain was considered necessary for activity [358]. The hydrophobic nature of the long chain apparently aids the piperidine compound in gaining access to the Na^+-K^+-ATPase enzyme in the synaptosomal membrane. The piperidine compound presumably proceeds to cause disruption of secondary bonding forces of the lipoprotein complex of the enzyme. This event presumably results in conformational changes, thus affecting the end product of enzyme inhibition.

Cryptophorine. The major alkaloid ($C_{17}H_{26}NO$) of *Bathiorhamnus crypto-phorus* [359] absorbed 4 mol of hydrogen to give a saturated piperidine alcohol. The original base had four conjugated double bonds, according to UV and IR spectroscopy. Dehydrogenation gave a 3-hydroxypyridine where NMR spectrum suggested a trisubstituted pyridine. NMR combined with shift reagent indicated four conjugated double bonds, an *N*-methyl group, a C=C—CH_2—CH_3 group, an ethyl group, and an all-*cis* stereochemistry of the piperidine substituents [360]. Hence, the structure of 1,2-dimethyl-3-hydroxy-6-($1'E,3'E,5'E,7'E$)-decatetraenyl piperidine (83) was assigned to the alkaloid.

3.5.2. Plant Alkaloids.

Carpaine. The major alkaloid of *Carica papaya* [361] has the molecular formula $C_{28}H_{50}N_2O_4$ and is a dimeric lactone based on its mass spactrum [362]. The previous [1] internal lactone structure, $C_{14}H_{25}NO_2$, has been rejected, as has the pyrrolidyl-lactone structure [363].

Carpaine (84) gave upon hydrolysis carpamic acid (85), a reaction reversed by thionyl chloride [364]. Carpamic acid was dehydrogenated by Pd to desoxycarpyrinic acid, that is 6-methylpyridine-2-octanoic acid. Under different conditions carpyrinic acid (3-hydroxy-6-methylpyridine)-2-octanoic acid) was obtained which was then synthesized. The most economic synthesis of the latter starts with furan and gives a 10% overall yield [365]. However, carpyrinic acid could not be reconverted into carpamic acid [366]. Hofmann methylation of (84) led ultimately to 12-keto-tetradecanoic acid [367]. The relative configuration of the methyl group with respect to the side chain of carpamic acid was established by Govindachari [366] to be *cis*. There is a strong intramolecular hydrogen bond between the piperidine nitrogen and the hydroxyl group at C(3); hence, Sicher [368] ascribed the all-*cis* configuration to methyl carpamate and, by inference, to carpaine itself. The absolute configuration of C(3) was deduced from the correlation with $R(-)$-3-tetradecanol by Coke and Rice [369]. Carpamic acid gave the dimeric lactone in a variety of ways [364], the best method being the recent use by Corey [370] of 2,2'-dipyridyl-disulfide and triphenylphosphine upon *N*-carbobenzoxy carpamic acid. The first total synthesis of (±)-carpamic acid was elaborated by Brown [371] as an intramolecular reductive amination process. We present a more recent synthesis [372] by Natsume and Ogawa (Scheme 35). The key intermediate is 1-carbobenzoxy (Cbz) 2-methyl-1,2-dihydropyridine, which is converted by singlet oxygen into an adduct. The latter is cleaved by ethyl vinyl ether and tin (2)chloride, and then ethanol to 3,6-*trans*-2-methyl-2-hydroxy-1,2,3,6-tetrahydropyridine-6-acetaldehyde acetal.

Epimerization at C(3) and hydrolysis of the acetal, followed by a Wittig reaction, gave the 2,3,6-all-*cis* aldehyde derivative. The aldehyde with the C_8-phosphorane led to the expected octenoic acid, and subsequent routine operation afforded (±)-85, carpamic acid. The last step, dilactone formation, was achieved according to Narasimhan [364]. By changing the phosphorane to a C_4 Wittig reagent, 2-methyl-3-hydroxypiperidine-6-hexanoic acid (±)-azimic acid (86) was obtained [372]. Another synthesis of 85 and 86 starts with D-glucose [373].

The biogenesis of the carpaine group most likely involves a linear combination of acetates [374].

Pharmacology of Carpaine. Extracts prepared from *Carica papaya* have long been used as a cardiac stimulant, an antihypertensive, and as a diuretic in

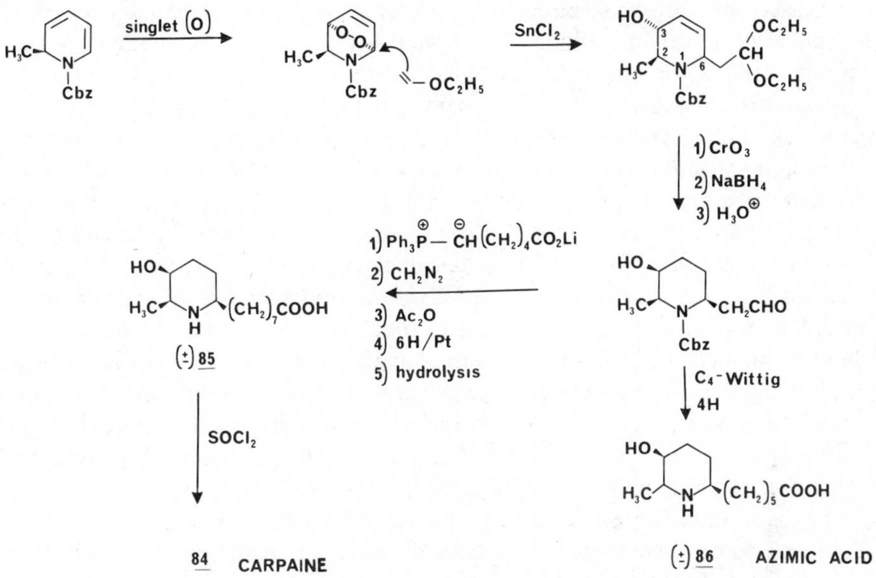

Scheme 35. Total synthesis of carpamic and azimic acids.

Eastern countries [375–377]. Because of these ascribed effects of the plant, in conjunction with the relatively high concentrations of carpaine contained in the leaves, the pharmacological properties of this alkaloid have been examined rather thoroughly.

One of the earliest reported effects of carpaine was its amoebicidal activity [378]. In subsequent studies relative to microbes, carpaine was found to inhibit *Mycobacterium tuberculosis* at quite low concentrations, that is, 1×10^{-4} M [379]. In more recent years this alkaloid has been observed to possess antitumor activity *in vitro* at low concentrations, that is, 0.1–0.2% [380]. The carcinomas investigated in the latter study included lymphoid leukemia L1210, Lymphocytic leukemia p388, and Ehrlich ascites tumor cells.

In an early report on the cardiovascular effects of carpaine this alkaloid was noted to produce a fall in blood pressure and some degree of tachycardia after administration to cats [381]. While contractile force of the myocardium, particularly that of the auricle, was directly depressed, conduction was not affected. Because of the similarity between the inotropic effects of carpaine and those of emetine, an alkaloid obtained from ipecac that is still used clinically for the treatment of amebiasis, the mechanism of action of these two alkaloids was presumed to be quite similar.

Other pharmacological effects observed to be produced by carpaine during the course of the latter study included marked relaxation of the uterus and bronchodilation in guinea pigs. In experiments with cats the therapeutic index of carpaine was shown to be relatively high; toxic doses were approximately 40 times greater in magnitude than doses evoking significant reduction of blood pressure. The frog was found to be relatively resistant to the effects of both carpaine and emetine. When effects were finally produced, changes in myocardial conduction and mild vasoconstriction occurred.

The cardiovascular effects of carpaine have quite recently been reassessed by Hornick and co-workers [382] in anesthetized rats. After intravenous administration of carpaine, a dose-related decrease in both systolic and diastolic blood pressure resulted. After administration of a dose at the high end of the dose–response curve, cardiac output, stroke volume, stroke work, and contractile force were likewise reduced, whereas calculated total peripheral resistance remained unchanged. Because neither atropine nor propranolol attenuated the circulatory effects of carpaine, a role of the autonomic nervous system in their mediation was ruled out. These results provide substantiation for earlier data [381] suggesting a direct effect of carpaine on the myocardium. Because macrocyclic dilactone structures such as carpaine are, in general, chelating agents, the authors concluded that carpaine possibly acts to produce its cardiovascular effects by antagonizing free calcium ions in plasma.

In the United States *Carica papaya* has been included as one of the plant sources known or suspected to produce dermatitis in man, with the likely agent being the sap [44].

Cassine.   The alkaloid ($C_{18}H_{35}NO_2$) of *Cassia excelsa* Shrad was isolated and its constitution and relative configuration determined by Highet [383, 384]. Dehydrogenation led to dehydrocassine, the ^1H NMR spectrum of which showed a pattern for a 2,3,6-trisubstituted pyridine. The alkaloid itself gave a positive iodoform test for a —$COCH_3$ group that was confirmed by the IR spectrum. The IR spectral data pointed to a secondary alcohol, and the ^1H NMR spectrum of a 2-methyl-3-hydroxypiperidyl-6-dodecan-11-one (**87**). The IR spectrum showed that the ring nitrogen is strongly H bonded to the alcoholic hydroxyl, an arrangement similar to that in carpamic acid [384]. The all-*cis* position of the three substituents in (−)-cassine was proven by an acetoacetate synthesis of its antipode (+)-cassine from (+)-carpaine via carpamyl chloride by Rice and Coke [385]. This correlation also established absolute configuration of the *Cassia* alkaloid. The total synthesis of (±)-cassine has been reported [386] by Brown. Condensation of 4,15-dioxo-hexadecanal with nitroethane gave 2-nitro-3-hydroxyoctadecane-6,17-dione, which upon reductive amination, gave (±)-cassine (**87**) and (±)-3-epicassine (**88**) (Scheme 36).

The biosynthesis of **87** most likely follows the acetate route, by analogy to carpaine.

Carnavaline (89).   The alkaloid (**89**) from *C. carnaval* [387] was correlated with (−)-cassine (**87**) by sodium borohydride reduction of the carbonyl group.

Julifloridine.   An alkaloid from the leaves of *Prosopis juliflora* DC [388] is a 12′-hydroxy regioisomer (**90**) of carnavaline, of unsettled stereochemistry.

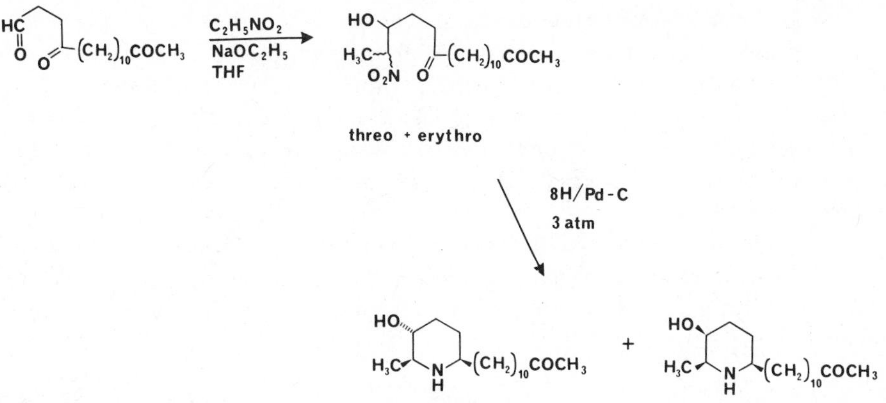

Scheme 36. Total synthesis of cassine and 3-epicassine.

Spectalinine. *C. spectabilis* DC seeds contain, based on optical rotation, NMR, and mass spectrometry [389], the 2,3,6-*cis* antipode (91) of carnavaline; however, it has a C_{14}-side chain, and the configuration of the 13'-hydroxyl is unknown.

Prosopine, Prosopinine, Prosophylline, Prosafrine, Prosafrinine, and Isoprosopinine. All these alkaloids have been isolated from leaves, stems, and roots of *P. africana* [390]. The structure elucidation was done mostly by spectral methods by Goutarel and his associates [391, 392].

Prosopinine. The molecular formula is $C_{18}H_{35}NO_3$. The IR spectrum shows hydroxyl and NH groups (3300 and 3400 cm^{-1}) and a carbonyl absorption (1720 cm^{-1}). The ^{1}H NMR spectrum gave a complex pattern that suggested assignment of a OC—CH_2—CH_3 group. Three protons could be exchanged by D_2O, and three others were adjacent to an alcoholic hydroxyl. The mass spectrum showed M$^+$ 313, and two major peaks at m/z 282 (M-31) and m/z 130, the latter corresponding to $C_6H_{12}NO_2$. There were three protons that could be exchanged by acetyl groups, giving triacetylprosopinone with an amide IR absorption at 1657 cm^{-1}. The nitrogen was methylated with formaldehyde, indicating a secondary amine. The presence of a ketone function was corroborated by oxime and semicarbazone formation. A modified Wolff–Kishner reduction eliminated the carbonyl function to give desoxoprosopinine (93). This compound gave the molecular peak of 299, the M-31 peak at 268, and the characteristic base peak at 130. Desoxoprosopinine reacted with benzaldehyde to give a cyclic acetal. Catalytic dehydrogenation of prosopinine with Pd/C gave dehydroprosopinine ($C_{18}H_{29}NO_2$) that showed a UV absorption typical of 3-hydroxypyridines. All these and other spectral data were consistent with a 2-methyl-3-hydroxypyridyl-dodecanone structure for dehydroprosopinine and, by inference a 2-hydroxy-methyl-3-hydroxy-6-(10'-oxododecyl)piperidine structure for prosopinine (92). The relative stereochemistry was established by double irradiation in NMR of the benzylidene acetal (94) of 10'-desoxoprosopinine [391, 392]. The coupling constant of H(2)/H(3) was 10 Hz, in agreement with a *trans*-arrangement of these hydrogens.

Periodic acid oxidation indicated the loss of one carbon from prosopinine and the formation of an *N*-formyl pyrrolidine. The formyl group, in turn, was reduced by lithium aluminum hydride to an *N*-methyl pyrrolidine that was analyzed by mass spectrometry.

Prosopine. Prosopine ($C_{18}H_{37}NO_3$) was converted into the same desoxo-prosopinine (93) as prosopinine via Oppenauer oxidation of an alcohol followed by deoxygenation of the ketone group. Therefore, the only differences were two additional hydrogens and the position of the oxygen function in the side chain. The mass spectrum and the ^{1}H NMR clearly indicated that prosopine is oxygenated at C(11') [391, 392], corresponding to structure 95. The absolute configuration at C(11') was established [392] by using Horeau's method.

(±)-*Prosophylline.* (±)-Prosophylline ($C_{18}H_{35}NO_3$) gives a triplet for a C-methyl group in the ^{1}H NMR spectrum, indicating an identical position of the carbonyl group as in prosopinine (92). However, the benzaldehyde cycloacetal (97) of 10'-desoxoprosophylline had a different coupling pattern on C(6) than (94) derived from desoxoprosopinine. A double irradiation study of the acetal led to the 2,6-*cis*-2,3-*trans* configurational assignment for this *racemic* alkaloid [391, 392].

α- and β-Isoprosopinines. α- and β-isoprosopinines could not be separated. These gave upon reduction of the ketone function the same desoxoprosopinine (93). The position of the carbonyl group was established by Baeyer–Villiger oxidation, leading to a 1:3 mixture of valeric and capronic acids that was characterized by GLC. Therefore, one of the compounds is the 7-oxododecyl-(98) and the other is 6-(8-oxododecyl)-2-hydroxymethyl-3-hydroxy piperidine (99) [391].

Prosafrine. Prosafrine ($C_{18}H_{37}NO_2$) has two hydroxyl functions corresponding to a desoxyprosopine structure. The mass spectrum shows a base peak m/z 114 in contrast to the aforementioned piperidines (m/z 130). Elimination of the carbonyl oxygen from prosafrinone and *N*-methylation gave *N*-methyl-desoxo-prosafrinone; the latter proved to be identical with *N*-methyl-desoxocassine of Highet. Therefore, there is a methyl group in position 2, instead of the hydroxymethyl as in prosopine (structure 100) [391].

Prosafrinine. Prosafrinine ($C_{18}H_{35}NO_2$) [391] is 2'-desoxyprosopinine (101).

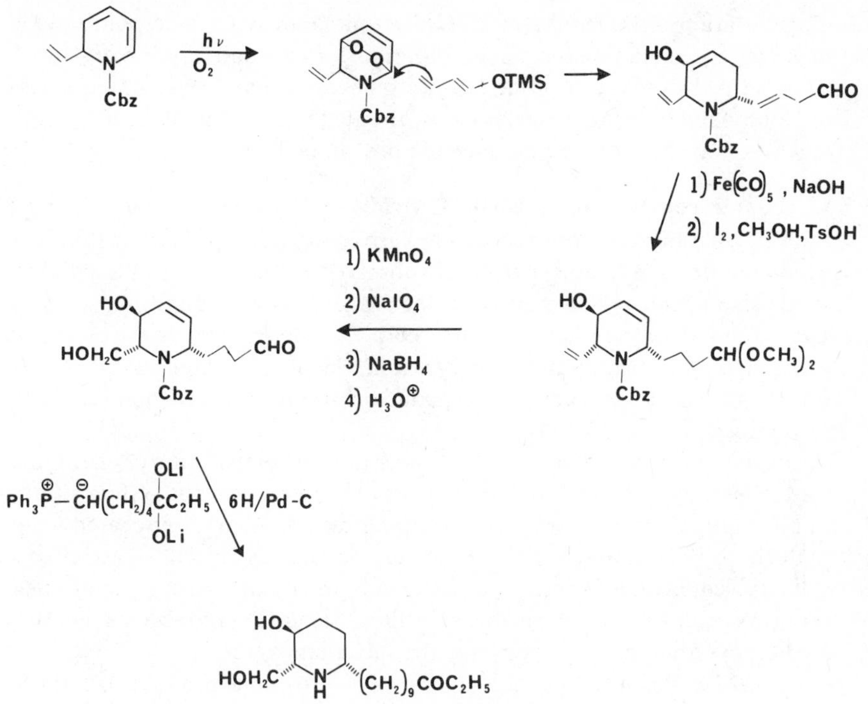

(±) **96** PROSOPHYLLINE

Scheme 37. Total synthesis of prosophylline.

3.5.3. Total Synthesis of the Prosopis Alkaloids.

Despite persistent efforts [393, 394] in the 2-hydroxymethyl class, only (±)-prosophylline has been synthesized [395] in addition to (−)-desoxoprosophylline, (−)-desoxoproso-pinine [396, 397], and (±)-all-*cis* prosopine [394]. Prosafrinine is also available by synthesis [372]. The total synthesis [395] of (±)-prosophylline is shown in Scheme 37. The novelty of this synthesis is the ring opening of the 3,6-photoperoxide of a 2-vinyl-*N*-Cbz-1,2-dihydropyridine by the TMS-enolether of 3-butenal. The vinyl side chain at C(2) that was introduced into pyridine at the outset is converted into the 1,2-diol and this, by sodium periodate, into the aldehyde, which is then reduced by metal hydride to the primary carbinol. The construction of the C_{12} side chain at C(6) is based on a Wittig reaction between a blocked 4,5-dehydropiperidyl-butanal and 6-oxo-octylphosphorane dilithio ketal followed by saturation of the double bonds. Prosafrinine was synthesized by a similar approach.

(±)-All-*cis* prosopine (**102**) was prepared, together with two other racemates, by synthesis of (±)-hexadehydroprosopine [365] and catalytic hydrogenation thereof, over rhodium on carbon [394]. The major produce of the hydrogenation was converted into a benzylidene acetal (**104**). The ^{1}HNMR, 250 MHz, using double irradiation, showed that the H2/H3 coupling constant corresponds to the *cis*-fusion of the metadioxane to the piperidine ring—unlike the *trans*-fused acetals of (+)-11-desoxyprosopine (10-desoxoprosopinine) **(94)** and also desoxoprosophylline **(96)** respectively. Thus, the methylol group at C(2) and the C(3) hydroxyl are *cis* **(103)** in the racemic prosopine isomer.

3.5.4.   Pharmacology of Long-Chain Piperidines: Prosopines.

Prosopis africana has historically been utilized by the natives of Africa for a wide diversity of purported effects. The most frequently observed usages include as a sedative, a hypotensive agent, a spasmolytic, a local anesthetic in conditions such as toothache, and an antiseptic agent [398, 399]. Presumably as a consequence of the variegated medicinal nature of *P. africana*, the pharmacological properties of two of the major alkaloids of the plant, prosopine and prosopinine, have been examined extensively [399].

The profile of pharmacological activity seen after administration of prosopine mandates classification of this alkaloid as a mild central nervous system stimulant. In mice hyperarousal and increased exploratory behavior appeared soon after injection. Locomotor activity, however, was not enhanced. These effects were short-lived, lasting for only 1 h. Although prosopine markedly shortened barbiturate sleeping time in mice, this piperidine did not antagonize the sedative effect of cresoxydiol, a drug acting via the medullary centers.

With regard to the autonomic nervous system, prosopine produced a mild to moderate elevation of blood pressure in anesthetized cats. In contrast with the usual case, the elevation was not dose related. Prosopine exerted no effect on the elevated blood pressure produced by carotid occlusion or administration of epinephrine and likewise did not affect the hypotension produced by administration of acetylcholine. It was accordingly concluded that the hypertensive effect of prosopine was mediated via direct vasoconstriction.

With the utilization of the rabbit cornea, prosopine was found to possess local anesthetic activity equal in magnitude to that of cocaine. Unlike cocaine, however, prosopine produced much local irritation.

Toxic doses of prosopine were found to be only 2–3 times greater than those producing mild central and peripheral effects. These doses produced marked central excitation, violent convulsions, and death due to respiratory arrest.

In the experiments on prosopinine, a wider battery of pharmacological tests was applied, and a greater range of pharmacological activity was uncovered. Like prosopine, prosopinine possessed central nervous system activity. The effects of this alkaloid in mice, however, were opposite to those of prosopine, with the production of prostration, decreased motility, and reduced exploratory behavior. Loss of the righting reflex was observed in 50% of the animals. Although barbiturate sleeping time was prolonged, the dose of strychnine required to produce seizures was reduced only if this convulsant was continuously infused rather than injected. No protection was likewise afforded by prosopinine against seizures produced by acute administration of pentylenetetrazol or by electroshock. On the other hand, hypermotility due to the administration of amphetamine was reduced, and lethality was likewise lowered. Tests for analgesia revealed that prosopinine lacked this property.

Like prosopine, prosopinine also affects the autonomic nervous system, but with the production of a fall in blood pressure. The hypotensive effect of prosopinine was not altered by the administration of sparteine, atropine, or mepyramine. The mild hypertension resulting from carotid occlusion as well as the fall in blood pressure after vagal stimulation or injection of acetylcholine, however, were reduced by prosopinine. These results indicate that prosopinine lowers blood pressure by exertion of a parasympatholytic effect.

Prosopinine also produced a negative inotropic effect and a slight reduction in cardiac rhythm in anesthetized dogs. Experiments in rabbits showed that, like the anticholinergics, prosopinine reduces capillary permeability.

With regard to the gastrointestinal tract, prosopinine decreased both resting tone and the force of contractions in isolated rat duodenum. Contractions induced by acetylcholine were also reduced by 50%. In this preparation prosopinine is 500-fold less potent than atropine.

Prosopinine did not demonstrate anti-inflammatory activity. While antimalarial and amebicidal activities were likewise lacking, moderate antibiotic activity against gram-positive organisms and slight antifungal activity were apparent.

Like prosopine, prosopinine exerted local anesthetic activity comparable to that of cocaine. Local irritation, however, was more pronounced, with the production additionally of blinking and tearing.

Toxic doses of prosopinine were, like prosopine, only 2–3 times greater in magnitude than those producing moderate central and peripheral effects. After initial prostration violent convulsions paradoxically ensued, and death occurred via respiratory arrest.

On the basis of the pharmacological profiles of prosopine and prosopinine, the authors concluded that most of the activity associated with extracts from *P. africana* was due to the prosopinine content.

In an effort to eliminate the irritant effects accompanying the local anesthetic activity of prosopinine, two derivatives, isoprosopinine and desoxoprosopinine, were prepared. Unfortunately, these compounds proved to be less effective as a local anesthetic while retaining activity as an irritant.

3.6. Piperidine-Chromone Alkaloids: Rohitukine

This *Meliaceae* alkaloid ($C_{16}H_{18}NO_5$), extracted from roots and stems of *Amoorha rohituka*, has been subject to X-ray crystal analysis [400] and proved to be a chromone alkaloid (105).

3.7. Spiropiperidines (Histrionicotoxin and Related Alkaloids)

3.7.1. Histrionicotoxin. From 1971 to 1974 Witkop and co-workers [401, 402] isolated several structurally unique alkaloids from the Colombian frog belonging to the genera *Dendrobates*. These alkaloids, which occur in the frog's defensive skin secretions, have been found to be highly active venoms as well as mucosal tissue irritants toward both mammals and reptiles.

Histrionicotoxin, which is isolated from *D. histrionicus*, has a unique spiropiperidine structure and two *cis*-enyne side chains (106). Both the relative and the absolute configuration were determined by X-ray crystallography by Karle [404]. Its action as a neurotoxin and the limited availability of natural material (HTX) has triggered considerable work. HTX itself has not been synthesized to date, but several routes were developed that led to the corresponding hydrogenated species perhydrohistrionicotoxin (107) (HTX-12), which was also obtained [403] by hydrogenation of histrionicotoxin over Pd/C in THF. HTC-12 has biological activity similar to HTX in the *in vitro* assay in nerve muscle preparations of frogs and therefore provides a biochemical standard.

3.7.2. Synthesis of Perhydrohistrionicotoxin.

Two different approaches to perhydrohistrionicotoxin are discussed here. One, described by Evans [405], pertains to the (±)-HTX-12, and the other, by Witkop and Brossi [406], relates to the naturally derived (−)-HTX-12.

Evans's approach is based on *a*-acyliminium ion-olefin cyclization. The iodomagnesium salt of glutaramide was treated with 4-nonenyl magnesium bromide to give a carbinolamide, which was transformed into a spirolactam after exposure to anhydrous formic acid. This process involves an acyliminium ion-olefin cyclization as the key step. The lactam thus formed was transformed over several steps into (±)-HTX-12 (Scheme 38).

Witkop and Brossi's approach [406] starts from one enantiomer of the bislactam. Its optical resolution was achieved by transformation of the alcohol moiety. Condensation of the (±)-lactam with (+)-α-methylbenzylisocyanate in toluene afforded a mixture of diastereomeric carbamates which were separated by preparative HPLC, to give the two esters in crystalline form. Hydrolysis led to the enantiomeric lactams, which were separately converted into (+)- and (−)-HTX-12 by the procedures followed previously for the racemic lactam. This synthesis was the first synthesis of natural (−)-perhydrohistrionicotoxin. Two syntheses of racemic perhydrohistrionicotoxin were devised recently [407, 408].

3.7.3. Pharmacology of Histrionicotoxins.

Because of the similarity of histrionicotoxin to acetylcholine with regard to the juxtaposition of the N and O atoms in the molecule, this animal alkaloid was strongly suspected shortly after elucidation of its structure [401] of interacting with cholinergic systems. Subsequent studies undertaken in laboratories around the country have confirmed this suspicion and have defined the nature of the interaction.

Utilization of histrionicotoxins has allowed further clarification of the events leading to postsynaptic changes after combination of the neurotransmitter, acetylcholine, with a receptor. In pioneering studies utilizing perhydrohistrionicotoxin, evidence was provided to suggest that two sets of binding sites participate at vertebrate neuromuscular junctions in the production of mem-

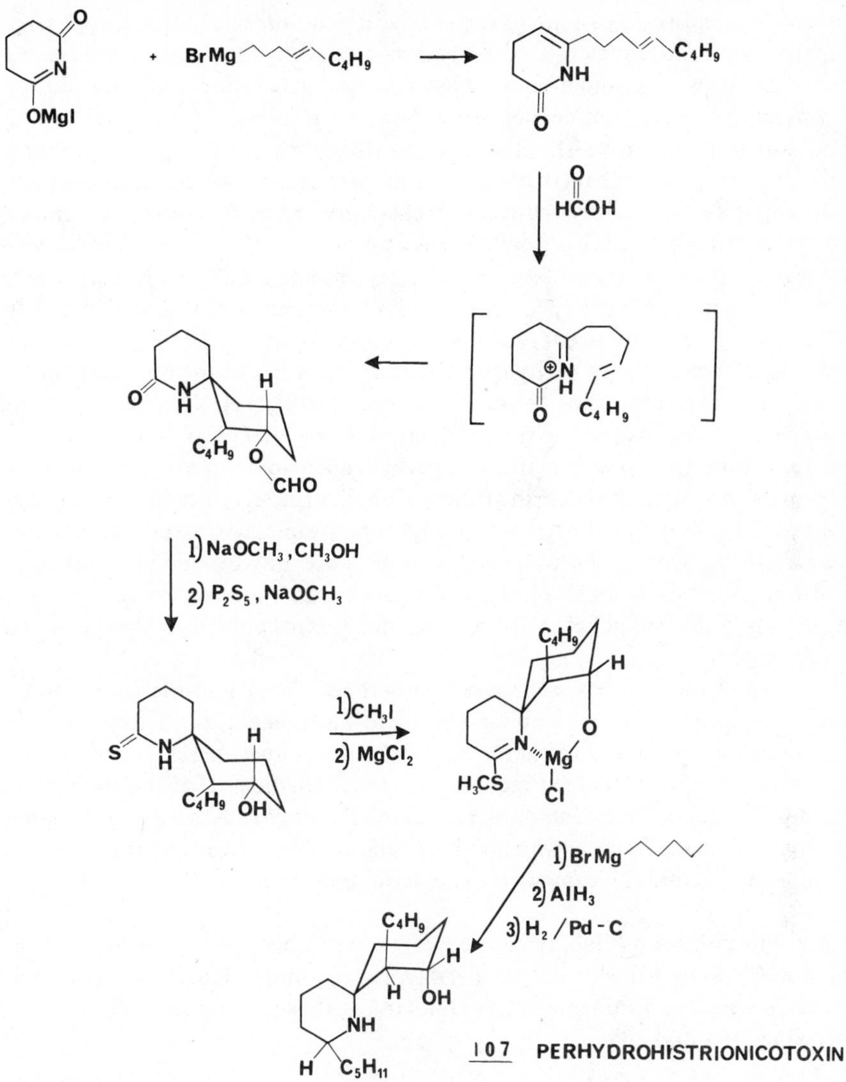

107 PERHYDROHISTRIONICOTOXIN

Scheme 38. Synthesis of perhydrohistrionicotoxin.

brane depolarization: the acetylcholine receptor sites and the ion conductance modulator sites [409]. Perhydrohistrionicotoxin apparently binds selectively at the ionic modulator sites. Whereas separately curare and perhydrohistrionicotoxin only partially block the binding of bungarotoxin (a specific antagonist at nicotinic cholinergic receptor sites in skeletal muscle), in combination these agents completely block bungarotoxin binding. As complete recovery of acetylcholine sensitivity occurs after administration of bungarotoxin in the presence of curare, this latter agent is presumably combining with acetylcholine receptors per se. In contrast, no recovery of acetylcholine sensitivity occurs when bungarotoxin is administered in the presence of perhydrohistrionicotoxin. As this semisynthetic alkaloid affects action potential generation and elicits failure of consecutive acetylcholine depolarizations, it is presumed to bind at ion modulator sites that are distinctly different from those acted upon by curare.

Histrionicotoxin appears to act quite similarly to its completely reduced form [410]. This alkaloid likewise reduces the sensitivity of denervated skeletal muscle to repetitive applications of acetylcholine and blocks the increase in ionic conductance resulting from combination of acetylcholine with its own receptor. In addition, histrionicotoxin specifically inhibits the increase of potassium conductance that normally occurs during an action potential.

In a more recent study perhydrohistrionicotoxin has been shown to inhibit muscarinic cholinergic sites in neuroblastoma cells [411]. The nature of the binding site for the alkaloid in the production of this effect was not determined. One dissimilarity from interactions of perhydrohistrionicotoxin with nicotinic systems was noted: no increase in affinity of cholinergic agonists for the receptor occurred. The binding properties of perhydrohistrionicotoxin and tetracaine, a local anesthetic, were found to be identical. Because dihydro-adaline, granatan-3α-ol and granatan-3β-ol had less affinity for the binding sites, it was suggested that the lipophilic side chains of the histrionicotoxins contribute to the actual binding process.

Similarities and differences between the actions of local anesthetics and those of histrionicotoxin are still under study. In experiments utilizing *Torpedo* membrane fragments and a fluorescent probe, local anesthetics were found to alter the binding kinetics of carbachol, while the toxin did not [412]. Binding and permeability measurements undertaken in clonal muscle cells, on the other hand, indicated that, in common with the local anesthetics, histrionicotoxin blocks activation of nicotinic receptors by allosteric inhibition.

3.7.4. Nitramine and Isonitramine.

Two alkaloids from *Nitraria* species ($C_{10}H_{19}NO$) were believed to be decahydroquinolines. Quite recently their structures have been reformulated [413, 414] as those of stereoisomeric spiro-piperidines **108** and **109**.

3.8. Fused-Ring Piperidines: Granatane Alkaloids

3.8.1. Pseudopelletierine (3-Granatanone).

Pseudopelletierine ($C_9H_{15}NO$), one of the *Punica granatum* alkaloids [288], is an optically inactive tertiary base. The formation of an oxime indicates the presence of a carbonyl group [415]. In addition, the compound forms a dibenzylidene derivative and a diisonitroso derivative, each of which indicate the presence of a —CH_2COCH_2— group in the molecule [416].

Reduction with sodium and alcohol gave an alcohol *N*-methyl granatoline, which, upon the action of hydriodic acid and phosphorus at 140°, is converted into N-methyl granatenine, an olefin with no oxygen function. Hofmann methylation of the latter, in analogy with Willstätter's previous studies on tropidine, gave rise to 1,3,5-cyclooctatriene [417]. This supported Piccinini's formula (110) for pseudopelletierine [418]. Based on this fact and on the similarity of tropinone and pseudopelletierine, the structure of pseudopelletierine was indeed assigned as 9-methyl-9-azabicylo[3.3.1]-nonane-3-one (110). An alternative name, granatan-3-one is also accepted.

Pseudopelletierine was synthesized by Robinson [419] using glutaraldehyde, methylamine, and calcium acetonedicarboxylate. The yield was greatly improved, to 70%, by Cope [420, 421].

3.8.2. 1-Methyl-9-nor-3-granatanone.
The 1-methyl homologue of 9-nor-pseudo-pelletierine was isolated from *Euphorbia atoto* in Australia [422]. Its structure (111) was determined by IR and NMR spectroscopy. The similarity of the spectral data with those of the synthetic 1-methyl-3-granatanone [423] was very helpful in ascertaining the structure of the natural alkaloid.

3.8.3. Adaline.
Adaline is a defensive alkaloid isolated [424] from the European ladybug, *Adalia bipunctata*. The molecular formula is $C_{13}H_{23}NO$. It is a secondary amine proved by chemical evidence and a ketone. The azabicyclononane skeleton of adaline is based on degradations. It differs from the structure of other defensive secretions of the *Coccinellidae*. Infrared spectral resemblance indicated that it is close to the familiar pomegranate alkaloid pseudopelletierine. Based on this, a 1-amyl-3-oxo-9-norgranatane structure (112) was assigned to adaline.

The synthesis of (±)-adaline was accomplished by several methods. The Robinson–Schöpf method gave racemic adaline in 17% yield. An intramolecular cycloaddition route, pursued by Gössinger and Witkop [425] starting from piperideine-N-oxide, furnished racemic adaline in an overall yield of 30% (Scheme 39).

Addition of amylmagnesium bromide was followed by oxidation by mercuric oxide to 6-amyl-2-piperideine and by a second Grignard addition of allyl-magnesium bromide. Intramolecular nitrone-olefine cycloaddition resulted in the formation of the tricyclic bridged hydroxylamine. Reductive cleavage to the 1-amyl-3β-norgranatanol and, in turn, oxidation to the ketone gave racemic adaline.

Recently Hill and Renbaum [426] have developed an alternative route for both racemic (±)-adaline and natural-(−) adaline (Scheme 40). A double conjugate addition of benzylamine to 3-amyl-2,7-cyclooctadienone, followed by debenzylation, resulted in racemic adaline in 39% overall yield. Asymmetric induction was achieved by using R(+)-α-methylbenzylamine in the conjugate addition step. The two diasteromers formed in unequal amounts and could be separated. Hydrogenolysis of the α-methylbenzyl group led to (−)-adaline or to

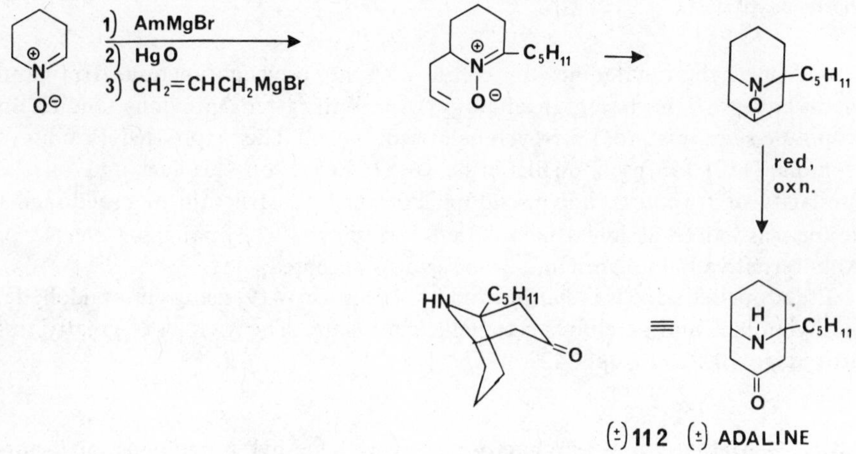

Scheme 39. Synthesis of (±)-adaline.

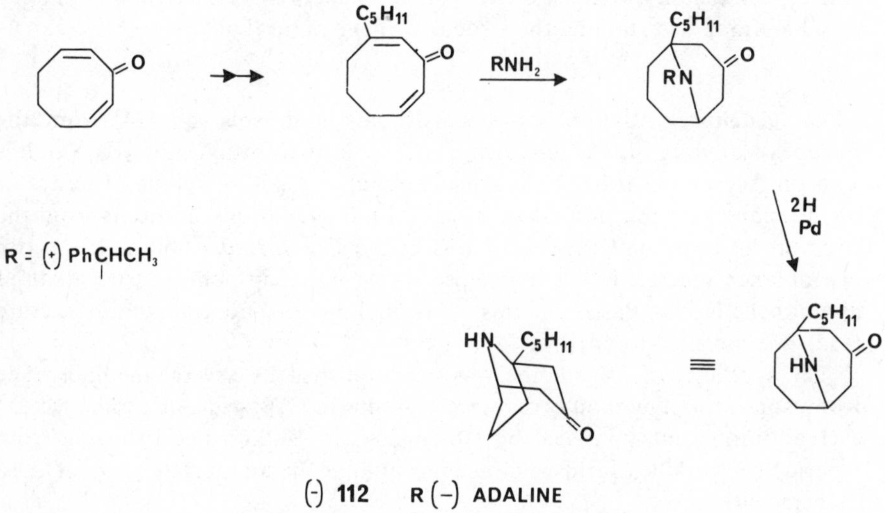

R = (+) PhCHCH₃

(-) **112**   R (−) ADALINE

Scheme 40. Synthesis of (−)-adaline.

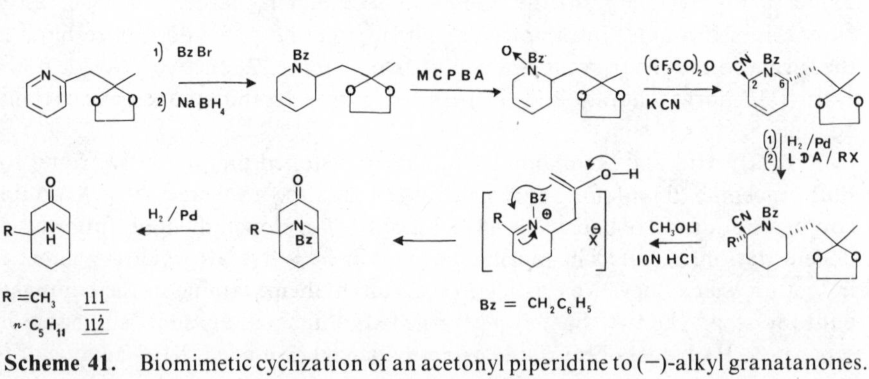

R = CH₃   <u>111</u>
n·C₅H₁₁   <u>112</u>

Bz = CH₂C₆H₅

Scheme 41. Biomimetic cyclization of an acetonyl piperidine to (−)-alkyl granatanones.

78

(+)-adaline, depending on which diastereomer was used. The levoratatory form was identical with the natural alkaloid.

The absolute configuration of dihydroadaline has been determined [427] as $1R,3S$ based on its ORD curve. The biosynthesis of adaline as 1-methyl-3-granatanone was supposed to involve [422] a pathway that is different from that of the formally related tropane. However, a very recent biomimetic approach [428](Scheme 41) to 1-methyl- and 1-amyl-3-norgranatranones, respectively, from a 2-alkyl-6-acetonyl-$\Delta^{1(2)}$-piperideine is reminiscent of Leete's well-established scheme of the biosynthesis of 3-tropanone via $\Delta^{1(2)}$-dehydrohygrine [429]. This idea is supported by the recent finding [430] of 6-methylpelletierine (113) in the defense secrete of the mealybug ladybug (*Cryptolaemus montronzoii*), which could lead via the piperideine, and in turn, internal cyclization, to a *tris*-nor-adaline.

ACKNOWLEDGMENTS

Thanks are due to Mr. R. Dharanipragada for literature search and help in editing this chapter, to Mrs. Duana Nacarate for her technical help with the manuscript, and to Miss Krystal Berry for preparing the formula tables.

REFERENCES

1. L. Marion, *Pyridine Alkaloids*, Vol. 1, *The Alkaloids*, R. H. F. Manske and H. L. Holmes, Eds., Academic, New York, 1950, pp. 167–258.

2. L. Marion, Vol. 6, 1960, pp. 123–144.

3. W. A. Ayer and T. E. Habgood, Ibid., Vol. 11, 1968, pp. 460–503.

4. D. Gross, *Naturstoffe mit Pyridinstruktur und ihre Biogenese*, in *Fortschr. Org. Naturst.*, Vol. 28, Springer, Wien–New York, 1970, pp. 109–161.

5. E. Leete, "Biosynthesis and Metabolism of the Tobacco Alkaloids," in *Alkaloids, Chemical and Biological Perspectives*, Vol. 1, S. William Pelletier, Ed. Wiley-Interscience, New York, 1983, pp. 85–152.

6. D. R. Dalton, *The Alkaloids, A Biogenetic Approach*, Dekker, New York, 1979.

7. G. A. Cordell, *Introduction to Alkaloids, A Biogenetic Approach*, Wiley, New York, 1981.

8. A. Hantzsch, *Chem. Ber.* 19, 31 (1886).

9. D. E. Danford and H. N. Munro, in *The Pharmacological Basis of Therapeutics*, A. G. Gilman, L. S. Goodman, and A. Gilman, Eds., Macmillan, New York, 1980, pp. 1560–1582.

10. R. I. Levy, in *The Pharmacological Basis of Therapeutics*, A. G. Gilman, L. S. Goodman, and A. Gilman, Eds., Macmillan Publishing Co., New York, 1980, pp. 834–847, Chap. 34.

11. P. Needleman and E. M. Johnson, in *The Pharmacological Basis of Therapeutics*, A. G. Gilman, L. S. Goodman, and A. Gilman, Eds., Macmillan, New York, 1980, Chap. 33.

12. S. Kety, *New Engl. J. Med.* 276, 325 (1967).

13. F. Fournier, *Plantes Medicinales et Veneneuses de France III*, Paris, 1948, p. 495.

14. J. Mishkinsky, B. Joseph, F. G. Sulman, and Al. Goldschmied, *Lancet* ii, 1311 (1967).

15. J. Mishkinsky, A. Goldschmied, B. Joseph, Z. Ahronson, and F. G. Sulman, *Arch. Int. Pharmacodyn.* **210**, 27 (1974).

16. J. J. Venit and P. Magnus, *Tetrahedron Lett.* **21**, 4815 (1980).

17. D. A. Evans and E. W. Thomas, *Tetrahedron Lett.*, 4841 (1979).

18. S. Ghosal and S. K. Dutta, *Phytochemistry* **10**, 195 (1971).

19. A. Wohl and A. Johnson, *Chem. Ber.* **40**, 4712 (1907).

20. R. Willstätter, *Chem. Ber.* **30**, 729 (1897); **35**, 615 (1902).

21. H. Gloge, H. Lullman, and E. Mutschler, *Br. J. Pharmacol. Chemother,* **27**, 185 (1966).

22. B. Kummer, H. Lullman, and E. Mutschler, *Naunyn-Schmiedebergs Arch. Exp. Path. Pharmak.* **254**, 159 (1966).

23. G. A. R. Johnston, P. Krogsgaard-Larsen, and A. L. Stephanson, *Nature (London)* **258**, 627 (1975).

24. D. Lodge, G. A. R. Johnston, D. R. Curtis, and S. J. Brand, *Brain Res.* **136**, 513 (1977).

25. O. Nieschulz, *Arzneim.-Forsch.* **20**, 218 (1970).

26. S. N. Pradhan and S. N. Dutta, *Psychopharmacologia* **17**, 49 (1970).

27. S. N. Pradhan and S. N. Dutta, *Int. Rev. Neurobiol.* **14**, 173 (1971).

28. G. G. Davis, *J. Am. Med. Assoc.* **64**, 711 (1915).

29. R. W. Mendelson and A. G. Ellis, *J. Trop. Med.* **27**, 274 (1924).

30. K. N. Arjungi, *Arzneim.-Forsch.* **26**, 951 (1976).

31. K. Suri, H. M. Goldman, and H. Wells, *Nature (London)* **230**, 383 (1971).

32. J. Ashby, J. A. Styles, and E. Boyland, *Lancet* **i**, 112 (1979).

33. H. F. Stich, W. Stich, and P. P. S. Lam, *Mutagen Res.* **90**, 355 (1981).

34. G. Schroeter and C. Seidler, *J. Prakt. Chem. (2)* **105**, 165 (1922).

35. G. Schroeter, C. Seidler, M. Salzbacher, and R. Kanitz, *Chem. Ber.* **65**, 432 (1932).

36. M. Dubeck and S. Kirkwood, *J. Biol. Chem.* **199**, 307 (1952).

37. E. Leete and F. H. B. Leitz, *Chem. Ind. (London),* 1572 (1957).

38. R. A. Hiles, R. U. Bjerrum, *Phytochemistry* **8**, 1927 (1969).

39. G. R. Waller, K. S. Yang, R. K. Gholson, and L. A. Hadwiger, *J. Biol. Chem.* **241**, 4411 (1966).

40. P. Fu, J. Kobus, and T. Robinson, *Phytochemistry* **11**, 105 (1972).

41. D. D. Bonnycastle, in Drill's *Pharmacology in Medicine,* J. R. Dipalma, Ed., 4th ed., McGraw-Hill, New York, 1971, p. 982.

42. P. G. Stecher, Ed., *The Merck Index,* 8th Ed., Merck, Rahway, NJ, 1968, p. 920.

43. J. M. Kingsbury, *Poisonous Plants of the United States and Canada,* Prentice-Hall, Englewood Cliffs, NJ, 1964, pp. 37, 194–197.

44. J. W. Hardin and J. M. Arena, *Human Poisoning from Native and Cultivated Plants*, Duke University Press, Durham, North Carolina, 1974, pp. 119–121.

45. S. N. Ganguli, *Phytochemistry* **9**, 1667 (1970); (CA **73**, 7744Oh).

46. E. Späth and G. Koller, *Chem. Ber.* **56**, 880 (1923).

47. R. Mukherjee and A. Chatterjee, *Chem. Ind.*, 1524 (1964).

48. R. Mukherjee and A. Chatterjee, *Tetrahedron* **22**, 1461 (1966).

49. N. F. Proskurnina, *J. Gen. Chem. (USSR)* **14**, 1148 (1944); *Chem. Abstr.* **40**, 7213 (1946).

50. M. S. Rabinovich and R. A. Konovalova, *J. Gen. Chem. (USSR)* **18**, 1510 (1948); *Chem. Abstr.* **43**, 2213 (1949).

51. T. R. Govindachari, K. Nagarajan, and S. Rajappa, *J. Chem. Soc.,* 551 (1957).

52. T. R. Govindachari, K. Nagarajan, and S. Rajappa, *J. Chem. Soc.*, 2725 (1957).

53. T. Kametani, M. Takeshima, M. Thara, and K. Fukumoto, *Heterocycles* **3**, 627 (1975).

54. T. Kametani, M. Takeshima, M. Thara, and K. Fukumoto, *J. Org. Chem.* **41**, 2542 (1976).

55. M. Plat, M. Koch, A. Bouquet, J. le Men and M.-M. Janot, *Bull. Soc. Chim. Fr.,* 781 (1963).

56. W. C. Wildman, J. Le Men, and K. Wiesner, "Monoterpene Alkaloids" in *Cyclopentanoid Terpene Derivatives,* W. I. Taylor and A. R. Battersby, Eds., Dekker, New York, 1969, Chap. 4, pp. 239-278.

57. S. K. Bhattacharya, S. Ghosal, R. K. Chaudhuri, and A. K. Sanyal, *J. Pharm. Sci.* **61,** 1838 (1972).

58. S. Ghosal, P. V. Sharma, R. K. Chaudhuri, and S. K. Bhattacharya, *J. Pharm. Sci.* **62,** 926 (1973).

59. E. Steinegger and T. Weibel, *Pharm. Acta Helv.* **26,** 333 (1951).

60. P. N. Natarajan, A. S. C. Wan, and V. Zaman, *Planta Med.* **25,** 258 (1974).

61. S. K. Bhattacharya, S. Ghosal, R. K. Chaudhuri, A. K. Singh, and P. V. Sharma, *J. Pharm. Sci.* **63,** 1341 (1974).

62. W. B. Mors, O. Gottlieb, and C. Djerassi, *J. Am. Chem. Soc.* **79,** 4507 (1957).

63. E. Ziegler and E. Nolken, *Mh. Chem.* **89,** 391 (1958).

64. D. de B. Correa and O. R. Gottlieb, *Phytochemistry* **14,** 271 (1973).

65. M. Beroza, *J. Am. Chem. Soc.* **74,** 1585 (1952).

66. M. Beroza, *J. Org. Chem.* **28,** 3562 (1963).

67. R. M. Smith, "Celastraceae Alkaloids," R. H. F. Manske, Ed., *The Alkaloids,* Vol. 16, Academic, New York, 1977, pp. 215–248.

68. W. T. Swingle, H. L. Haller, E. H. Siegler, and M. C. Swingle, *Science* 93, 60 (1941).

69. F. Acree and H. L. Haller, *J. Am. Chem. Soc.* **72,** 1608 (1950).

70. M. Beroza, *J. Am. Chem. Soc.* **73,** 3656 (1951).

71. M. Beroza, *J. Am. Chem. Soc.* **74,** 1585(1952).

72. S. M. Kupchan, W. A. Court, R. G. Dailey, C. J. Gilmore, and R. F. Bryan, *J. Am. Chem. Soc.* **94,** 7194 (1972).

73. G. Juling, Y. Shixiang, W. Xichun, X. Shixi, and L. Deda, *Chin. Med. J.* **94,** 405 (1981).

74. Q. Wanzhang, L. Chenghuang, Y. Shumei, Z. Guangdou, H. Kunyuan, F. Li, F. Shufang, T. Guantian, G. Zhiming, W. Hongtu, L. Chengzhu, J. Huiming, Z. Gusheng, S. Yongde, D. Jiahe, L. Peng, Z. Caiyi, Z. Kegang, and Q. Guowei, *Chin. Med. J.* **94,** 827 (1981).

75. W. Posselt and L. Reimann, *Geigers Mag. Pharmac.* **24,** 138 (1828).

76. L. H. F. Melsens, *Ann. Chim.* (3), **9,** 465 (1843).

77. H. Weidel, *Ann. Chem.* **165,** 328 (1873).

78. A. Pinner, *Chem. Ber.* **25,** 2807 (1892).

79. R. Laiblin, *Ann. Chem.* **196,** 173 (1879).

80. A. Pictet and P. Crépieux, *Chem. Ber.* **28,** 1904 (1895); **31,** 2018 (1898).

81. A. Pictet and A. Rotschy, *Chem. Ber.* **37,** 1225 (1904).

82. M. Nakane and C. R. Hutchinson, *J. Org. Chem.* **43,** 3972 (1978).

83. E. Leete, *J. Chem. Soc. Chem. Commun.,* 1091 (1972).

84. M. Shibagaki, H. Matsuta, S. Shibata, A. Saito, Y. Tsujino, and H. Kaneko, *Heterocycles* **19,** 1641 (1982).

85. P. Karrer and R. Widmer, *Helv. Chim. Acta* **8,** 364 (1925).

86. J. F. Whidby and J. I. Seeman, *J. Org. Chem.* **41,** 1585 (1976).

87. T. B. Pitner, W. B. Edwards, R. L. Bassfield, and J. F. Whidby, *J. Am. Chem. Soc.* **100,** 246 (1978).

88. J. D. Phillipson and S. S. Handa, *Phytochemistry* **14,** 2683 (1975).

89. A. H. Beckett, P. Jenner, and J. W. Gorrod, *Xenobiotica* **3,** 557 (1973).

90. P. Jenner and J. W. Gorrod, *Res. Commun. Chem. Pathol. Pharmacol.* **6,** 829 (1973).

91. J. Cossey and J.-P. Pete, *Tetrahedron Lett.,* 4941 (1978).

92. M. Ehrenstein, *Arch. Pharm.* **269,** 627 (1931).

93. E. Späth and E. Zajic, *Chem. Ber.* **68,** 1667 (1935).

94. E. Späth, L. Marion, and E. Zajic, *Chem. Ber.* **69,** 251 (1936).

95. C. S. Hicks and H. Le Meswuier, *Aust. J. Exptl. Biol. Med. Sci.* **13,** 175 (1935).

96. E. Späth and F. Kesztler, *Chem. Ber.* **70,** 704 (1937).

97. E. Leete, M. R. Chedekel, and G. B. Bodem, *J. Org. Chem.* **37,** 4465 (1972).

98. A. H. Warfield, W. D. Galloway, and A. G. Kallianos, *Phytochemistry* **11,** 3371 (1972).

99. A. J. N. Bolt, *Phytochemistry* **11,** 2341 (1972).

100. S. S. Hecht, C. B. Chen, and D. Hoffmann, *Acc. Chem. Res.* **12,** 92 (1979).

101. E. Späth and E. Zajic, *Chem. Ber.* **68,** 1667 (1935).

102. E. Späth and F. Kesztler, *Chem. Ber.* **70,** 2450 (1937).

103. F. Blau, *Chem. Ber.* **27,** 2535 (1894).

104. H. S. Ryang and H. Sakurai, *J. Chem. Soc. Chem. Commun.,* 594 (1972).

105. E. Späth, A. Wenusch, and E. Zajic, *Chem. Ber.* **69,** 393 (1936).

106. T. Kisaki and E. Tamaki, *Phytochemistry* **5,** 293 (1966).

107. O. Fejer-Kossey, *Phytochemistry* **11,** 415 (1972).

108. E. Leete and M. R. Chedekel, *Phytochemistry* **13,** 1853 (1974).

109. C. R. Eddy and A. Eisner, *Anal. Chem.* **26,** 1428 (1954).

110. B. Witkop, *J. Am. Chem. Soc.* **76,** 5597 (1954).

111. D. Spitzner, *Synthesis,* 242 (1977).

112. S. Brandange and L. Lindblom, *Acta Chem. Scand.* **30B,** 576 (1976).

113. E. Noga, *Fachl. Mitteil. österr. Tabakregie* **14,** 1 (1914) (*Chem. Centralblatt,* **1915,** I, 434).

114. E. Späth, J. P. Wibaut, and F. Kesztler, *Chem. Ber.* **71,** 100 (1938).

115. A. Wenusch, *Z. Unters. Lebensm.* **84,** 498 (1942) [*Chem. Abtr.* **37,** 6405 (1943)].

116. E. Werle and K. Koekbe, *Ann. Chem.* **562,** 60 (1949).

117. E. Späth and F. Kesztler, *Chem. Ber.* **70,** 704 (1937).

118. A. Pictet and A. Rotschy, *Chem. Ber.* **34,** 696 (1901).

119. A. Orechoff and G. Menschikoff, *Chem. Ber.* **64,** 266 (1931).

120. A. Orechoff, *C. R. Acad. Sci.* **189,** 945 (1929); *Arch. Pharm.* **272,** 673 (1934).

121. W. C. Evans and A. Somanabandhu, *Phytochemistry* **16,** 1859 (1977).

122. A. Orechoff and G. Menschikoff, *Chem. Ber.* **65,** 232 (1932).

123. G. Menschikoff, A. Grigorovitsch, and A. Orechoff, *Chem. Ber.* **67,** 1398 (1934).

124. F. E. Scully, Jr., *J. Org. Chem.* **45,** 1515 (1980).

125. R. Lukeš, A. A. Arojan, J. Kovař, and K. Blaha, *Coll. Czech. Chem. Commun.* **27,** 751 (1962).

126. A. Orechoff and S. Norkina, *Chem. Ber.* **65,** 1126 (1932).

127. Ya L. Gol'dfarb, R. M. Ispiryan, and I. L. Belenkii, *Izvest. Akad. Nauk SSR Ser. Khim.,* 923 (1969).

128. J. W. Wheeler, O. Olubajo, C. B. Storm, and R. M. Duffield, *Science* **211,** 1051 (1981).

129. P. M. Quan, T. K. B. Karns and L. D. Quin, *J. Org. Chem.* **30,** 2769 (1965).

130. A. Orechoff and N. Proskurnina, *Chem. Ber.* **68,** 1807 (1935); *Bull. Soc. Chim. Fr.* (5) **5,** 29 (1938).

131. C. Schöpf and F. Braun, *Naturwiss.* **36,** 377 (1945).

132. C. Schöpf, F. Braun, and F. Kreibich, *Ann. Chem.* **674,** 87 (1964).

133. I. Ribas and M. del Rosario Mendez, *Anales Real Soc. Espan. Fis. y Quim. Ser. B.* **51,** 55 (1955); *Chem. Abstr.* **49,** 8316e (1955).

134. I. Ribas and E. Rivera, *Anales Real Soc. Espan. Fis. y. Quim.* **B49,** 707 (1953); *Chem. Abstr.* **49,** 4681e (1955).

135. C. Schöpf and W. Merkel, *Naturwiss.* **53,** 274 (1966).

136. L. Costa and I. Ribas, *Anales Real Soc. Espan. Fis. y. Quim.* **B48,** 699 (1952); *Chem. Abstr.* **48,** 2721C (1954).

137. J. Dominguez, M. R. Mendez, and I. Ribas, *Anales. Real Soc. Espan. Fis. y. Quim.* (Madrid) **B52,** 133 (1956); *Chem. Abstr.* **51,** 1213 (1957).

138. I. Murakoshi, E. Kidoguchi, J. Haginiwa, S. Ohmiya, K. Higashi-Yama, and H. Otomasu, *Phytochemistry* **20,** 1407 (1981).

139. L. D. Quin, *J. Org. Chem.* **24,** 914 (1959).

140. E. Späth and F. Kuffner Tabakalkaloide", in *Fortschr. Chem. Organ. Naturst.* Vol. 2, Springer, Wien, 1939, pp. 248–300.

141. E. Späth and E. Zajic, *Chem. Ber.* **69,** 2448 (1936).

142. F. Blau, *Chem. Ber.* **24,** 327 (1892).

143. F. Kuffner and E. Kaiser, *Mh. Chem.* **85,** 896 (1954).

144. F. Kuffner and N. Faderl, *Mh. Chem.* **87,** 71 (1956).

145. T. Kisaki, S. Misuzaki, and E. Tamaki, *Phytochemistry* **7,** 323 (1968).

146. A. Pinner, *Arch. Pharm.* **231,** 378 (1893).

147. W. G. Frankenburg and A. A. Vaitekunas, *J. Am. Chem. Soc.* **79,** 149 (1957).

148. W. L. Alworth, *Diss. Abstr.* **29,** 482B (1969).

149. B. Testa and P. Jenner, *Mol. Pharmacol.* **9,** 10 (1973).

150. M. Noguchi, H. Sakuma, and E. Tamaki, *Arch. Biochem. Biophys.* **125,** 1017 (1968); *Phytochemistry* **7,** 1861 (1968).

151. M. Frankel and Y. Knobler, *J. Am. Chem. Soc.* **80,** 3147 (1980).

152. P. S. Larson, H. B. Haag, and H. Silvette, *Tobacco: Experimental and Clinical Studies,* The Williams and Wilkins Company, Baltimore, 1961.

153. M. S. G. Clark, M. J. Rand, and S. Vanov, *Arch. Int. Pharmacodyn.* **156,** 363 (1965).

154. H. Nishimura and K. Nakai, *Science* **127,** 877 (1958).

155. M. W. Crowe, in *Effects of Poisonous Plants on Livestock,* R. F. Keeler, K. R. Van Kampen and L. F. James, Eds., Academic Press, New York, 1978.

156. R. F. Keeler and M. W. Crowe, unpublished observation.

157. M. Rindl and M. L. Sapiro, *Onderstepoort J. Vet. Sci. Anim. Ind.* **22,** 301 (1949).

158. C. R. Smith, *J. Am. Chem. Soc.* **57,** 959 (1935).

159. R. F. Keeler, *Clin. Tox.* **15,** 417 (1979).

160. R. F. Keeler and L. D. Balls, *Clin. Tox.* **12,** 49 (1978).

161. M. Masore, *C. R. Acad. Sci.* **204,** 890 (1937).

162. R. Adams and V. Jones, *J. Am. Chem. Soc.* **69,** 1803 (1947).

163. J. L. Johnson, *J. Am. Chem. Soc.* **69,** 1810 (1947).

164. M. Mascré, *C. R. Acad. Sci.* **204,** 890 (1937).

165. J. P. Wibaut and J. P. Schuhmacher, *Rec. Trav. Chim.* **71,** 1017 (1952).

166. S. L. Bonting and F. R. Schepman, *Rec. Trav. Chim.* **69,** 1007 (1950).

167. H. P. Tiwari, W. R. Penrose, and I. D. Spenser, *Phytochemistry* **6,** 1245 (1967).

168. J. P. F. D'Mello and D. E. Taplin, *World Rev. Anim. Prod.* **14,** 41 (1978).

169. R. J. Jones, *World Anim. Rev.* **31,** 13 (1979).

170. E. M. Hutton and R. A. Bray, *Aust. CSIRO Division of Tropical Crops and Pastures Divisional Report* 1976–77, 1977.

171. Morris (1896). Cited in Tropenplanzen, p. 200, L. Zoo., 1897.

172. D. Kostermans, *Rec. Trav. Chim.* **65,** 319 (1946).

173. F. C. Kraneveld and R. Djaenoedin, *Hemera Zoa.* **57,** 623 (1950).

174. M. P. Hegarty, P. G. Schinckel, and R. D. Court, *Aust. J. Agric. Res.* **15,** 153 (1964).

175. W.-C. Tsai and K.-H. Ling, *Toxicon* **9,** 241 (1971).

176. M. P. Hegarty, C. P. Lee, G. S. Christie, F. G. DeMunk, and R. D. Court, *Aust. J. Biol. Sci.* **31,** 115 (1978).

177. A. H. Reisner, C. A. Bucholtz, and K. A. Ward, *Mol. Pharmacol.* **16,** 278 (1979).

178. W. D. DeWys and T. C. Hall, *Eur. J. Cancer* **9,** 281 (1973).

179. E. L. Willet, J. H. Quisenberry, L. A. Henke, and C. Maruyama, *Hawaii Agric. Exp. Stat. Bien. Rep.* 1944–1946, pp. 46–47, 1947.

180. O. Wayman and I. I. Iwanaga, *Proc. West. Sect. Am. Soc. Anim. Prod.* **8,** 1 (1957).

181. J. R. Sandoval, *Phillipp. Agric.* **38,** 374 (1955).

182. J. W. Hylin and I. J. Lichton, *Biochem. Pharmacol.* **14,** 1167 (1965).

183. H. S. Joshi, *Aust. J. Agric. Res.* **19,** 341 (1968).

184. S. Dewreede and O. Wayman, *Teratology* **3,** 21 (1970).

185. R. K. Yoshida, Ph.D. thesis, University of Minnesota, 1944.

186. H. Matsumoto, E. G. Smith, and G. D. Sherman, *Arch. Biochem. Biophys.* **33,** 201 (1951).

187. L. von Sallmann, P. Grimes, and E. Collins, *Am. J. Ophthalmol.* **47,** 107 (1959).

188. R. I. Hamilton, L. E. Donaldson, and L. J. Lambourne, *Aust. Vet. J.* **44,** 484 (1968).

189. J. H. G. Holmes, *Proc. Aust. Soc. Anim. Prod.* **11,** 453 (1976).

190. R. J. Jones, C. G. Blunt, and J. H. G. Holmes, *Trop. Grassl.* **10,** 113 (1976).

191. M. P. Hegarty, C. P. Lee, G. S. Christie, R. D. Court, and K. P. Haydock, *Aust. J. Biol. Sci.* **32,** 27 (1979).

192. J. F. Thompson, C. J. Morris, and I. K. Smith, *Ann. Rev. Biochem.* **38,** 137 (1969).

193. M. P. Hegarty and P. J. Peterson, in *Chemistry and Biochemistry of Herbage,* Vol. 1, G. W. Butler and R. W. Bailey, Eds., Academic, London, 1973, pp. 1–62.

194. C. G. Casinovi, G. Grandolini, R. Mercantini, N. Oddo, R. Oliviari, and A. Tonolo, *Tetrahedron Lett.,* 3175 (1968).

195. P. S. White, J. A. Findlay, and W. H. J. Tam, *Can. J. Chem.* **56,** 1904 (1978).

196. N. N. Girotra, Z. S. Zelawski, and N. L. Wendler, *J. Chem. Soc. Chem. Commun.,* 566 (1976).

197. N. N. Girotra and N. L. Wendler, *Heterocycles* **9,** 417 (1978).

198. N. N. Girotra, A. A. Patchett, S. B. Zimmerman, D. L. Achimov, and N. L. Wendler, *J. Med. Chem.* **23,** 209 (1980).

199. E. Medina and G. Spitteler, *Chem. Ber.* **114,** 814 (1981).

200. E. Medina and G. Spitteler, *Ann. Chem.,* 538 (1981).

201. G. Voss and H. Gerlach, *Ann. Chem.,* 1466 (1982).

202. T. Terashima, Y. Kuroda, and Y. Kaneko, *Tetrahedron Lett.,* 2535 (1969).

203. M. Pailer and E. Haslinger, *Mh. Chem.* **101,** 508 (1970).

204. H. W. Gschwend, *Tetrahedron Lett.,* 271 (1970).

205. M. Onda, Y. Konda, Y. Narimatsu, H. Tanaka, J. Awaya, and S. Omura, *Chem. Pharm. Bull. (Japan)* **23,** 2462 (1975).

206. H. L. Sleeper and W. Femical, *J. Am. Chem. Soc.* **99,** 2367 (1977).

207. T. Sakan, A. Fujino, F. Murai, A. Suzui, Y. Butsugan, and Y. Terashima, *Bull. Chem. Soc., Japan* **53,** 712 (1960).

208. D. Gross, G. Edner, and H. R. Schütte, *Arch. Pharm.* **304,** 19 (1971).

209. G. W. Cavill, P. L. Robertson, D. V. Clark, R. Duke, C. J. Orton, and W. D. Plant, *Tetrahedron* **38,** 1931 (1982).

210. M.-M. Janot, J. Guilhem, O. Contz, G. Venera, and E. Ciong, *Ann. Pharm. Fr.* **37**, 413 (1979).

211. L. B. Davies, S. G. Greenburg, and P. G. Sammes, *J. Chem. Soc. Perkin Trans. I,* 1909 (1981).

212. M. Nitta, A. Sekiguchi, and H. Koba, *Chem. Lett.,* 933 (1982).

213. G. W. K. Cavill and A. Zeitlin, *Aust. J. Chem.* **20**, 349 (1967).

214. J. Wolinsky and D. Chan, *J. Org. Chem.* **30**, 41, (1965).

215. G. G. Ayerst and K. Schofield, *J. Chem. Soc.,* 4097 (1958).

216. K. J. Clark, G. E. Fray, R. H. Jaeger, and R. Robinson, *Tetrahedron* **6**, 217 (1959).

217. D. Gross, W. Berg, and H. R. Schütte, *Biochem. Physiol. Pflanz.* **163**, 576 (1972).

218. H. Auda, H. R. Juneja, E. J. Eisenbraun, G. R. Waller, W. R. Kays, and H. H. Appel, *J. Am. Chem. Soc.* **89**, 2476 (1967).

219. T. Sakan, F. Murai, Y. Hayashi, Y. Honda, T. Shono, M. Nakajima, and M. Kato, *Tetrahedron* **23**, 4635 (1967).

220. H. Ripperger, *Pharmazie* **34**, 557 (1979).

221. A. V. Danilova and R. A. Konovalova, *Zh. Obshch. Khim.* **22**, 2237 (1952); *Chem. Abstr.* **48**, 691a (1954).

222. A. V. Danilova, *Zh. Obsch. Khim.* **26**, 2069 (1956); *Chem. Abstr.* **51**, 5098d (1957).

223. K. Torssell, *Acta. Chem. Scand.* **22**, 2715 (1968).

224. J. Koyama, T. Sugita, Y. Sujuta, and H. Irie, *Heterocycles* **12**, 1017 (1979).

225. V. Regnault, *Ann. Chim. Phys.* (2), **68**, 158 (1838).

226. T. Wertheim, *Ann. Chem.* **70**, 58 (1849).

227. R. Fittig and W. H. Mielch, *Ann. Chem.* **172**, 134 (1874).

228. L. Rügheimer, *Chem. Ber.* **15**, 1390 (1882).

229. S. Tsuboi and A. Takeda, *Tetrahedon Lett.,* 1043 (1979).

230. A. Schulze and H. Oediger, *Ann. Chem.,* 1725 (1981).

231. H. Lohaus and H. Gall, *Ann. Chem.* **517**, 278 (1935).

232. E. Ott and F. Eichler, *Chem. Ber.* **55**, 2563 (1922).

233. D. Dwuma-Badu, J. Jadot, G. Dardenne, M. Marlier, and J. Casimir, *Phytochemistry* **15**, 747 (1976).

234. H. D. Scharf and J. Janus, *Tetrahedron* **35**, 385 (1979).

235. I. Addae-Mensah, F. G. Torto, and I. Baxter, *Tetrahedron Lett.,* 3049 (1976).

236. L. Addae-Mensah, F. G. Torto, B. Torto, and H. Achenbach, *Planta Med.* **41**, 200 (1981).

237. S. Linke, J. Kurz, and H.-J. Zeiler, *Ann. Chem.,* 1142 (1982).

238. C. K. Atal and S. S. Banga, *Current Sci. (India)* **32**, 354 (1963).

239. A. Chatterjee and C. P. Dutta, *Sci. Cult. (Calcutta)* **29**, 568 (1963).

240. C. K. Sehgal, P. L. Kachroo, R. L. Sharma, S. C. Taneja, K. L. Dhar, and C. K. Atal, *Phytochemistry* **18**, 1865 (1979).

241. S. Prakash, M. A. Khan, H. Khan, and A. Zaman, *Phytochemistry* **22**, 1836 (1983).

242. S. Dasgupta and A. B. Ray, *Indian J. Chem.* (B) **17**, 538 (1979).

243. R. M. Smith, *Tetrahedron* **35**, 437 (1979).

244. H. Greger, M. Grenz, and F. Bohlmann, *Phytochemistry* **20**, 2579 (1981).

245. H. Greger, M. Grenz, and F. Bohlmann, *Phytochemistry* **21**, 1071 (1982).

246. L. Mikhailova, *Izv. Mikrobiol. Inst.* **21** (a), 277, (b) 291 (1970); *Chem. Abstr.* **75**, 45926w (1971).

247. Y. Q. Pei, *Epilepsia* **24**, 177 (1983).

248. M. R. Srinivasan and M. N. Satyanarayana, *Nut. Reports Int.* **23**, 871 (1981).

249. E. Ajitey-Smith and I. Addae-Mensah, *Proc. West African Soc. Pharmacol.,* 7th Scientific Meeting, Zaria, 1977.

250. I. Addae-Mensah, F. G. Torto, C. I. Dimonyeka, I. Baxter, and J. K. M. Sanders, *Phytochemistry* **16,** 757 (1977).

251. A. Ladenburg, *Chem. Ber.* **19,** 439 (1886).

252. W. Leithe, *Chem. Ber.* **65,** 927 (1932).

253. E. Leete, *J. Am. Chem. Soc.* **85,** 3523 (1963).

254. E. Leete, *J. Am. Chem. Soc.* **86,** 2509 (1964).

255. E. Leete and N. Audityachaudhury, *Phytochemistry* **6,** 219 (1967).

256. M. F. Roberts, *Phytochemistry* **10,** 3057 (1971).

257. O. Diels and K. Alder, *Ann. Chem.* **498,** 1 (1932).

258. G. Fodor and G. A. Cooke, *Tetrahedron Suppl.* **8,** 113 (1966).

259. A. Kolliker, *Arch. Path. Anat.* **10,** 235 (1856).

260. A. R. Cushny, *J. Exp. Med.* **1,** 202 (1896).

261. B. Moore and R. Row, *J. Physiol.* **22,** 273 (1898).

262. J. N. Langley and W. C. Dickenson, *J. Physiol.* **11,** 509 (1890).

263. A. R. Cushny, *A Textbook of Pharmacology and Therapeutics,* Lea Brothers, Philadelphia, 1899.

264. D. M. R. Culbreth, *A Manual of Materia Medica and Pharmacology,* Lea & Febiger, Philadelphia and New York, 1917, pp. 411–415.

265. J. W. Fairbairn and S. B. Challen, *Biochem. J.* **72,** 556 (1959).

266. N. V. Mody, R. Henson, P. A. Hedin, U. Kokpol, and D. H. Miles, *Experientia* **32,** 829 (1976).

267. J. De Boer, *Arch. Int. Pharmacodyn.* **83,** 473 (1950).

268. W. C. Bowman and I. S. Sanghvi, *J. Pharm. Pharmacol.* **15,** 1 (1963).

269. S. R. Sampson, D. W. Esplin, and B. Zablocka, *Proc. West. Pharmacol. Soc.* **6,** 4 (1963).

270. S. R. Sampson, D. W. Esplin, and B. Zablocka, *J. Pharmacol. Exp. Ther.* **152,** 313 (1966).

271. S. R. Sampson, *Int. J. Neuropharmacol.* **5,** 171 (1967).

272. D. R. Curtis and G. A. R. Johnston, in *Neuropoisons,* Vol. 2, L. L. Simpson and D. R. Curtis, Eds., Plenum, New York, 1974, Chap. 6.

273. W. F. von Oettingen, *The Therapeutic Agents of the Pyrrole and Pyridine Group,* Edwards Brothers, Ann Arbor, MI, 1936, p. 85.

274. E. J. Church, *The Trial and Death of Socrates.* Macmillan, London, 1896, p. 211–213.

275. E. P. Claus, *Gathercoal, and Wirth, Pharmacognosy.* Lea & Febiger, Philadelphia, 1956, pp. 434–436.

276. R. F. Keeler, *Clin. Tox.* **7,** 195 (1974).

277. R. F. Keeler and L. D. Balls, *Clin. Tox.* **12,** 49 (1978).

278. R. Willstätter, *Chem. Ber.* **34,** 3166 (1901).

279. E. Späth and E. Adler, *Mh. Chem.* **63,** 127 (1933).

280. R. K. Hill, *J. Am. Chem. Soc.* **80,** 1609 (1958).

281. E. Galinowsky and H. Mulley, *Mh. Chem.* **79,** 426 (1948).

282. G. Fodor and E. Bauerschmidt, *J. Heterocyclic Chem.* **5,** 205 (1968).

283. E. Leete and J. O. Olson, *J. Am. Chem. Soc.* **94,** 5472 (1972).

284. E. Späth, F. Kuffner, and L. Ensfellner, *Chem. Ber.* **66,** 591 (1933).

285. K. Balenović and N. Stimač, *Croat. Chim. Acta* **29,** 153 (1957).

286. R. K. Hill, *J. Am. Chem. Soc.* **80,** 1611 (1958).

287. K. E. Harding and S. R. Burks, Abstract, ORGN 122, 183rd ACS National Meeting, Washington, DC, August 1983; *J. Org. Chem.* **49,** 40 (1984).

288. C. Tanret, *C. R. Acad. Sci.* **86**, 1270 (1878); **90**, 695 (1880).

289. K. Hess and A. Eichel, *Chem. Ber.* **50**, 1386 (1917).

290. K. Hess and A. Eichel, *Chem. Ber.* **51**, 741 (1918).

291. K. Hess, *Chem. Ber.* **52**, 1005 (1919).

292. R. Gilman and L. Marion, *Bull. Soc. Chim. Fr.,* 1993 (1961).

293. J. Quick and R. Oterson, *Synthesis,* 745 (1976).

294. E. Leete, *J. Am. Chem. Soc.* **91**, 1697 (1969).

295. R. N. Gupta and I. D. Spenser, *Phytochemistry* **8**, 1937 (1969).

296. M. Keogh and D. G. O'Donovan, *J. Chem. Soc.* (C), 1792 (1970).

297. E. Leistner and I. D. Spenser, *J. Am. Chem. Soc.* **95**, 4715 (1973).

298. J. E. Braekman, R. N. Gupta, D. B. MacLean, and I. D. Spenser, *Can. J. Chem.* **50**, 2591 (1972).

299. B. Holmstedt and G. Liljestrand, *Readings in Pharmacology,* MacMillan, New York, 1963, p. 2.

300. R. C. Wren, *Potter's New Cyclopaedia of Botanical Drugs and Preparations,* Sir Isaac Pitman & Sons, London, 1956, pp. 243, 244.

301. E. P. Claus, *Gathercoal and Wirth Pharmacognosy,* Lea & Febiger, Phildelphia, 1956, p. 432.

302. W. T. Salter, *A Textbook of Pharmacology,* W. B. Saunders, Philadelphia and London, 1952, pp. 1128–1143.

303. J. U. Lloyd and G. D. Lloyd, *Drugs and Medicine of North America 2,* 1886, p. 65.

304. B. Franck, *Chem. Ber.* **91**, 2803 (1958).

305. H. C. Beyerman, L. Maat, A. van Veen, A. Zweistra, and E. v. Philipsborn, *Rec. Trav. Chim.* **84**, 1367 (1965).

306. G. Fodor and D. Butruille, *Chem. Ind. (London),* 1437 (1968).

307. D. Butruille, G. Fodor, C. Saunderson-Huber, and F. Létourneau, *Tetrahedron* **27**, 2055 (1971).

308. H. C. Beyerman, L. Maat, J. P. Visser, J. Cymerman-Craig, R. P. Chan, and S. K. Roy, *Rec. Trav. Chim.* **88**, 1012 (1969).

309. C. Schöpf, G. Dummer, and W. Wüst, *Ann. Chem.* **626**, 134 (1959).

310. J. J. Tufariello and S. A. Ali, *Tetrahedron Lett.,* 4647 (1978).

311. H. C. Beyerman, J. Eenshvistra, and W. Eveleens, *Rec. Trav. Chim.* **76**, 415 (1957); H. C. Beyerman, J. Eenshvistra, W. Eveleens, and E. Zweistra, *Rec. Trav. Chim.* **78**, 43 (1959).

312. C. Hootelé, B. Colan, F. Halin, J. P. Declercq, G. Germain, and M. Van Meersch, *Tetrahedron Lett.,* 5063 (1980).

313. C. Houtelé, J. P. Etienne, and B. Colau, *Bull. Soc. Chim. Belg.* **88**, 111 (1979).

314. R. N. Gupta and I. D. Spenser, *Can. J. Chem.* **45**, 1275 (1967).

315. A. Nordal, *A Pharmacognostical Study of Sedum acre L.,* Johan Grundt Tanum Forlag, Oslo, 1946.

316. L. P. S. Francis and G. W. Francis, *Planta Med.* **32**, 268 (1977).

317. G. Büchi, W. Broger, and G. Schmidt, *Helv. Chim. Acta* **48**, 275 (1965).

318. W. D. Erdmann, H. J. Ruff, and G. Schmidt, *Arzneim.-Forsch* **11**, 835 (1961).

319. G. Büchi, F. Fracher, W. Broger, and H. Tiedemann, *Pharm. Acta Helv.* **50**, 337 (1975).

320. H. Wieland, *Chem. Ber.* **54**, 1784 (1921).

321. H. Wieland, C. Schöpf, and W. Hermsen, *Ann. Chem.* **444**, 40 (1925).

322. H. Wieland and O. Dragendorff, *Ann. Chem.* **473**, 83 (1929).

323. C. Schöpf and G. Lehman, *Ann. Chem.* **518**, 1 (1935).

324. C. Schöpf, E. Müller, and E. Schenkenberger, *Ann. Chem.* **687**, 241 (1965).

325. A. Ebnöther, *Helv. Chim. Acta* **41**, 386 (1958).

326. R. N. Gupta and I. D. Spenser, *Can. J. Chem.* **49**, 384 (1971).

327. M. F. Keogh, and D. G. O'Donovan, *J. Chem. Soc.* (C), 247 (1970).

328. A. Hellwig, *Zentralbl. Chir.* **48**, 731 (1921).

329. H. Wieland, A. Eckstein, and E. Rominger, *Ztschr. Kinderh.* **28**, 218 (1921).

330. F. R. Curtis and S. Wright, *Lancet* **ii**, 1255 (1926).

331. W. J. R. Camp, *J. Pharmacol. Exp. Ther.* **31**, 393 (1927).

332. C. W. Edmunds, *Am. J. Physiol.* **11**, 79 (1904).

333. M. Nisisita, *Okayama-Igakkai Zasshi* **39**, 1985 (1927); *Chem. Abstr.* **24**, 5876 (1930).

334. S. M. Mansuri, V. V. Kelkar, and M. N. Jindal, *Arzneim.-Forsch.* **23**, 1721 (1973).

335. H. Wieland and R. Mayer, *Arch. Exp. Path. Pharmakol.* **92**, 195 (1922).

336. M. I. Steinberg and R. L. Volle, *Naunyn-Schmiedeberg's Arch. Pharmacol.* **272**, 16 (1972).

337. J. L. Dorsey, *Ann. Int. Med.* **10**, 628 (1936).

338. I. S. Wright and D. Littauer, *J. Am. Med. Assoc.* **109**, 649 (1937).

339. P. W. Bradshaw, *Int. J. Addict.* **8**, 353 (1973).

340. W. H. Tallent, V. L. Stromberg, and E. C. Horning, *J. Am. Chem. Soc.* **77**, 3361 (1955).

341. E. Leete and R. A. Carver, *J. Org. Chem.* **40**, 2151 (1975).

342. R. K. Hill, T. H. Chan, and J. A. Joule, *Tetrahedron* **21**, 147 (1965).

343. E. Leete and K. N. Juneau, *J. Am. Chem. Soc.* **91**, 5614 (1969).

344. E. Leete, J. C. Lechleiter, and R. A. Carver, *Tetrahedron Lett.*, 3779 (1975).

345. M. Marlier, G. A. Dardenne, and J. Cassimir, *Phytochemistry* **11**, 2597 (1972).

346. L. E. Fellows, E. A. Bell, D. G. Lynn, F. Pilkiewicz, I. Miura, and K. Nakanishi, *J. Chem. Soc., Chem. Commun.*, 977 (1979).

347. D. P. Rainey, E. R. Smalley, M. H. Crump, and F. M. Strong, *Nature*, **205**, 203 (1965).

348. S. D. Aust, H. P. Broquist, and K. L. Rinehart, Jr., *J. Am. Chem. Soc.* **88**, 2879 (1966).

349. R. A. Gardiner, K. L. Rinehart, Jr., J. J. Snyder, and H. P. Broquist, *J. Am. Chem. Soc.* **90**, 5639 (1968).

350. W. J. Gensler and N. W. Hu, *J. Org. Chem.* **38**, 3848 (1977).

351. E. Clevestine, H. P. Broquist, and T. M. Harris, *Biochemistry* **18**, 3663 (1979).

352. H. Schildknecht, D. Krauss, J. Connert, H. Essenbreis, and N. Orfanides, *Angew. Chem. Int. Ed.* **14**, 427 (1974).

353. H. Schildknecht, D. Berger, D. Krauss, J. Connert, J. Gehlhaus, and H. Essenbreis, *J. Chem. Ecol.* **2**, 1 (1976).

354. K. Fuji, K. Ichikawa, and E. Fujita, *Chem. Pharm. Bull. (Japan)* **27**, 3183 (1979).

355. J. G. MacConnell, M. S. Blum, and H. M. Fales, *Tetrahedron* **26**, 1129 (1971).

356. J. M. Brand, M. S. Blum, H. M. Fales, and J. G. MacConnell, *Toxicon* **10**, 259 (1972).

357. R. B. Koch, D. Desaiah, D. Foster, and K. Ahmed, *Biochem. Pharmacol.* **26**, 983 (1977).

358. A.-S. Abdelfattah and R. B. Koch, *Biochem. Pharmacol.* **30**, 3195 (1981).

359. J. Bruneton, A. Cavé, and R. R. Paris, *Plant Med. Phytother.* **9**, 21 (1975); *Chem. Abstr.* **83**, 283992 (1975).

360. J. Bruneton and A. Cavé, *Tetrahedron Lett.*, 739 (1975).

361. J. L. van Ryn, *Arch. Pharm.* **231**, 184 (1893); **235**, 332 (1897).

362. M. Spitteler-Friedmann and G. Spitteler, *Mh. Chem.* **95**, 1234 (1964).

363. G. Barger, R. Robinson, and T. S. Work, *J. Chem. Soc.*, 711 (1937).

364. M. S. Narasimhan, *Chem. Ind. (London)* 1526 (1956).

365. G. Fodor, V. Sankaran, and J.-P. Fumeaux, *Synthesis*, 464 (1972).

366. T. R. Govindachari and N. S. Narasimhan, *J. Chem. Soc.,* 1563 (1955).

367. H. Rappoport and H. D. Baldridge, Jr., *J. Am. Chem. Soc.* **74,** 5365 (1952).

368. M. Tichy and J. Sicher, *Tetrahedron Lett.,* 511 (1962).

369. J. L. Coke and W. Y. Rice, *J. Org. Chem.* **30,** 3420 (1965).

370. E. J. Corey, K. C. Nicolaou, and L. S. Melvin, *J. Am. Chem. Soc.* **97,** 654 (1975).

371. E. Brown and S. A. Bourgouin, *Tetrahedron* **31,** 1047 (1975).

372. M. Natsume and M. Ogawa, *Heterocycles* **12,** 159 (1979); **14,** 169 (1980).

373. S. Hannesian and R. Frenette, *Tetrahedron Lett.,* 3391 (1979).

374. C. W. L. Bevan and A. V. Ogan, *Phytochemistry* **3,** 591 (1964).

375. M. Greshoff, *Buitenzorg.* **7,** 5 (1890).

376. D. G. Fairchild, *Garden Islands of the Great East,* Schribner's, New York, 1943.

377. I. G. Noble, *Habana* **85,** 198 (1946).

378. S. To and C. Kyu, *J. Med. Sci.* **8,** 52 (1934).

379. A. M. Ramaswamy and M. Sirsi, *Indian J. Pharm.* **22,** 34 (1960).

380. L. Oliveros-Balardo, V. A. Masilungan, V. Cardeno, L. Luna, F. De Vera, E. de la Cruz, and E. Valmonte, *Possible Anti-Tumor Constituent of Carica papaya L.* Scientific Session, Third Asian Congress of Pharmaceutical Sciences, Manila, Philippines, Nov. 16–21, 1970.

381. B. J. M. Tuffley and C. H. Williams, *Aust. J. Pharm.* **52,** 796 (1951).

382. C. A. Hornick, L. I. Sanders, and Y. C. Lin, *Res. Commun. Chem. Path. Pharmacol.* **22,** 277 (1978).

383. R. J. Highet, *J. Org. Chem.* **29,** 471 (1964).

384. R. J. Highet and P. F. Highet, *J. Org. Chem.* **31,** 1275 (1966).

385. W. Y. Rice and J. L. Coke, *J. Org. Chem.* **31,** 1010 (1966).

386. E. Brown and A. Bonte, *Tetrahedron Lett.,* 2881 (1975).

387. D. Lythgoe and M. J. Vernengo, *Tetrahedron Lett.,* 1133 (1967).

388. V. U. Ahmad, A. Basha, and W. Haque, *Z. Naturforsch.* **33b,** 347 (1978).

389. I. Christofidis, A. Weller, and J. Jadot, *Tetrahedron* **33,** 3005 (1977).

390. G. Ratle, X. Monseur, B. C. Das, J. Yassi, Q. Khuong-Huu, and R. Goutarel, *Bull. Soc. Chim. Fr.,* 2945 (1966).

391. G. Ratle, thesis, Faculté des Sciences Université de Paris, Orlay, 1968.

392. Q. Khuong-Huu, G. Ratle, X. Monseur, and R. Goutarel, *Bull. Soc. Chim. Belg.* **81,** 425, 443 (1972).

393. A. J. G. Baxter and A. B. Holmes, *J. Chem. Soc. Perkin I,* 2343 (1977).

394. G. Fodor, Abstract ORGN 069, Lecture at the 169th ACS National Meeting, Philadelphia, 1975; V. Sankaran, Thesis, West Virginia University, 1972.

395. M. Natsume and M. Ogawa, *Heterocycles* **16,** 973 (1981).

396. Y. Saitoh, Y. Moriyama, and Q. Khuong-Huu, *Tetrahedron Lett.* **21,** 75 (1980).

397. Y. Saitoh, Y. Moriyama, H. Hirota, T. Takahashi, and Q. Khoung-Huu, *Bull. Chem. Soc. Japan* **54,** 283 (1981).

398. J. M. Watt and M. G. Breyer-Brandwijk, *Medicinal and Poisonous Plants of Southern and Eastern Africa,* Livingstone, Edinburgh-London, 1962.

399. P. Bourrinet and A. Quevauviller, *Ann. Pharm. Franc.* **26,** 787 (1968).

400. A. D. Harmon, U. Weiss, and J. V. Silverton, *Tetrahedron Lett.,* 721 (1979).

401. J. W. Daly, I. Karle, C. W. Myers, T. Tokuyama, J. A. Waters, and B. Witkop, *Proc. Nat. Acad. Sci. USA* **68,** 1870 (1971).

402. T. Tokiyama, K. Uenoyama, G. Brown, J. W. Daley, and B. Witkop, *Helv. Chim. Acta* **57,** 2597 (1974).

403. J. W. Daly, B. Witkop, T. Tokuyama, T. Nishikawa, and I. Karle, *Helv. Chim. Acta* **60**, 1128 (1977).

404. I. L. Karle, *J. Am. Chem. Soc.* **95**, 4036 (1973).

405. D. Evans, E. Thomas, and R. Cherpeck, *J. Am. Chem. Soc.* **104**, 3694 (1982).

406. K. Takahashi, B. Witkop, A. Brossi, M. A. Maleque, and E. X. Albuquerque, *Helv. Chim. Acta* **65**, 252 (1982).

407. T. Ibuka, H. Minakata, Y. Mitsui, E. Tabushi, T. Taga, and Y. Inubushi, *Chem. Lett.,* 1409 (1981).

408. T. Ibuka, Y. Mitsui, K. Hayashi, H. Minakata, and Y. Inubushi, *Tetrahedron Lett.* **22**, 4425 (1981).

409. E. X. Albuquerque, E. A. Barnard, T. H. Chiu, A. J. Lapa, J. O. Dolly, S.-E. Jansson, J. W. Daly, and B. Witkop, *Proc. Nat. Acad. Sci.* **70**, 949 (1973).

410. A. J. Lapa, W. X. Albuquerque, J. M. Starvey, J. W. Daly, and B. Witkop, *Exp. Neurol.* **47**, 558 (1975).

411. W. Bürgermeister, W. L. Klein, M. Nirenberg, and B. Witkop, *Mol. Pharmacol.* **14**, 751 (1978).

412. S. M. J. Dunn, S. G. Blanchard, and M. A. Raftery, *Biochemistry* **20**, 5617 (1981).

413. A. A. Ibragimov, Z. Osmanov, B. Tashchodzhaev, N. D. Abudullaev, M. R. Yagudaev, and S. Yu. Yunusov, *Khim. Prir. Soedin.,* 623 (1981).

414. Z. Osmanov, A. A. Ibragimov, and S. Yu. Yunusov, *Khim. Prir. Soedin* 126 (1982); *Chem. Abstr.* **97**, 20680c (1982).

415. G. Ciamician and P. Silber, *Chem. Ber.* **29**, 490 (1896).

416. A. Piccinini, *Gazz. Chim. Ital.* **29**, II, 311 (1899).

417. R. Willstätter and H. Veraguth, *Chem. Ber.* **38**, 1975 (1905).

418. A. Piccinini, *Atti Accad. Lincei V* **8**, (ii), 219 (1899).

419. R. C. Menzies and R. Robinson, *J. Chem. Soc.* **125**, 2163 (1924).

420. A. C. Cope, H. L. Dryden, Jr., C. G. Overgerger, and A. A. D'Addieco, *J. Am. Chem. Soc.* **73**, 3416 (1951).

421. A. C. Cope, H. L. Dryden, and C. F. Howell, *Org. Synth. Coll. IV,* 816 (1963).

422. N. K. Hart, S. R. Johns, and J. A. Lamberton, *Aust. J. Chem.* **20**, 561 (1967).

423. K. Alder, H. Betzing, and R. Kuth, *Ann. Chem.* **620**, 73 (1959).

424. B. Tursch, J. C. Braekman, D. Daloze, C. Hootele, D. Losman, R. Karlsson, and J. M. Pasteels, *Tetrahedron Lett.* 201 (1973).

425. E. Gössinger and B. Witkop, *Mh. Chem.* **111**, 803 (1980).

426. R. K. Hill and L. Renbaum, *Tetrahedron* **38**, 1959 (1982).

427. B. Tursch, C. Chome, J. C. Brackman, and D. Daloze, *Bull. Soc. Chim. Belg.* **82**, 699 (1973).

428. D. H. Gnecco Medina, D. S. Grierson, and H.-P. Husson, *Tetrahedron Lett.* **24**, 2099 (1983).

429. E. Leete "Alkaloids Derived from Ornithine, Lysine and Nicotinic Acid," in *Encyclopedia of Plant Physiology, Secondary Plant Products,* E. A. Bell and B. V. Charlwood, Eds., vol. 8, Springer, Berlin, Heidelberg, 1980, p. 69.

430. W. V. Brown and B. P. Moore, *Aust. J. Chem.* **35**, 1255 (1982).

431. G. B. Fodor, "Alkaloids Derived from HISTIDINE and other Precursors" in *Encyclopedia of Plant Physiology, Secondary Plant Products,* E. A. Bell and B. V. Charlwood, Eds., vol. 8, Springer-Verlag, Berlin, Heidelberg, 1980.

Chapter Two

The Indolosesquiterpene Alkaloids of the Annonaceae

Peter G. Waterman
Phytochemistry Research Laboratory
University of Strathclyde
Glasgow, Scotland, U.K.

CONTENTS

1. INTRODUCTION

The term *indolosesquiterpene* was coined by Leboeuf et al. [1] to describe polyalthenol **(1)**, an alkaloid characterized by its ready division into indole and sesquiterpene (C_{15}) subunits. Since the initial report of **(1)** in 1976, the isolation and structure elucidation of 11 other indolosesquiterpenes have been recorded. The structures of several of these alkaloids have later been revised or have yet to be confirmed.

To date, all of the indolosesquiterpenes that have been characterized have been isolated from two closely allied West African species of Annonaceae: *Greenwayodendron (Polyalthia) oliveri* (Engl.) Verdc. and *Greenwayodendron (Polyalthia) suaveolens* (Engl. & Diels) Verdc. Until recently *G. oliveri* and *G. suaveolens* were considered to be part of the large and heterogeneous pantropical genus *Polyalthia*. However, the floral characters of these two species differ markedly from those of *Polyalthia sensu strictu* and, according to Verdcourt [2] there are no grounds for their retention within the larger genus; in fact they are not even closely related [3]. Unfortunately much of the work carried out on these two *Greenwayodendron* species have been under the former generic name of *Polyalthia* and as a result many of the alkaloids have trivial names relating to the latter, for example, polyalthenol, polyveoline, polyavolensin.

The purpose of this chapter is to review the literature relating to the identification of the indolosesquiterpenes and then to speculate on their possible mode of biogenesis. The alkaloids have been divided into two groups depending on whether there is a single link between the hetero ring of indole and the sesquiterpene (the tetracyclic group) or two links with the formation of an additional ring (the pentacyclic group). For both groups the numbering system adopted in this review is the same, giving priority to the sesquiterpene system (see **1**).

The following two sections concentrate on reviewing the data presently available on the alkaloids of these two groups in relation to the proposed structures. As the reader will see, there is still a considerable amount of work required.

The tetracyclic group of polyalthenol, isopolyalthenol, and neopolyalthenol have been assigned structures entirely on the basis of arguments developed from spectral data. As far as can be ascertained from available literature there has been no attempt to carry out chemical modifications among this group of alkaloids even though polyalthenol has been isolated in reasonable quantity. However, work on this group of compounds continues at the Université de Paris Sud and that data included for iso- and neopolyalthenol must be considered of a preliminary nature. Furthermore, while chemical manipulation should be able to confirm some features such as C(3)-stereochemistry and position of the double bond in the decalene system, other less tractable problems such as decalene conformation and configuration of the C(8)-methyl may only yield to X-ray analysis.

Among the pentacyclic alkaloids the structure and absolute stereochemistry of polyveoline **(5)** has been established by X-ray crystallography. Two series of compounds (in each case made up of related alcohol, acetate, and ketone) were reported from *G. suaveolens,* the polyavolensins **(11–13)** and the greenwayodendrines **(7–10)**. X-ray crystallography, reported independently for one member of each of these groups, led to the revision of the structures initially assigned to the polyavolensins and established that they are identical to the greenwayodendrines so that in reality only nine indolosesquiterpenes are known rather than twelve. This confusion regarding identity has led to an excess of trivial names in this group of pentacyclic alkaloids (see Section 3.2).

2. TETRACYCLIC INDOLOSESQUITERPENES

2.1. Polyalthenol and Isopolyalthenol

Polyalthenol was first isolated, together with the unusual triterpene polycarpol, from petroleum extracts of the leaf and the stem bark of *G. oliveri* [4, 5]. Further extraction of these plant materials, after basification with ammonia, yielded 11 aporphine alkaloids [5]. Polyalthenol was purified from the crude petrol concentrates of the leaves by column chromatography, initially over Sephadex LH20 eluting with chloroform–methanol (7 : 3) and subsequently over silica gel eluting with chloroform–benzene (1 : 1), giving a final yield of about 0.2% [5]. Similar treatment of the stem bark extract, but omitting the Sephadex chromatography, also gave polyalthenol. The structure assignment of polyalthenol as **1** rests largely on the interpretation of spectral data reported by Leboeuf et al. [1].

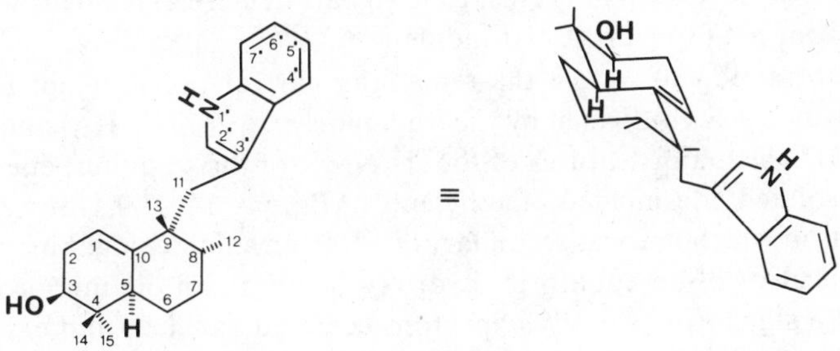

Polyalthenol crystallized from hexane as beige needles melting at 149–150°C and with $[\alpha]_D$ of +50°. Accurate mass measurement showed a molecular ion at m/z 337 for $C_{23}H_{31}NO$ with a base peak at m/z 130 (C_9H_8N) and a UV spectrum (Table 1) that are both indicative of an indole nucleus. The ^1H NMR spectrum,

Table 1.   UV and $Log_{10}\epsilon^a$ Values for Indolosesquiterpene

Alkaloids	
Polyalthenol (**1**)	226 (4.52), 285 (3.76), 292 (3.73)
Isopolyalthenol (**2**)	224 (4.61), 285 (3.97), 292 (3.95)
Neopolyalthenol (**4**)	224 (4.51), 285 (3.89), 292 (3.83)
Polyveoline (**5**)	249 (3.76), 301 (3.37)
Greenwayodendrin-3β-ol (**7**)	227, 277, 282, 293
Greenwayodendrin-3β-yl-acetate (**8**)	227, 277, 282, 292
Greenwayodendrin-3-one (**9**)	225 (4.63), 276 (3.88), 282 (3.89), 293 (3.76)
Polyavolensin (**11**)	238 (3.25), 278 (3.88), 285 (3.89), 295 (3.78)
Polyavolensinone (**13**)	237 (3.21), 279 (3.77), 285 (3.78), 294 (3.65)
Polyavolinamide (**15**)	217 (4.10), 255 (4.02), 285 (3.42), 393 (3.34)

^{a}In parentheses.

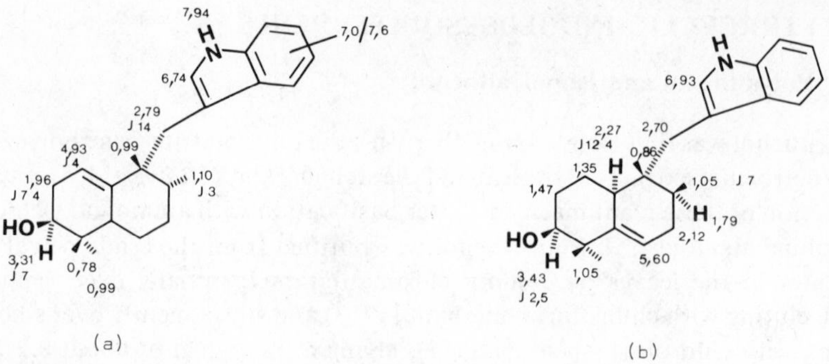

Figure 1. ¹H assignments for polyalthenol (a) and Isopolyalthenol (b).

run in deuterochloroform (Fig. 1a), gave signals in close agreement with those anticipated for a C(3')-substituted indole.

After subtraction of indole the remaining part of polyalthenol had to be $C_{15}H_{25}O$, which was confirmed by the fragments m/z 207 ($C_{15}H_{25}$) and m/z 189 (207—H_2O). The major features of the ¹H NMR of the sesquiterpene (Fig. 1a) were an isolated, deshielded, methylene (ABq, J 14 Hz), four C-methyl substituents of which one was secondary (J 3 Hz), an axial oxymethine proton (t, J 7Hz) indicative of an equatorial hydroxy function, an olefinic proton (t, J 4 Hz), and a signal for two allylic protons centered at δ 1.96 and exhibiting J values of 7 and 4 Hz. Irradiation of the δ 1.96 resonance caused the collapse of both olefinic and oxymethine signals to singlets, and formation of polyalthenol-O-acetate caused large shifts in two of the methyl resonances, thus requiring their close proximity to the oxymethine group. As a result of these observations the partial structure $RR'C{=}CHCH_2CH_{ax}(OH)C(Me)_2R''$ could be deduced for the sesquiterpene unit.

The ¹³C NMR spectrum (Table 2) showed resonances that could be assigned to both the indole and to the part structure proposed for the sesquiterpene. It further indicated the occurrence of one more quaternary center (43.6 ppm), which must in turn bear the third tertiary methyl substituent. The partial structure for the terpene was expanded by observing in the SFORD spectra a residual "3-bond" coupling of the quaternary carbon at 43.6 ppm to the olefinic H(1) and a further long-range interaction of the former to the allylic proton at C(5). These findings were used to expand the partial structure by assigning the allylic C(5)-carbon and the quaternary center to R and R'.

At this stage, assuming a 3-hydroxydecalin nucleus for the sesquiterpene, the remaining unassigned ¹³C resonances (1 × CH, 3 × CH₂, 1 × CH₃) could be fitted to the proposed structure (1), which also seemed to be the most plausible on biogenetic grounds. Structure 1 was further substantiated by an interesting ¹³C experiment measuring the shifts induced by the lanthanide shift reagent Yb(fod)₃ on binding to the hydroxyl. Assuming that the decalin skeleton maintains the

Table 2. ^{13}C NMR of Tetracyclic Indolosesquiterpenes [1, 7]^a

Carbon	Polyalthenol (1)	Isopolyalthenol (2)	Neopolyalthenol (4)
C(1)	116.4	31.6	31.5
C(2)	31.2	29.0	29.0
C(3)	74.8	76.7	76.5
C(4)	37.0	37.6	37.4
C(5)	44.4	142.6	142.6
C(6)	27.8	128.2	128.2
C(7)	30.7	31.6	31.5
C(8)	42.6	32.6	32.4
C(9)	43.6	42.3	42.3
C(10)	142.8	39.4	39.8
C(11)	26.0	29.5	29.0
C(12)	16.9	14.8	14.7
C(13)	23.4	23.4	23.1
C(14)	16.3	24.6	24.5
C(15)	25.5	26.3	26.1
C(2′)	122.5	123.4	135.5
C(3′)	112.4	112.9	102.4
C(3a′)	129.0	129.0	128.6
C(4′)	118.9	119.8	119.4
C(5′)	118.5	121.5	120.6
C(6′)	121.0	118.8	119.4
C(7′)	110.7	110.8	110.1
C(7a′)	135.3	135.8	136.8

^{a}Spectra were run in CDCl$_3$.

anticipated half chair–chair conformation, the observed shifts (Fig. 2) for a 1:1 molar ratio of Yb(fod)$_3$ and polyalthenol were in close agreement with values anticipated for **(1)**.

From these data the sesquiterpene part of polyathenol was designated 3β-hydroxy-4α,4β,8α,9β-tetramethyl-5Hα-decal-1(10)-ene with a 9α substitution to indole through a methylene bridge originating from the terpene. The proposals made for ring-A assignments are well supported by NMR and by

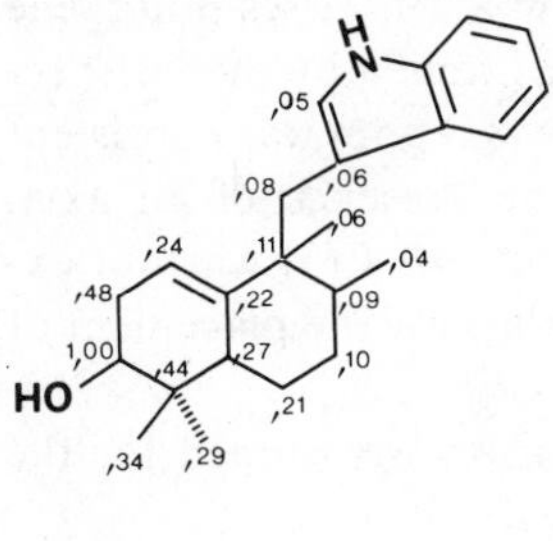

Figure 2. Shift values (relative to a shift of 1.00 for C(3)) for a 1:1 molar ratio of polyalthenol and Yb(fod)$_3$.

analogy with published data, but those for ring B are more tenuous with respect to configuration. Direct evidence for placing the C(8)-methyl in the α-(equatorial) position is found only in the low shift produced in the Yb(fod)$_3$ experiment; its presence in the axial configuration would be expected to lead to shifts in the same order as those observed for C(9)-substituents. No direct evidence is proposed to support the C(9)-methyl being in the equatorial position other than the biogenetic analogy (Section 4), although the relatively deshielded position of C(13) when compared with the axial C(20)-methyl of kolavane diterpenes [6] may be relevant.

Subsequently polyalthenol has been isolated from *G. suaveolens,* in this case in company with two other tetracyclic indolosesquiterpenes [7]. One of the additional compounds, which was given the trivial name isopolyalthenol, crystallized from petroleum ether and gave melting point 129° C and $[\alpha]_D$ of $-49°$. Accurate mass measurement again showed the molecular ion to be $C_{23}H_{31}NO$, and neither the fragmentation pattern nor the UV spectrum (Table 1) was significantly different from polyalthenol. The ^{1}H NMR spectrum (Fig. 1b) showed the presence of the same basic structure and substituents as polyalthenol, but varied in a number of ways, notably:

1. The oxymethine proton at δ 3.43 occurred as a broad singlet or small triplet (J 2.5 Hz) indicative of an equatorial proton and therefore an axial hydroxyl function.

2. The olefinic proton at δ 5.60 also appeared as a singlet.

3. The signals for the C(4)-methyls were deshielded to δ 1.16, consistent with the presence of an axial hydroxyl at C(3).

4. A 2H-singlet at δ 2.70 for the C(11)-methylene.

5. A signal at δ 2.27 with J of 12 and 4 Hz for the allylic H(10) which must therefore be axial.

In the light of an n.o.e. experiment in which the irradiation of the δ 1.16 methyl signal caused a 20% enhancement in the olefinic proton at δ 5.60, the gross structure of the sesquiterpene of isopolyalthenol was defined as decal-5(6)-ene with substituents in the same positions as in polyalthenol. The ^{13}C NMR spectrum of isopolyalthenol (Table 2) agreed closely with that of polyalthenol for the 3-substituted indole portion of the molecule and signals for the sesquiterpene part were compatible with the decal-5(6)-ene structure implied by the ^{1}H spectrum. The deshielding of the C(14)-methyl resonance by 8.3 ppm was consistent with that anticipated for the axial C(4)-methyl in the presence of an axial hydroxyl [8]. The shielding of 10 ppm for the resonance for C(9) was not explained by Dubois [7] but can be taken as strong evidence for the placement of the C(13)-methyl axial.

If, as in polyalthenol, a chair–half chair conformation is assumed for the decalin moiety, then the two isomeric structures **2** and **3** are possible for isopolyalthenol: **2** has the hydroxyl axial and β, and as a consequence, the axial C(10)-proton must be α and *trans* to the axial methyl group on C(9), whereas in **3** the hydroxyl is axial and α with the C(10)-proton β and *cis* to the methyl on C(9) which has become equatorial. The evidence available at present appears to favor **2**.

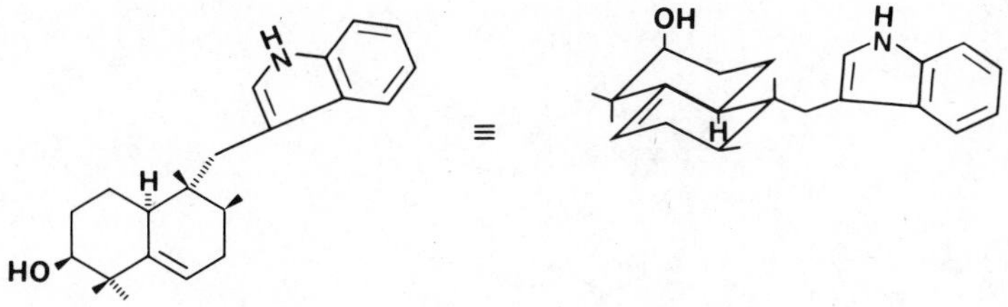

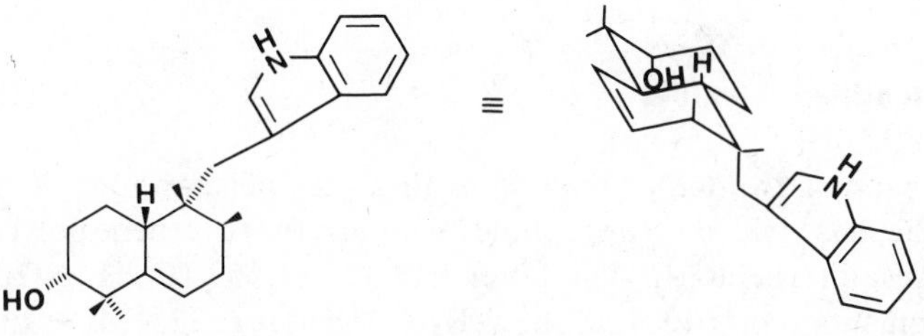

2.2. Neopolyalthenol

Neopolyalthenol, isolated together with isopolyalthenol and polyalthenol from *G. suaveolens,* crystallized from petroleum ether with a melting point of 89° C and $[\alpha]_D$ of $-19°$ [7]. Accurate mass measurement once more indicated $C_{23}H_{31}NO$ and both UV (Table 1) and mass spectra were very similar to those of isopolyalthenol. In the ^{1}H NMR spectrum the appearance of a singlet at δ 6.16 and the absence of the δ 6.93 signal of isopolyalthenol (Fig. 1b) suggested attachment of the sesquiterpene to C(2′) of the indole rather than C(3′). This arrangement was confirmed by the ^{13}C NMR spectrum in which the shielded doublet at 102.4 ppm (Table 2) can be assigned to the unsubstituted C(3′)-position of indole. This change in substitution position causes slight shielding of the C(11)-protons (from δ 2.70 to δ 2.56) but otherwise both ^{1}H and ^{13}C spectra agree closely with those of isopolyathenol. The full ^{13}C spectrum is given in Table 2.

Neopolyalthenol is therefore the C(2′)-substituted isomer of isopolyalthenol and can provisionally be given structure **4**, although following the arguments made for isopolyalthenol, the α-hydroxy form (as in **3**) cannot be discarded.

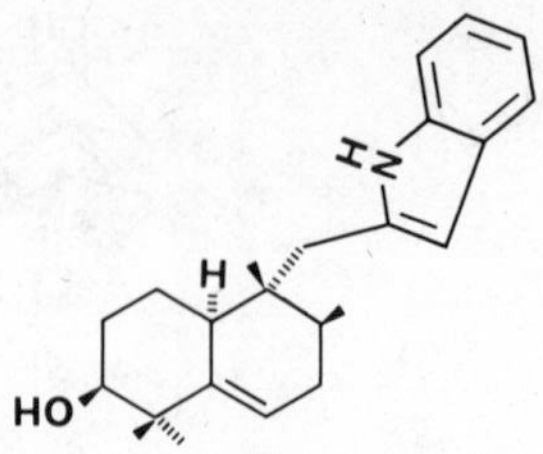

4

2.3. Unidentified Alkaloid

The stem bark of *Isolona campanulata* Engl. & Diels, another West African Annonaceae, has yielded six aporphine alkaloids, the triterpene polycarpol, and an indolosesquiterpene [9]. The latter had M^+ of 353 ($C_{23}H_{31}NO_2$) with the typical fragments at m/z 130 (indole, C_9H_8N) and at m/z 223 ($M^+ - 130$) and 205 (223—H_2O). The IR spectrum showed an absorption at 1725 cm^{-1} for a carbonyl, while the ^{1}H NMR spectrum did not reveal the presence of a C(3)-oxymethine proton. Because of lack of material and the sensitive nature of this compound, structural identification was not achieved. This isolate appears to be an indolosesquiterpene, probably a 3-oxo compound, and represents the only example of this class of alkaloids so far recorded from any taxa other than *Greenwayodendron*.

3.   PENTACYCLIC INDOLOSESQUITERPENES

3.1. Polyveoline

Polyveoline was the first indolosesquiterpene to be isolated from the stem bark of *G. suaveolens,* being obtained from a chloroform extract of the ammoniated plant material by a process involving column chromatography over both alumina and silica gel and, finally, preparative TLC over silica gel [10]. It made up approximately 25% of the total tertiary alkaloid fraction, the remainder being 7-oxygenated aporphines. Polyveoline recrystallized from methanol to give melting point 172°C and $[\alpha]_D$ of −23°. Unlike the previously discussed indolosesquiterpenes, polyveoline is basic, readily forming a hydrochloride (melting at 297° with decomposition and with $[\alpha]_D$ of −110°).

Accurate mass measurement of polyveoline gave M^+ 339 ($C_{23}H_{33}NO$), with 2H more than polyalthenol and its congeners. The fragmentation pattern indicated loss of CH$_3$ and H$_2$O and major ions at m/z 131 (C_9H_9N), 130, 118, and 117 with, in this instance, m/z 131 as the base peak. The UV spectrum (Table 1) lacked the typical indole pattern seen in **1,** being more like that of dihydroindoles.

Table 3. ^{1}H Resonances for H(2′) and H(3′) of
N-substituted Polyveoline [11]

N-substituent	δ Value for	
	H(2′)	H(3′)
H	4.29	3.43
Me	3.79	3.33
COMe	4.50	3.75

The ^{1}H NMR spectrum revealed [11] the four indole aromatic protons but not those for either H(2′) or H(3′) of indole. Replacement of the indole nucleus by dihydroindole was also suggested by the occurrence of 1H resonances at δ 4.29 (multiplet) and δ 3.43 (*d, J* 11 Hz) which could be attributed to H(2′), adjacent to a proton-containing center, and H(3′), adjacent to a quaternary carbon. This hypothesis was sustained by synthesis of the *N*-methyl and *N*-acetyl derivatives which produced the anticipated shifts for the H(2′) and H(3′) protons (Table 3). In the sesquiterpene the resonance for the H(3) oxymethine occurred as a triplet at δ 3.41 (*J* 3 Hz), indicative of an axial configuration for the hydroxyl. Four quaternary methyl resonances were observed at δ 0.82, 0.89, 1.14, and 1.34.

At this point the structure and stereochemistry of polyveoline (5) was established by X-ray crystallography performed on the hydrochloride [11, 12]. This result confirmed the presence of the dihydroindole moiety linked through C(2′)–C(11) and through C(3′)–C(8). In the sesquiterpene the C(3)-hydroxy function was axial as were methyl substituents at C(8) and C(10) and the proton at C(5), while the C(9)-proton was equatorial.

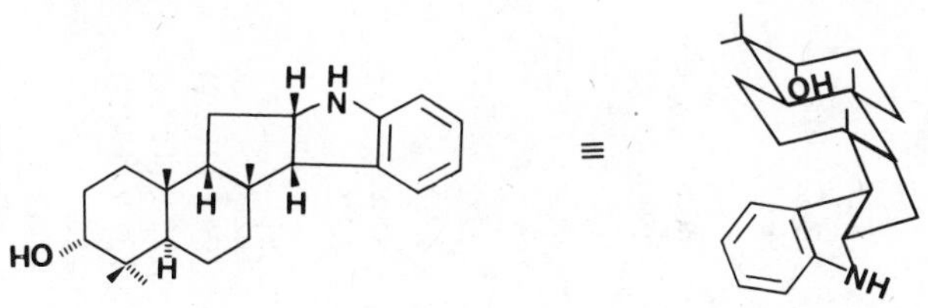

5

With hindsight the assignments of the four methyl resonances can now be made. The most deshielded (δ 1.34) may be assigned to C(12), which will be influenced both by its proximity to the indole and by 1,3-diaxial interaction with C(13) in the same manner as noted for the greenwayodendrines (Section 3.2). The latter argument requires that the δ 1.14 resonance be assigned to C(13). The ^{13}C NMR spectrum of polyveoline was assigned by Hocquemillar et al. [11] by analogy with the spectra for dihydroindole and 3β-hydroxyl-4α,4β,10β-tri-methyl-5Hα-decalin. The assigned values for polyveoline are given in Table 4.

Table 4. ^{13}C NMR Spectra of a Number of Pentacyclic Indolosesquiterpenes[a]

Carbon Number	Polyveoline (5) [11]	Polyavolinamide (15) [18]	Greenwayodendrin-3β-ol (7) [13]	Polyavolensin (11) [16]	Greenwayodendrin-3-one (9) [13]	Polyavolensinone[b] (13) [16]
C(1)	37.1 (31.2)	37.3	38.4[e]	38.5	38.2[e]	38.1
C(2)	31.2[e] (25.0)	(25.5)	27.0	22.8	33.8	33.7
C(3)	76.4	76.4	78.8	80.7	216.7	216.1
C(4)	44.3 (36.3)	37.6	38.4	37.9	47.4	47.3
C(5)	39.6	(59.0)	55.9	56.1	55.0	55.1
C(6)	18.6	19.1	19.8	19.9	21.2	21.2
C(7)	31.7[e] (31.2)		37.9[e]	37.7	37.8[e]	37.8
C(8)	61.6	60.8	63.0	63.1 (65.1)[c]	63.0	62.8
C(9)	61.1	59.0	65.2	65.1	64.4	64.3
C(10)	36.3 (44.3)	36.7	36.4	36.2[d]	36.2	36.2
C(11)	25.0 (37.1)		22.6		22.8	22.8
C(12)	28.1		19.7[f]	20.2	19.7	19.6
C(13)	21.9	22.1	15.0	16.2	15.5	15.4
C(14)	23.2	23.3	20.0[f]	21.5 (20.2)	20.8	20.8
C(15)	28.2	28.9	28.0	28.1	26.9	26.9
C(2')	61.1	63.5	131.8[g]	131.9	132.1[f]	131.9[e]
C(3')	37.2	(60.8)	94.3	94.6	94.8	94.6
C(3a')	127.5	131.4	132.6[g]	132.8	133.0[f]	132.9[e]
C(4')	124.3	124.3 (122.9)	120.1	120.3	120.7	120.4
C(5')	117.6	122.9 (127.9)	118.8	119.0	119.4	119.2
C(6')	127.5	127.9 (124.3)	120.5	120.7	121.0	121.8
C(7')	108.8	116.7	109.5	109.6	109.6	109.6
C(7a')	152.9	144.3	142.8	142.9	142.8	142.5

[a] The assignments are those of the reviewer; where they differ from original assignments the latter are given in parentheses.
[b] No assignments were made by Okorie [16]; those given are the reviewer's.
[c] 65.1 is a doublet, 63.1 a singlet.
[d] The only remaining value reported, 33.7 ppm, does not fit the requirements for C(11).
[e-g] Values in the same column with the same superscript are interchangeable.

3.2. The Greenwayodendrines (Polyavolensins)

The co-identity of the separately reported greenwayodendrines and polyavolensins is now established. In this section arguments used to support originally proposed structures will be discussed and in Table 4, 5, and 6. The numbers assigned to the original (now known to be incorrect) polyavolensin structures will be used to differentiate data reported for them by Okorie [16] from data reported for the greenwayodendrines by Hasan et al. [13].

Polyveoline and four novel nonbasic pentacyclic indolosesquiterpenes were isolated from the stem bark of *G. suaveolens* collected from the rain forest of west Cameroun [13]. These compounds, known collectively as the greenwayodendrines, were obtained by elution from a silica gel column with petroleum ether containing increasing amounts of ethyl acetate and, where necessary, by subsequent TLC. Their interrelationship was established by simple chemical conversions which showed the three major compounds to be an acetate, the corresponding alcohol and the derived ketone. The epimeric alcohol was a minor component. The UV spectra of these alkaloids (Table 1) were typical of the indole system and all four compounds showed a major fragment in the mass spectrum, usually the base peak, at m/z 130 (C_9H_8N).

Structure elucidation of the greenwayodendrines was achieved by detailed studies of ^{1}H NMR at 360 MHz (Table 5) and ^{13}C NMR (Table 4) [13]. The major alcohol ($C_{23}H_{31}NO$, melting point 167–169°C, $[\alpha]_D$ +11°) showed the typical proton resonances for the aromatic ring and H(3') of indole, for four quaternary methyl substituents, and for an axial oxymethine proton. A feature of the ^{1}H spectrum was the presence of an ABX system that could be assigned to the isolated moiety $=C(R)CH_2CH_{ax}R'R''$. Assuming that this compound was similar to polyveoline but with indole rather than dihydroindole and with indole and sesquiterpene linked through N(1') and C(2') rather than C(2') and C(3'), the ^{13}C resonances could be assigned, with the deshielded tertiary signal at 65.2 ppm placed at C(9), as in related decalins with an analogous substitution pattern [8, 14], and a quaternary signal at 63.0 ppm at C(8), deshielded by the adjacent N(1'). On this basis the general structure (6) could be drawn for the greenwayodendrines. High-field NMR spectra (Tables 4 and 5) were also obtained for the corresponding ketone ($C_{23}H_{29}NO$, melting point 191–194° C $[\alpha]_D$ +47°) and showed those changes anticipated for oxidation at C(3). In the case of the ketone decoupling experiments allowed unambiguous assignment of the ^{1}H resonances and splitting patterns for all the protons of the sesquiterpene. The acetate ($C_{25}H_{33}NO_2$, melting point 211–214° C, $[\alpha]_D$ −1°) was not subjected to the same rigorous spectral analysis (for details of 90-MHz ^{1}H spectrum see Table 5), but direct chemical interconversion to and from the major alcohol confirmed the relationship.

For the greenwayodendrines, stereochemical problems at C(3), C(5), C(8), and C(10) had to be resolved. For C(3) the hydroxy and acetoxy substituents were equatorial (β) because of the large J value obtained for H(3) (10 and 5 Hz). The

Table 5. ¹H NMR of the Greenwayodendrines and Polyavolensins [13, 16, 18]ᵃ

	Greenwayodendrines			Polyavolensins		
	3β-acetate (8)	3β-ol (7)	3-one (9)	3β-acetate (11)	3β-ol (12)	3-one (13)
H(3)	4.54 dd(10,6)	3.25 dd(10,5)	—	4.55 t(7.5)	3.20 t(7.5)	—
H(5)		1.27 dd(12,5)	1.63 dd(12,4)			
H(7$_{ax}$)		2.04 dt(12,4)	2.05 dt(12,4)			
H(7$_{eq}$)		2.72 td(12,5)	2.74 td(12,4)			
H(8)		—	—	2.80 m	2.80 m	2.80 m
H(9)	2.22 dd(12,7)	2.22 dd(13,7)	2.26 dd(13,6)	—	—	—
H(11$_{ax}$)	2.74 m	2.68 dd(15,13)	2.72 dd(15,13)			
H(11$_{eq}$)	2.74 m	2.82 dd(15,7)	2.85 dd(15,6)			
H(3')	6.18 s	6.18 s	6.20 s	6.18 s	6.17 s	6.20 s
H(4')–H(7')	7.00–7.60 m	7.02–7.55 m	7.02–7.55 m	6.94–7.63 m	6.93–7.63 m	6.93–7.63 m
12(Me)	1.21 s	1.22 s	1.24 s	1.18 sᶜ	1.18 sᶜ	1.18 sᶜ
13(Me)	1.10 s	1.06 sᵇ	1.17 s	1.07 sᶜ	1.00 sᶜ	1.12 sᶜ
14(Me), 15(Me)	0.95	1.05,ᵇ 0.88	1.13, 1.15	0.93ᶜ	1.00,ᶜ 0.82ᶜ	1.08
OAc	2.07			2.05		

ᵃ J values are in parentheses.
ᵇInterchangeable.
ᶜUnassigned to specific methyl groups.

signal for H(5) could be seen in the spectra of both the alcohol and ketone and exhibited a large J (12 Hz) requiring it to be axial (α). The two methyl substituents, C(12) and C(13), on C(8) and C(10), respectively, were assigned axial configurations because of their pronounced deshielding because of 1,3-diaxial interaction with one another. The axial positioning of C(12) was also evident from the deshielding of H(7_{eq}) [$\sim \delta$ 2.70, against δ 2.05 for H(7_{ax})]. This deshielding was attributed [13] to H(7_{eq}) lying in the plane of the indole nucleus. This situation would only be true if the C(12)-methyl was axial. If the C(12)-methyl were equatorial and the normal chair–chair form were retained, then H(9) would be forced into an equatorial configuration with a dihedral angle of about 45° to each of the C(11)-protons; whereas its larger J value of 12.5 Hz clearly placed it in an axial position. On the basis of the above arguments the sesquiterpene unit was defined as 3β-hydroxy-4α,4β,8β,10β-tetramethyl-5Hα,9Hα-decalin with the three major isolates being greenwayodendrin-3β-ol (7), greenwayodendrin-3β-yl acetate (8), and greenwayodendrin-3-one (9). These assignments were subsequently confirmed by X-ray crystallography of 9 [13].

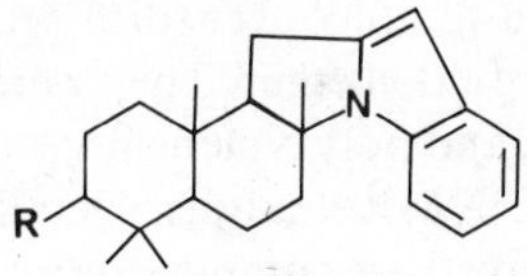

6

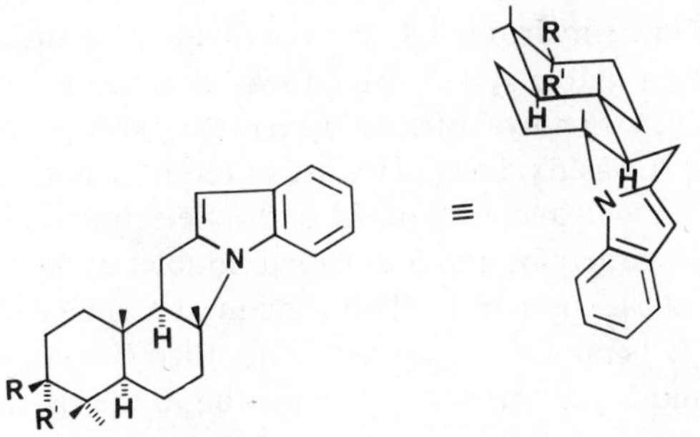

7   R = OH   R' = H

8   R = OAc   R' = H

9   R , R' = O

10   R = H   R' = OH

The fourth alkaloid was obtained only in trace amounts and as a gum (M$^+$ 337, $C_{23}H_{31}NO$). It was tentatively identified as greenwayodendrin-3α-ol (10) by reduction of greenwayodendrin-3-one with NaBH$_4$ to the α- and β-hydroxy epimers, which were identical with 7 and 10 by TLC. The β-epimer made up about 90% of the synthetic mixture.

11   R = OAc   R' = H

12   R = OH   R' = H

13   R = R' = O

14

15

The greenwayodendrines have recently been isolated again, from *G. suaveolens* collected in Congo Brazzaville [15].

In an earlier investigation of the constituents of stems and stem bark of *G. suaveolens* collected in the Apomu Forest Reserve in Nigeria, Okorie [16] isolated three hexane-soluble indolosesquiterpenes. These alkaloids were separated by column chromatography over silica gel, eluting with hexane containing increasing amounts of diethyl ether. They were identified as polyavolensin **(11)**, polyavolensinol **(12)**, and polyavolensinone **(13)** by simple chemical interconversions and ^{1}H and ^{13}C NMR analysis. Details of the published UV, ^{13}C NMR, and ^{1}H NMR data for these compounds are listed in Tables 1, 4 and 5, respectively.

Using arguments applied to other indolosesquiterpenes discussed previously, the polyavolensins were established as pentacyclic and as possessing an equatorial hydroxyl in a 3-hydroxy-4,4,10-trimethyldecalin nucleus with another tertiary methyl and an indole system which was attached to the sesquiterpene through N(1′) and C(2′). In polyavolensin the hydroxy group was acetylated and in polyavolensinone it was oxidized. This set of requirements [16] allowed two possible structures for polyavolensinol, **12** or **14**. Selection of **12** as the correct structures was based largely on the deshielded doublet at 65.1 ppm in the ^{13}C spectrum of **11** and at 64.3 ppm in **13**. This signal was assigned to a CH adjacent to N(1′) with the CH being further linked to another H-bearing carbon as indicated from the coupling observed for this proton in the ^{1}H spectrum (δ 2.80, Table 5, given as multiplet, J values not quoted).

The structural arguments used by Okorie [16] were not conclusive. In addition to the failure to take into account the alternate C(8)/C(9) to N(1′)/C(2′) pattern of linkage found in the greenwayodendrines **(6)**, they exhibit the following specific weaknesses. (1) In addition to the tertiary carbon resonating at above 60 ppm the ^{13}C spectra of both **11** and **13** show a second deshielded resonance (this time quaternary) that must arise from the sesquiterpene. In neither case was this signal assigned. In the greenwayodendrines these two deshielded carbons were placed at C(8) and C(9) by analogy with diterpenes of related structure (see above and ref. [13]), but in the case of the polyavolensin structure, with a methyl at C(9) and C(8) tertiary, it is to be anticipated that C(9) would resonate at about 45 ppm, as is demonstrated in the polyalthenol group (Table 2). (2) In the

Table 6. Comparison of Reported Melting Points and
Optical Rotations for the Greenwayodendrines and
Polyavolensins [13, 16, 18]

Compound	Optical Rotation	Melting Point
Greenwayodendrin-3-yl acetate (8)	−1.2°	211–214° C
Polyavolensin (11)	−3.8°	210–212° C
Greenwayodendrin-3β-ol (7)	+11°	167–169° C
Polyavolensinol (12)	+6.2°	163–165° C
Greenwayodendrin-3-one (9)	+46.8°	191–194° C
Polyavolensinone (13)	+40°	185–187° C

^{1}H spectra of the polyavolensins (Table 5) there should be a deshielded signal, either a singlet or ABq, for the C(11)-methylene protons. In the greenwayodendrines these were found at about δ 2.80, where Okorie [16] assigned the C(8)-proton of his compounds.

In general the similarities between published data for the greenwayodendrines and polyavolensins extended to ^{13}C spectra (Table 4), ^{1}H spectra (Table 5), melting points and optical rotations (Table 6). In particular a comparison of the ^{13}C spectra of greenwayodendrin-3-one and polyavolensinone showed the biggest discrepancy between the published values for these two to be only 0.6 ppm. Attention was drawn to these similarities by Hasan et al. [13], and their doubts regarding the proposed structures 11–13 were confirmed by the appearance, before publication of the greenwayodendrine paper [13], of a revision of the structure of the polyavolensins based on an X-ray analysis [17]. This confirms that in reality polyavolensin = greenwayodendrin acetate, polyavolensinol = greenwayodendrin-3-ol, and polyavolensinone = greenwayodendrin-3-one. While Hasan et al. [13] recognized that the greenwayodendrines were likely to represent the true structures of the polyavolensins they argued for the assignment of the new trivial name on two counts; (a) they were not certain of co-identity when their paper was written, (b) they argued that whether the polyavolensins were correctly named or not, the structures proposed for them were biogenetically feasible (see Section 4) and that the name polyavolensin should be allowed to reside with structures 11–13. This argument agrees in philosophy with the rules used in botanical nomenclature, where a published but taxonomically incorrect name remains irrevocably bound to the taxon it initially, though incorrectly, described.

3.3. Polyavolinamide

As a result of further treatment of stem and root bark of Nigerian *G. suaveolens,* Okorie [18] was able to isolate a fourth alkaloid for which he assigned the trivial name polyavolinamide and structure 15. This compound had an analysis corresponding to $C_{25}H_{35}NO_2$, with melting point 249–251° and a UV spectrum (Table 1) indicative of a 2,3-dihydroindole. The presence of an N-acetyl group was indicated by a methyl resonance at δ 2.24 (3H), ^{13}C signals at 169.3 and 24.3 ppm, and a significant ion at m/z 339 (M⁺, CH_2CO). Major features of the ^{1}H spectrum were the deshielded signal for H(7') (δ 8.29, d, J 8Hz), a doublet at δ 3.57 (1H, J 11 Hz) and multiplet at δ 4.58 (1H) similar to H(3') and H(2') of polyveoline (5) [11], a 1H multiplet at δ 3.40 with a small coupling indicative of an equatorial oxymethine proton at H(3), and four methyl resonances as singlets at δ 1.40, 1.13, 0.81, and 0.80.

Polyavolinamide was assigned structure 15 primarily on the basis of its nonidentity with polyveoline. Nonidentity was established [18] mainly by differences in the melting points of deacetylpolyavolinamide and polyveoline (150–152° C vs. 172° C), of deacetylpolyavolinamide HCl and polyveoline HCl (210° C vs. 297° C), and of O-acetylpolyavolinamide and O,N-diacetylpolyveoline (246–248° C vs. 239° C).

However the ^{1}H spectrum for deacetylpolyavolinamide was almost identical with that of polyveoline in all published resonances—for example, values for polyveoline in parentheses, H(2'): δ 4.20 (4.29); H(3'): δ 3.42 (3.43); H(3): δ 3.42 (3.41); methyls: δ 1.33, 1.13, 0.90, 0.83 (1.35, 1.13, 0.87, 0.81). Okorie [18] partially assigned the ^{13}C resonances for polyavolinamide (Table 4) suggesting that doublets at 60.8 and 63.5 ppm were attributable to C(3') and C(2'), respectively. This assertion is at odds with the earlier assignments made for polyveoline [11] and for dihydroindole alkaloids such as criophylline [8] where a tertiary C(3') has been assigned a chemical shift of less than 45 ppm. Furthermore, a doublet at 59.0 ppm was assigned to C(5), whereas with an axial hydroxy at C(3) the signal of C(5) would be expected to be shielded in comparison with compounds in which the C(3) carries an equatorial hydroxyl (~ 56 ppm, Table 4). This is demonstrated by polyveoline, where the hydroxyl at C(3) is axial and C(5) resonates at 39.6 ppm [11].

It is somewhat surprising that when the structures of the polyavolensins were revised [17] there was no mention of polyavolinamide despite its obvious affinity to them. From the above discussion it is clear that the proposed structure 15 has not been fully substantiated, not even to the extent where the reviewer feels able to discount the possibility that polyavolinamide is actually N-acetylpolyveoline.

4. BIOSYNTHESIS

No experimental work has yet been reported about the biosynthesis of the indolosesquiterpenes and, in view of the nature and distribution of the taxa so far known to produce them, it is unlikely that anything will be done in the forseeable future. However, the processes involved do not appear to be in any way obscure, certainly involving the interaction of indole with a suitably activated sesquiterpene.

The sesquiterpene part of the indolosesquiterpene is clearly akin to the rare

16

group of sesquiterpenes typified by driman-3β-ol (16)[19]. Compound 16 can be envisaged as a product of the cyclization of *trans-trans*-farnesyl pyrophosphate (17) through the carbocation intermediate (18) (Scheme 1). The transformation of 18 to the sesquiterpene of polyalthenol (1) can be explained by the series of reactions depicted in Scheme 1: (a) movement of the charge from C(8) to C(9) with the C(8)-methyl taking the equatorial configuration 20, which could occur through the corresponding 8(9)-ene (19); (b) methyl transfer from C(10) to C(9) with the methyl adopting the equatorial configuration 21; (c) loss of a C(1)-proton to remove the charge and form the required 1(10)-ene (22). Alternatively the initial transfer of H(9) to C(8) with the C(8)-methyl taking an axial configuration (23) followed by the same C(10)-methyl transfer (24), but then H-transfer from C(5) to C(10) (25), and finally loss of charge by removal of a C(6)-proton and inversion of the nucleus could lead to the sesquiterpene form (26) considered to occur in isopolyalthenol (2) and neopolyalthenol (4) (see Scheme 1). It is recognized that the sequence from 18 to 26 should be considered as one concerted step rather than the discrete series shown in Scheme 1, but the latter is done in order to indicate clearly the resulting configuration of 22 and 26.

The linkage between the pyrophosphate activated C(11)-position of the sesquiterpene and C(2′) of the indole precursor, presumably tryptophan (27), is easy to rationalize (Scheme 2A). Linkage of the sesquiterpene to C(3′) rather than C(2′) can be envisaged through the quaternary form of 27 (Scheme 2B). Probably this combination can take place at different stages in the formation of the sesquiterpene. Where the carbocation has been discharged prior to linkage, tetracyclic products are formed; where not, there is scope for further interactions to form pentacyclic products.

Polyveoline and the greenwayodendrines could be formed through the hypothetical intermediate 28 (Scheme 3). To form the greenwayodendrines, all that is required is loss of the N(1′)-proton with ring closure between N(1′) and C(8) with the C(8)-methyl group taking the axial configuration. Polyveoline requires inversion of 28 at both C(3) and C(9), plausibly through the formation of 29 and its subsequent reduction to 30. Ring closure of C(8) to C(3′), again with the C(8)-methyl taking the axial position, would give 31, which on reduction yields polyveoline.

The polyavolensin structure could be derived from the alternative carbocation (32) (from indole and 20) via a methyl transfer from C(8) to C(9) to give 33 and subsequent ring closure as in the greenwayodendrine series (Scheme 4). Polyavolinamide would require a different sesquiterpene precursor (34) the biogenesis of which is obscure.

Finally, note that while the indolosesquiterpenes are, biosynthetically speak-

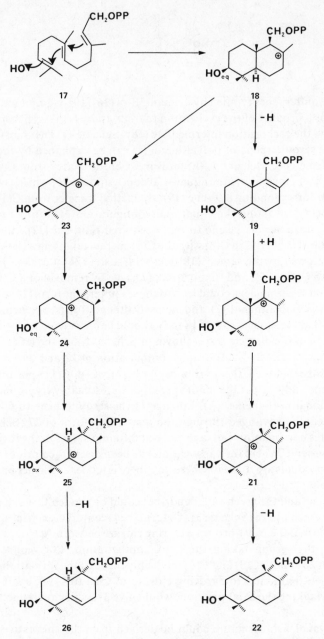

Scheme 1. Possible biogenetic routes to the sesquiterpene precursors of the tetracyclic indolosesquiterpenes.

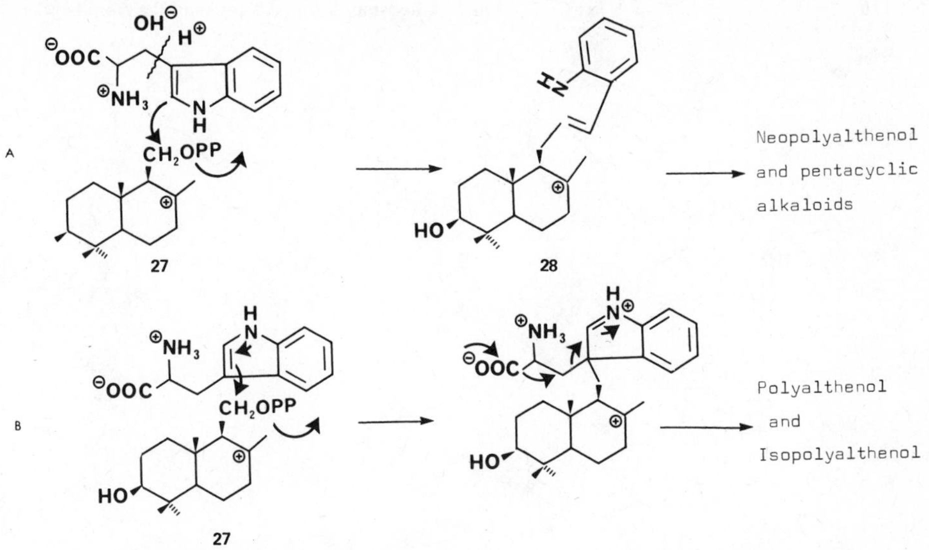

A

B

Scheme 2. Linkage of sesquiterpene and indole precursors.

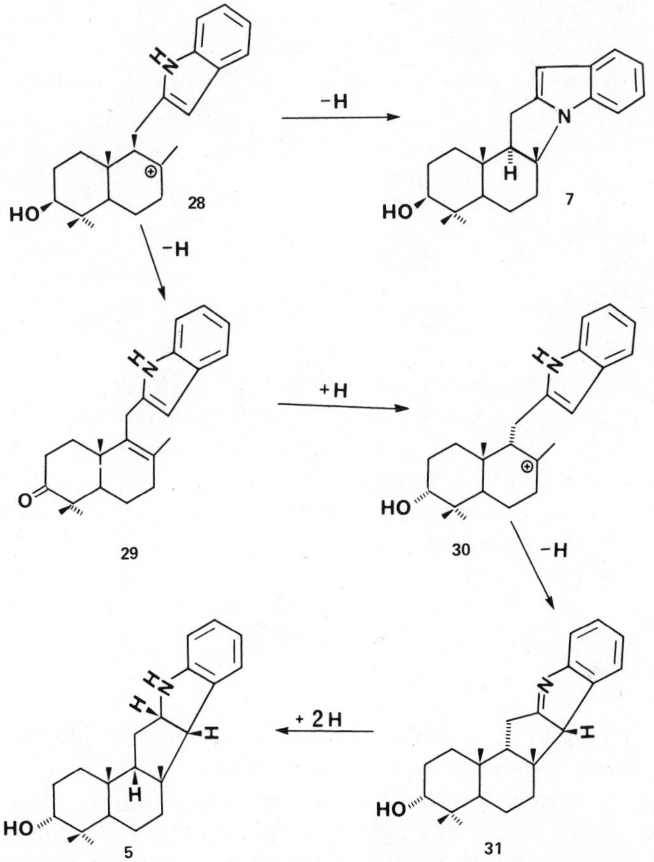

Scheme 3. Possible biogenetic routes to polyveoline (**5**) and greenwayodendrin-3β-ol (**7**).

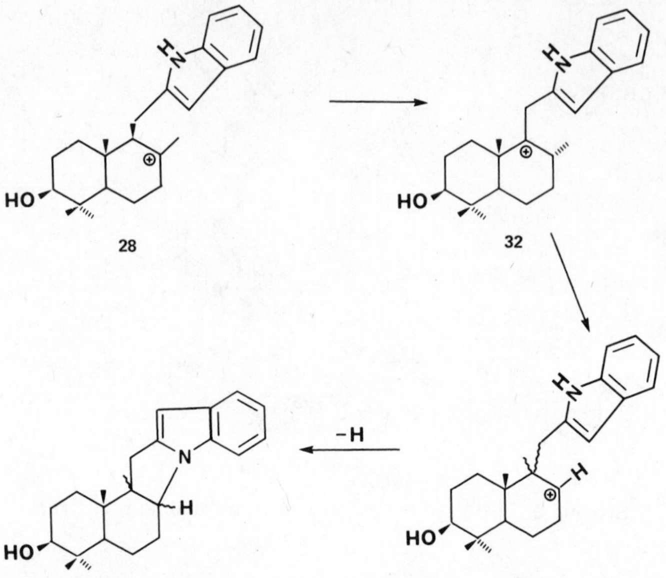

Scheme 4. Possible biogenetic route to polyavolensinol **(12)**.

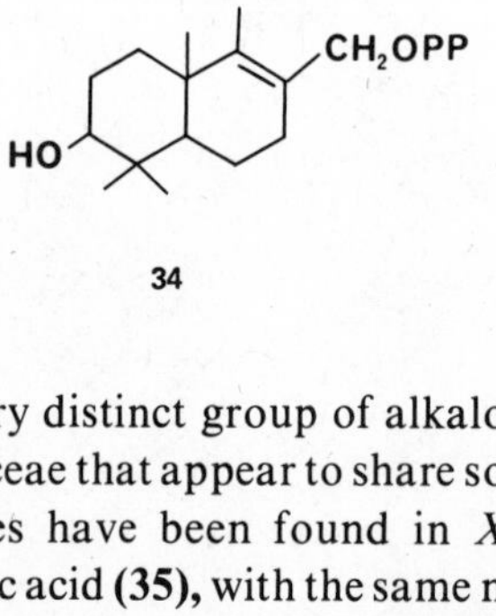

ing, a very distinct group of alkaloids, there are other metabolites found in the Annonaceae that appear to share some of their biogenetic origins. Kolavane-type diterpenes have been found in *Xylopia* [6] and *Annona* [20] species while polyalthic acid **(35),** with the same methylation pattern as in the sesquiterpenes, is reported from the Asian species *Polyalthia fragrans* [21]. Likewise simple 3-prenylated and 3,6-diprenylated indoles have been recorded from species of *Uvaria* [22] and *Monodora* [23].

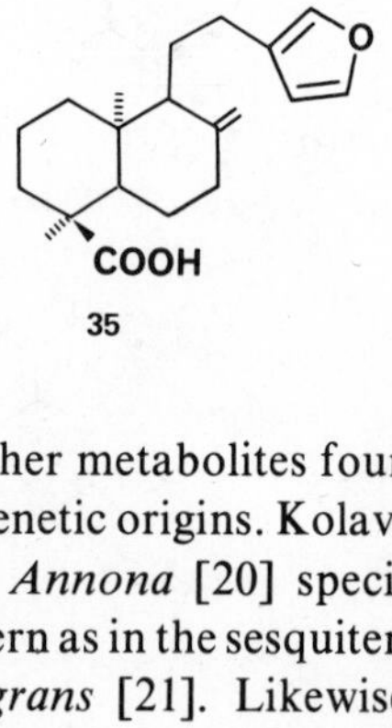

5. SUMMARY

Since the first report of polyalthenol a considerable number of indolosesquiter-penes have been recorded in the two species of *Greenwayodendron*, particularly *G. suaveolens.* Inasmuch as they have not been found in species still retained in *Polyalthia*, despite studies on a number of species of the latter, this result would tend to support Verdcourt's assertion [2] concerning the taxonomy of these taxa.

Furthermore, while the report of an unidentified alkaloid of this type from an *Isolona* does confirm that they do occur elsewhere in the family, the absence of more definite sources, despite the increasing research effort being made on this family, suggests that they are a very rare alkaloid type.

ACKNOWLEDGMENTS

The author wishes to extend his thanks to Professor André Cavé, Université de Paris Sud, for making available unpublished information on some indolosesquiterpenes.

REFERENCES

1. M. Leboeuf, M. Hamonniere, A. Cavé, H. E. Gottlieb, N. Kunesch, and E. Wenkert, *Tetrahedron Lett.,* 3559 (1976).

2. B. Verdcourt, *Adansonia ser. II* **9,** 87 (1969).

3. B. Verdcourt, "Annonaceae," in *Flora of Tropical East Africa,* Crown Agents, U.K., 1971, pp. 66–69.

4. M. Hamonniere, *These de Doctorat de 3eme Cycle,* Paris Sud Orsay, 1976.

5. M. Hamonniere, M. Leboeuf, and A. Cavé, *Phytochemistry* **16,** 1029 (1977).

6. C. M. Hasan, T. M. Healey, and P. G. Waterman, *Phytochemistry* **21,** 1365 (1982).

7. G. Dubois, *These de Doctorat de 3eme Cycle,* Paris Sud Orsay, 1981.

8. F. W. Wehrli and T. Nishida, *Fortschritte d. Chem. org. Naturst.* **36,** 71 (1979).

9. R. Hocquemiller, P. Cabalion, J. Bruneton, and A. Cavé, *Plantes médicinales et phytothérapie* **12,** 230 (1978).

10. A. Cavé, H. Guinaudeau, M. Leboeuf, A. Ramahatra, and J. Razafindrazaka, *Planta Med.* **33,** 243 (1978).

11. R. Hocquemiller, G. Dubois, M. Leboeuf, A. Cavé, N. Kunesch, C. Riche, and A. Chiaroni, *Tetrahedron Lett.* **22,** 5057 (1981).

12. C. Riche, A. Chiaroni, G. Dubois, R. Hocquemiller, M. Leboeuf, and A. Cavé, *Planta Med.* **39,** 206 (1980).

13. C. M. Hasan, T. M. Healey, P. G. Waterman, and C. H. Schwalbe, *J. Chem. Soc., Perkin Trans., I* 2807 (1982).

14. J. R. Hansen, M. Silverns, F. Piozzi, and G. Savona, *J. Chem. Soc., Perkin Trans., I* 114 (1976).

15. A. Cavé, private communication, June 29 1983.

16. D. A. Okorie, *Tetrahedron* **36,** 2005 (1980).

17. C. P. Falshaw, T. J. King, and D. A. Okorie, *Tetrahedron* **38,** 2311 (1982).

18. D. A. Okorie, *Phytochemistry* **20,** 2575 (1981).

19. H. H. Appel, C. J. W. Brooks, and K. H. Overton, *J. Chem. Soc.* 3322 (1959).

20. M. Leboeuf, A. Cavé, P. K. Bhaumik, ,B. Mukherjee, and R. Mukherjee, *Phytochemistry* **21,** 2783 (1982).

21. K. W. Gopinath, T. R. Govindachari, P. C. Parthasarathy, and N. O. Viswanathan, *Helv. Chim. Acta* **44,** 1040 (1961).

22. H. Achenbach and B. Raffelsberger, *Tetrahedron Lett.,* 2571 (1979).

23. M. N. Nwaji, S. O. Onyiriuka, and D. A. H. Taylor, *J. Chem. Soc., Chem. Comm.,* 327 (1972).

Chapter Three

Cyclopeptide Alkaloids

Madeleine M. Joullié

Department of Chemistry
University of Pennsylvania
Philadelphia, Pennsylvania 19104

Ruth F. Nutt

Merck Sharp & Dohme Research Laboratories
Medicinal Chemistry, 26A-4044
West Point, Pennsylvania 19486

CONTENTS

ABBREVIATIONS

Boc   *t*-Butyloxycarbonyl

Z   Benzyloxycarbonyl

DCC   *N,N'*-Dicyclohexylcarbodiimide

OSu   *N*-Succinimidoyl ester

OBn   Benzyl ester

DIPEA   Diisopropylethylamine

CDI   *N,N'*-Carbonyldiimidazole

DMAP   4-Dimethylaminopyridine

TFA   Trifluoroacetic acid

HOBT   1-Hydroxybenzotriazole

DBU   1,8-Diazabicyclo[5,4,0]undec-7-ene

DEAD   Diethyl azodicarboxylate

Phth   Phthalimido

DNP   Dinitrophenyl

NOE   Nuclear Overhauser enhancement

MIKES   Metastable ion kinetic energy spectra

1. INTRODUCTION

Cyclopeptide alkaloids are a rapidly growing group of closely related polyamide bases of plant origin. They are principally composed of simple amino acids; their basicity is attributable to an *N*-terminal amino acid residue, the amine function being either an *N*-methyl or an *N,N*-dimethyl group. They are further distinguished by an additional fragment that is either a hydroxyphenethylamine moiety or its respective dehydration or oxidation derivative. With one exception, all cyclopeptide alkaloids known to date contain at least one β-hydroxy-α-amino acid whose oxygen is involved in the ring. A general structure for these compounds may be represented by **1**. The term *peptide alkaloid* was first proposed by Goutarel and co-workers [1]; however, since these alkaloids, with the exception

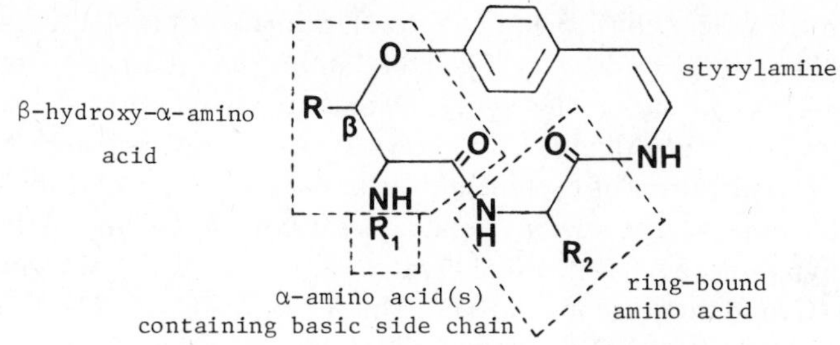

β-hydroxy-α-amino acid

styrylamine

α-amino acid(s) containing basic side chain

ring-bound amino acid

1

of lasiodine-A, are all cyclic, the term *cyclopeptide alkaloid* was deemed more appropriate. The terms *ansapeptide* and *phencyclopeptine* have also been proposed for these macrocyclic peptide alkaloids [2a].

Since the structure of pandamine was confirmed in 1966 [3], over 100 cyclopeptide alkaloids have been isolated and their structures elucidated. Cyclopeptide alkaloids are particularly common in plants of the Rhamnaceae family but are also found in over 25 other species of plants. The widespread occurrence of these compounds makes them an important class of natural products. They occur in the leaves, bark, and other parts of the plants as well. Yields from dried plant material vary from 0.01 to 1% and depend not only on the plant source but also on the method of isolation. The function of these alkaloids in the plants in which they are produced is still unknown and pharmacological studies are few. The low natural availability of cyclopeptide alkaloids and the lack of adequate synthetic methodology to prepare the macrolide ring have prevented thorough studies of their biological and pharmacological properties. There are several excellent reviews of cyclopeptide alkaloids [4–8], and the present article will primarily be concerned with the most recent developments in this rapidly expanding field.

2. DISCOVERY AND CLASSIFICATION

The presence of alkaloids in a plant from the Rhamnaceae family (*Ceanothus americanus*), long used in folk medicine, was noted as early as 1884 [9]. Goutarel and co-workers isolated adouetines -X, -Y, and -Z from *Waltheria americana* [10], and Swiss investigators isolated zizyphine from *Zizyphus oenoplia* [11]. The latter authors recognized proline and isoleucine as the amino acid components. Two years later the same authors proposed a complete structure for the alkaloid [12], but this structure was revised in 1973 by Tschesche and co-workers [13]. In 1964 the occurrence of alkaloids in *Scutia buxifolia* was reported [14]. The same year the structure of pandamine, extracted from *Panda oleosa*, was shown to have a characteristic peptide structure whose components were phenylalanine, *N,N*-dimethylleucine, β-hydroxyleucine, and 2-(*p*-hydroxyphenyl) ethylamine

[1]. These results were confirmed in 1966 [3]. Shortly after, alkaloids from *C. americanus* [15, 16], *S. buxifolia* [17], and *W. americana* [18] were isolated and recognized to have structures closely related to that of pandamine. Since then the number of cyclopeptide alkaloids of known structure has increased steadily.

The first classification of cyclopeptide alkaloids was proposed by Païs and Jarreau [5], based on the various residues that constituted the molecule: the *N,N*-dimethylamino acid (A), the β-hydroxyamino acid (B) the ring-bound amino acid (C) and the aminophenol (D). The A—B, B—C, and C—D bonds are amide linkages while B—D is an aryl alkyl ether. Such an arrangement was termed "type 4 cyclic" since it included the B, C, and D fragments of a cycle to which was bonded a fourth fragment A, as represented by structure **2**.

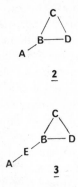

The "type 4 cyclic" group encompasses many of the known cyclopeptide alkaloids. Alternatively, there are a number of compounds that incorporate between A and B, an additional amino acid (E). These contain four amide bonds A—E, E—B, B—C, and C—D, and the same ether linkage B—D. This arrangement was called "type 5 cyclic" and was represented by **3**.

At the time this classification was proposed, only two cyclopeptide alkaloids could not be included in this classification: lasiodine-A and zizyphine. Thus far, lasiodine-A is the only naturally occurring acyclic peptide alkaloid in which a free phenolic hydroxyl and an α,β-unsaturated amide unit are present. However, under suitable conditions, lasiodine-A could presumably undergo ring closure. Zizyphine, to which an open chain structure was assigned originally [12], was later found to be a cyclopeptide. The revised structure was subsequently assigned the name zizyphine-A, since further examination of *Z. oenoplia* afforded other cyclopeptide alkaloids [13]. With a few modifications, the structure of zizyphine-A can be fitted into the Païs–Jarreau "type 5 cyclic" classification (**3**). In spite of the potential of this classification, it does not appear to have been used widely nor has it been modified to suit our present knowledge of this field.

Tschesche [7] used a different classification based on structural components. As all peptide alkaloids of the type discussed are cyclic, except for lasiodine-A, the first division was based on the size of the cycle, 13-, 14-, or 15-membered rings. Further distinctions were based on the β-hydroxyamino acid present in the ring. For instance, in the largest group, the 14-membered ring compounds, those containing β-hydroxyleucine were included in the frangulanine class. β-Hydroxyphenylalanine and *trans*-3-hydroxyproline formed the integerrine and amphibine-B classes, respectively. Only rarely were β-hydroxyvaline (hymenocardine) or β-hydroxyisoleucine (ceanothine-D) found to be part of the ring. Although this classification includes a large number of compounds, it does not encompass all known cyclopeptide alkaloids.

The Païs–Jarreau classification, with some modifications, could serve as a general method of classifying cyclopeptide alkaloids. The terms "type 4" or "type 5" cyclic used by these authors could be modified to "type 4" or "type 5," describing the number of units making up the cyclopeptide: A, B, C, and D or A, B, C, D, and E. The word *cyclic* could be omitted as acyclic molecules such as lasiodine-A are properly not class members. It would be useful to know the total number of ring atoms and these could be indicated in parentheses. For instance, "type 4 (14)" would designate a 14-membered ring cyclopeptide alkaloid having four units.

To eliminate the large amount of space required for the Païs–Jarreau structures, the ring units would be inserted in parentheses. The letters used in the Païs–Jarreau classification could be retained: A, for the basic amino acid, B for the β-hydroxyamino acid, and D for the amino alkyl phenol unit. A subscript could be used to indicate the substitution or degree of saturation of the amino chain. The chain could be saturated (D_{C-C}) or unsaturated ($D_{C=C}$) or have substituents such as hydroxyl (D_{OH}) or keto group ($D_{C=O}$). For those instances in which the β-hydroxyamino acid (B) involved in bond formation with the aromatic ring is missing and the ring linkage occurs instead via a β-carbon, the subscript A_C would indicate that the connection occurs through a carbon atom. Attachments of the B and D portions through the aromatic ring could be designated by their relative positions and shown by the letters *m*- (meta) or *p*-(para). Any other substituents in the aromatic ring could be indicated by a superscript such as D^R. The descriptive value of such a system is shown in Table 1.

The proposed system could be extended to lasiodine-A by the description "type 4 acyclic," $AEBD_{C=C}$-*p*-OH. Such a system emphasizes the basic relationships among the various cyclopeptide alkaloids. They are either type 4 or 5 and may form 13-, 14-, or 15-membered rings.

The simple amino acids found in these compounds are relatively few and usually belong to the L-family. The basic amino acids may be mono- or dimethylated. Typical amino acids for A, B, C, and E are shown in Table 2.

3. NATURAL OCCURRENCE AND ISOLATION

Cyclopeptide alkaloids are common in plants of the Rhamnaceae family but have also been found in plants of other families such as Sterculiaceae, Pandaceae, Rubiaceae, Urticaceae, Hymenocardiaceae, and Celastraceae.

TABLE 1. Cyclopeptide Alkaloids

Structure	Common Name	Description
	Frangulanine	Type 4 (14) A(B-p-D$_{C=C}$C)
	Amphibine-B	Type 5 (14) AE (B-p-D$_{C=C}$C)
	Zizyphine-A	Type 5 (13) AE (B-m-D$_{C=C}^{p-OMe}$C)
	Pandamine	Type 4 (14) A (B-p-D$_{OH}$C)
	Integerrine	Type 4 (14) A (B-p-D$_{C=C}$C)

TABLE 1. (*Continued*)

Structure	Common Name	Description
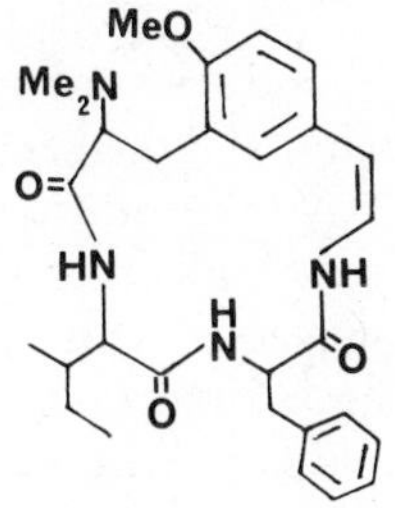	Mucronine-A	Type 4 (15) $A_C m\text{-}D_{C=C}^{o\text{-}OMe}EC$
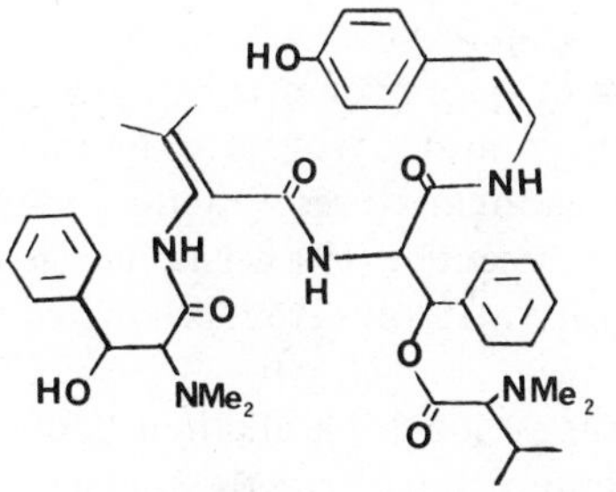	Lasiodine-A	Type 4 acyclic $AEBD_{C=C}p\text{-}OH$

Although cyclopeptide alkaloids are widespread in plants, being found in the leaves, bark, root bark, and seeds, they are generally difficult to isolate as they are often present in small amounts in very complex mixtures. The relative amounts of the various constituents of such mixtures depend on many factors, including the region of growth and the maturity of the plant. Therefore, accounts of the percentages of main and secondary alkaloids may vary considerably. The yields of cyclopeptide alkaloids from dried plant material range from 0.01 to 1%. The yields are very much dependent on the method of isolation, in addition to other factors related to their source.

TABLE 2. Common Amino Acids in Cyclopeptide Alkaloids

A	B	C	E
N-Me-Ala	β-OH-Ile	Ile	Ile
N,N-Me$_2$-Ala	β-OH-Leu	Leu	Leu
N,N-Me$_2$-Ile	β-OH-Phe	Phe	Pro
N,N-Me$_2$-Leu	β-OH-Pro	Pro	
N-Me-Phe	β-OH-Val	Trp	
N,N-Me$_2$-Phe		Val	
N-Me-Pro			
N,N-Me$_2$-Trp			
N-Me-Val			
N,N-Me$_2$-Val			

The first step in the isolation of cyclopeptide alkaloids is an extraction procedure which is modified according to the plant source and the bases present. These techniques have been described and referenced in previous review articles [4–7].

The dried ground plants are sometimes treated with a dilute basic solution (10% aqueous ammonia or 1% aqueous sodium carbonate) and then extracted with an organic solvent such as ether. Conversely, they may be treated directly with a solvent such as methanol. The resulting solution is then basified with ammonia and extracted with benzene. In some instances the dried material is simply heated with benzene–concentrated aqueous ammonia–methanol (100:1:1). The bases are usually separated from the extracts by treatment with 5% aqueous citric acid. Further purification and separation of the individual bases are accomplished by standard chromatographic methods. Chromatographic separations may be effected on alumina or, more commonly, on silica gel columns using solvents such as chloroform, acetone, ethyl acetate, dioxane, acetonitrile, and ether as well as chloroform–methanol mixtures. Fractions may be detected with an ultraviolet light source. More recently, HPLC has emerged as a valuable tool for the separation of cyclopeptide alkaloids from crude plant extracts [2].

The isolation and structural identification of the cyclopeptide alkaloids found in *S. buxifolia,* a member of the Rhamnaceae family, have been the subject of many investigations. The first member of the scutianines was isolated by Tschesche et al. in 1967 and named scutianine-A, an example of a type 4 (14), A(B-*p*-DC) cyclopeptide alkaloid. As the B unit is β-hydroxyleucine, these compounds are related to frangulanine. The structure elucidation of scutianine-A was an important contribution to the field of cyclopeptide alkaloids [17]. In 1971 scutianine-B was isolated [19]. In 1974, another research group reported the isolation of two new alkaloids, scutianine-C and -D and described structural studies on scutianine-A, -C and -D [20–22]. Simultaneously, Tschesche and co-workers published work on the isolation of three alkaloids, scutianine-C and two diastereomers, scutianine-D and -E [23]. The previously reported scutianines-C and -D [21, 22] were identical to the scutianines-D and -C, respectively, reported by the German group [23]. Subsequently, from the bark of *S. buxifolia* scutianines-G, -F, and -H were isolated [24]. While scutianine-G is a diastereomer of scutianines-D and -E, scutianine-F has structural resemblance to scutianine-B. The bark of another plant from the Rhamnaceae family, *Z. sativa,* yielded two compounds, sativanine-A and sativanine-B, in addition to other known cyclopeptides. Recently sativanines-C, -D, -E, and -F have been isolated from *Berberis* and *Corydalis* species [25a–c]. These compounds are of type 4 (14) A(B-*p*-DC). The B portion is β-hydroxyphenylalanine and therefore these compounds are related to integerrine [25a].

The bark of the African tree *Z. mauritiana Lam,* afforded two additional compounds, mauritine-A and mauritine-B, as well as the known cyclopeptide alkaloids frangulanine and amphibine-D. The mauritines were isolated, characterized, and found to be related to amphibine-D [26], type 5 (14) AE(B-*p*-DC), in which B is β-hydroxyproline. The stem bark of *Z. oenoplia Mill* afforded nine alkaloids. Four of these were known: zizyphine-A and -B and abyssinine-A and -B [27]. Among the others were zizyphine-C [type 4 (13) A(B-*m*-DC), B = β-hydroxyproline], zizyphine-D and zizyphine-E [type 4 (15) (AB-*m*-D^{R}C)]. The known alkaloids mauritine-A, mucronine-D, amphibine-H, nummularine-A and -B, as well as two novel compounds, jubanine-A and -B, were isolated from the stem bark of *Z. jujuba* [28a]. Recently, nummularine has been isolated from *Z. nummularia* [28b]. The new cyclopeptide alkaloids were representatives of type 5 (13) AE(B-*m*-D^{R}C), in which B is 3-hydroxyproline. Seven cyclopeptide alkaloids were found distributed among three forms of the shrub *C. integerrimus* (Deer Brush), a polymorphic species of the family Rhamnaceae occuring from southern Washington through California into western New Mexico [2]. These cyclopeptide alkaloids correspond to type 4 (14) A(B-*p*-DC) where A = *N*-Me-Val, *N,N*-Me$_2$-Val, *N*-Me-Ile, *N,N*-Me$_2$-Ile, B = β-OH-Phe, β-OH-Val, C = Phe, Trp, Leu.

An extract from the roots of *Melochia tomentosa* was found to contain three cyclopeptide alkaloids: scutianine-B and melonovines-A and -B [29]. These compounds are also type 4 (14) A(B-*p*-DC) systems. The methanolic extract of leaves of *Cocculus villosus*, a plant belonging to the family Menispermaceae, yielded a number of 13-membered ring cyclopeptide alkaloids [30, 31]. *C. villosus* is a climber located in the subtropical region of Indo-Pakistan. The sap of the leaves forms a mucilage used as a cooling, soothing lotion in prurigo, eczema, or impetigo [30]. The *Cocculus* alkaloids are type 5 (13) AE(B-*m*-DC); the B unit is a 3-hydroxyproline, classifying these compounds as relatives of amphibine.

4. GENERAL PROPERTIES

Cyclopeptide alkaloids, as the free bases, are generally crystalline and crystallize easily; the amphibines and mauritines are amorphous. They have high melting points, primarily in excess of 200° C. Reduction of the *p*-hydroxystyrylamine unit usually causes an increase in melting point of 20–80° C. With few exceptions (aralionine-A, +82°, methanol) and lasiodine-A (+38°, chloroform), most of the alkaloids are levorotatory and have fairly large negative rotations (−200° to −400° in either chloroform or methanol). The dihydro derivatives have lesser negative rotations. In 15-membered cyclopeptide alkaloids the optical activity depends both on the solvent and on the degree of methylation of the amino group [7]. Cyclopeptide alkaloids are usually sparingly soluble in water but readily soluble in alcohols, chloroform, and several other organic solvents.

The pK_a values of cyclopeptide alkaloids have been determined in methyl cellosolve-water. The pK_a range is between 5.39 for adouetine-Z and 6.55 for zizyphine-A. They are very weak bases and often crystallize from aqueous solutions of weak acids. Many salts of cyclopeptide alkaloids are known, including hydrochlorides, oxalates, hydroiodides, and perchlorates. Several of these salts are appreciably soluble in nonpolar solvents. The use of ion exchange resins in the isolation and purification of cyclopeptide alkaloids has been investigated [32].

5. BIOLOGICAL ACTIVITY

Plants containing cyclopeptide alkaloids were originally investigated because of their therapeutic reputations. For example, Z. mucronata was employed by the natives of central and southern Africa in the treatment of diarrhea and dysentery. Unfortunately, the cyclopeptide alkaloids isolated from these sources exhibited no significant biological activity. Pharmacological investigations, however, have been hampered by the lack of large quantities of pure products [5].

Cyclopeptide alkaloids are reported to have some antibiotic properties and to be weakly active against lower fungi and gram-positive bacteria. Ring size does not appear to affect the biological activity, but the degree of methylation of the amino group is reported to be important. Desmethyl bases are active only against fungi, and antibacterial activity is only observed after methylation of the amino function [7]. Adouetine-Z has been studied most thoroughly and has not shown any specific activity [33]. The effect of frangulanine on mitochondrial swelling has been investigated [34]. Frangulanine-induced mitochondrial swelling in 0.15 M KCl solution at a 6.5 M concentration. The alkaloid showed ion selectivity on the induction of mitochondrial swelling. Mitochondria underwent swelling in 0.15 M KCl or RbCl solution but not in either NaCl or LiCl solution. The ion selectivity might be caused by the formation of a complex with K^+ or Rb^+, which would act as ionophore in the mitochondrial inner membranes in a manner similar to valinomycin. Such a complex could have biological significance in plants, perhaps being involved in absorption of nutrients from the soil, especially alkali metals [34].

Photophosphorylation in isolated spinach chloroplasts was inhibited by 21 peptide alkaloids [35]. Zizyphine-A and -B, adouetine-Z, amphibine-B, -C, and -D, and scutianine-A inhibited the coupled but not the uncoupled electron transport. The other alkaloids stimulated nonphosphorylating electron flow, behaving like uncouplers: arabonine-A, lasiodine-A, and mucronine-B were the strongest inhibitors and uncouplers. Lasiodine-A and homaline enhanced several-fold the proton gradient induced by light across the thylakoid membranes. All of the cyclopeptide alkaloids assayed inhibited photophosphorylation. Some of them specifically affected ATP synthesis while others behaved like uncouplers. Cyclopeptide alkaloids may become useful tools in the study of energy conservation in chloroplasts. The sensitivity of the photosynthetic energy conservation machinery to cyclopeptide alkaloids may be related to their still unknown biological role in plants.

6. SYNTHESIS

6.1. General

Although the synthesis of cyclic peptides has been the subject of many investigations [36], no 14-membered ring cyclopeptide alkaloid has been prepared to date. In 1981 the syntheses of two 13-membered [39] and two 14-membered [37, 38] dihydrocyclopeptide alkaloids were reported. The first syntheses of a 13- and 15-membered cyclopeptide alkaloid have been recently accomplished [55, 57].

Cyclic peptides are divided into two classes, *homodetic* and *heterodetic*, depending on whether the peptide backbone is formed exclusively from amide linkages or contains other linkages as well. Heterodetic peptides usually possess disulfide bonds or ester functions. Cyclic peptides containing ester groups in the ring are also known as *depsipeptides* and comprise many substances of microbial origin. Several of these have generated interest because of their ability to transport cations across membranes.

Cyclopeptide alkaloids are a new class of heterodetic cyclic peptides that contain aryl alkyl ether linkages. They also have the ability to transport cations across membranes. Since it is believed that the biological activity of a cyclic peptide will be retained in analogs of different sequence and configuration, provided their structures are similarly arranged in space, it may be expected that cyclopeptide alkaloids will show a biological similarity to related depsipeptides. Therefore, practical synthetic methods for cyclopeptides are of utmost importance if this potential similarity is to be explored.

The steps required for the synthesis of heterodetic cyclic peptides are (1) formation of peptide bonds, (2) formation of nonpeptide bonds, and (3) cyclization, the latter step being most crucial.

As in any cyclization, the formation of a peptide ring requires the generation of mutually reactive chain ends under conditions that favor an intramolecular process. The formation of a peptide bond requires a free amino group and a carboxyl function activated toward nucleophilic attack. The intramolecular process is favored by allowing the coupling of the ends to proceed at high dilution (10^{-3}–10^{-4} M). In these circumstances the carboxyl activation and the cyclization step must be separable.

Cyclization is always a slow process and it is important that the activated carboxyl group should not undergo unimolecular or solvent-induced decomposition nor racemize before reaction with the terminal amino group.

Cyclization may be accomplished by methods that cleanly separate the activation and cyclization steps or methods that effect direct cyclization of unprotected peptides.

The most widely used methods of carboxyl activation are the active ester, the azide, and mixed anhydride procedures. Direct cyclization involves protection of the *N*- and *C*-terminal groups by functions that can be removed simultaneously, followed by ring closure of the terminally unblocked chain by means of carboxyl activating agents.

Carboxyl Activation: Active Ester Method. Activated esters are stable intermediates and do not decompose easily, though they may racemize if the cyclization step is slow.

Conversion of peptide acid to the corresponding ester may be carried out with *p*-nitrophenyl sulfite or *p*-nitrophenyl trifluoroacetate [36]. Reaction of *p*-nitrophenol, a carbodiimide, and the acid has also been used. The *N*-terminal residue is usually blocked as the *tert*-butoxycarbonyl or benzyloxycarbonyl derivative. Acid cleavage removes these groups and reaction of the ester and free amino group occurs in dilute basic solution. Cyclization has often been carried out by adding the active ester hydrohalide to a large volume of pyridine at 60–100° C.

p-Nitrophenyl esters have been employed most commonly, although recently more reactive pentafluorophenyl esters have been used. *N*-Hydroxysuccinimide- or 2,4,5-trichlorophenyl esters have also shown utility.

When the active ester is *p*-nitrothiophenyl and the *N*-terminal blocking group is *o*-nitrophenylsulfenyl, cyclization does not require a separate step to deblock the amino function. With nitrothiophenol, the cyclization can be effected in pyridine either in the presence or absence of imidazole.

Acyl Azide Method. Although the acyl azide function is thermally labile, the aminoacyl azide does not have a tendency to racemize; its precursor may be a methyl or ethyl ester function. The ester is treated with hydrazine to afford an acyl hydrazide which can then be diazotized. Coupling is carried out in basic solution in a large volume of cold solvent (water, dimethylformamide, or pyridine) so that the reactive peptide is generated in 10^{-3}–10^{-4} M solutions. Diazotization may be carried out in an organic solvent with an alkyl nitrite; in this case hydrolysis is eliminated as a side reaction.

Mixed Anhydride Method. An alternative carboxyl activation method utilizes a mixed carbonic anhydride. As noted with the azides, these active intermediates are more thermally labile than esters. In this procedure the amino group to be coupled is protonated. Complete protonation is assured by the presence of a buffer during the formation of the anhydride [36]. The activated peptide is cyclized by addition to dilute base.

Direct Method. Cyclizations of unprotected peptides have been carried out with *N,N'*-dialkylcarbodiimides, such as *N,N'*-dicyclohexylcarbodiimide, or *N*-ethyl-*N'*-(3-dimethylaminopropyl)-carbodiimide. Although these reagents are convenient to use, the activation step requires a high concentration of carbodiimide and peptide while cyclization requires low concentrations. During slow cyclizations, rearrangement of an active *O*-acylisourea to an inactive *N*-acylurea may occur, but there are ways to prevent this side reaction [36].

Other procedures have been successfully employed: terminally unblocked open-chain peptides have been cyclized using isoxazolium salts such as Woodward reagent K; cyclizations of unblocked open-chain peptides have also been accomplished with derivatives of pyrophosphorous acid in diethyl phosphite at 140° C and diphenyl phosphoryl azide at 0° C [36]. The different methods of activiation and cyclization, as applied to cyclopeptide alkaloids, will be discussed in more detail when individual syntheses are considered.

6.2. Synthetic Strategy

The main challenges to the synthesis of cyclopeptide alkaloids are (1) the preparation of the β-aryloxyamino acid, (2) the formation of an enamide, and (3) the generation of the macrocycle. In the most commonly found 14-membered cyclopeptides, this involves creation of a cycle with considerable strain. Synthetic studies have principally focused on the dihydro derivatives of the naturally occurring cyclopeptide alkaloids. These derivatives present similar synthetic problems since reduction of the vinylic double bond does not relieve ring strain to an appreciable extent.

As the generation of the macrocycle is the key step in the synthesis of any cyclopeptide alkaloid, we might first consider the different positions for ring closure in a 14-membered dihydrocyclopeptide alkaloid, assuming that introduction of the double bond could be accomplished after cyclization. These sites are shown in Figure 1 and are indicated by the numbers 1–4.

Cyclization at sites 1 and 2 requires a high-dilution procedure using any of the standard peptide bond formation methods, described in section A. Cyclization at position 3 requires an intramolecular Michael addition of the phenolic function to a dehydroamino acid residue. As such reactions are implicated in the biosynthesis of several natural products, this approach is a biomimetic one. A desirable feature of the Michael cyclization is that it allows the use of the intact alkaloid structure, including the styrylamide moiety, as a linear precursor. Cyclization at site 3 could also be accomplished via nucleophilic attack by a phenolate ion on a suitably substituted aziridine. Finally, cyclization at position 4 would require aldol-type reactions. Such reactions might be incompatible with the groups already present in the linear precursors.

The various attempts, both successful and unsuccessful, to synthesize cyclopeptide alkaloids will be reviewed according to the cyclization site chosen by the various investigators.

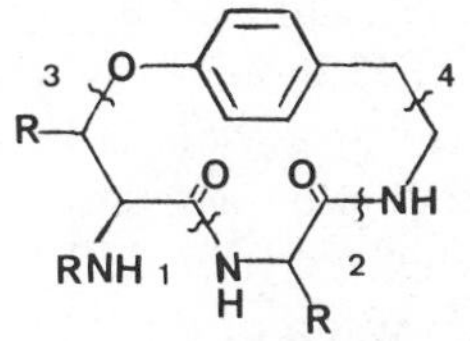

Figure 1.

6.3.   Cyclization at Sites 1 and 2

6.3.1.   Frangulanine Class.   The first attempts to synthesize cyclopeptide alkaloids were carried out by French investigators [40, 41]. Early isolations and structural determinations of cyclopeptide alkaloids were also carried out by this group [10].

The target molecule was a dihydro derivative (4) of frangulanine, type 4 (14) A(B-*p*-D$_{C=C}$C). This 14-membered macrocycle is made up of a β-hydroxyleucine (B), a free amine blocked with a *tert*-butoxycarbonyl group (Boc), phenylalanine (C), and a saturated tyramine (D$_{C-C}$). The synthesis of 4 was never accomplished, although larger ring systems such as 5 were prepared. Cyclization

4   5

was attempted at both sites 1 and 2, using two different linear precursors (14) and (16) (Scheme 1). Both precursors required a β-aryloxyleucine derivative of *erythro* configuration as a common intermediate. Compound 13 was prepared [40, 42] by an eight-step synthesis beginning with methyl *erythro*-β-hydroxyleucinate (6), whose amino function was protected with a benzyloxycarbonyl group (Z). The ester function of 6 was reduced with lithium borohydride to the corresponding alcohol (7). Tosylation of the alcohol, followed by heating, afforded the oxazolidinone derivative 8. Tosylation of 8 afforded 9, which was then treated with the sodium salt of *N-tert*-butyloxycarbonyltyramine, prepared, in turn, from the protected tyramine and sodium hydride in dimethylformamide. *N*-Benzyloxycarbonyltyramine may also be used. The crucial conversion of 9 to 11 proceeded via aziridine intermediate 10. The formation of 10 was inferred from the stereochemical course of the S$_N$2 displacement. This nucleophilic displacement produced the ether linkage in the correct configuration but racemic at the α-position (11). The *erythro*-configuration of 8 was retained in derivative 11. In the *threo*-series the intermediate aziridine (10) could even be isolated. Basic hydrolysis of the oxazolidinone ring of 11, followed by protection of the free amino group with carbobenzoxy chloride, afforded 12. Oxidation with Jones reagent gave 13. Elaboration of 13 to linear intermediate 14, useful for cyclization at site 2, was accomplished by acylation of 13 with phenylalanine methyl ester using dicyclohexylcarbodiimide activation. Compound 16, desired for cyclization at site 1, was synthesized from 12 by selective removal of the *tert*-butoxycarbonyl group and acylation of the tyramine amino group of 15 with *tert*-butoxycarbonylphenylalanine *N*-hydroxysuccinimide ester [41].

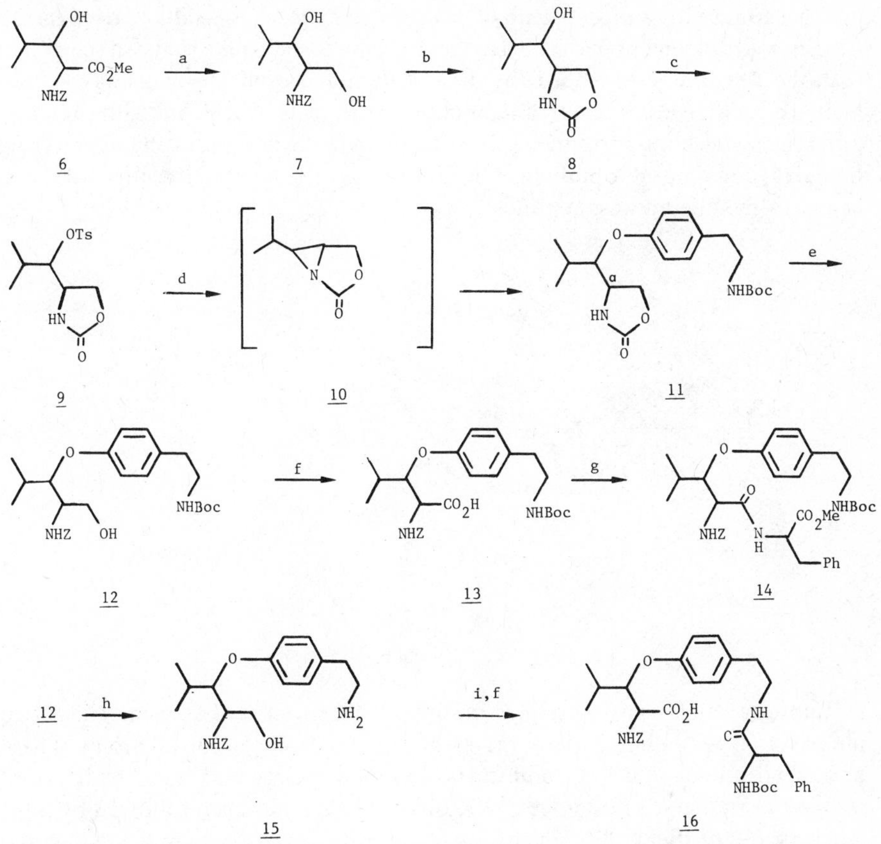

^{a}LiBH$_4$; ^{b}TsCl, Δ; ^{c}TsCl; ^{d}N-Boc-tyramine, NaH, DMF; ^{e}NaOH, Z-Cl; ^{f}Jones reagent; ^{g}Phe-OMe, DCC, ^{h}H$^+$; ^{i}Boc-Phe-OSu

Scheme 1.

Cyclization attempts at sites 1 and 2, designed to form a 14-membered ring, involved standard peptide bond formation methods such as azide or active ester activation.

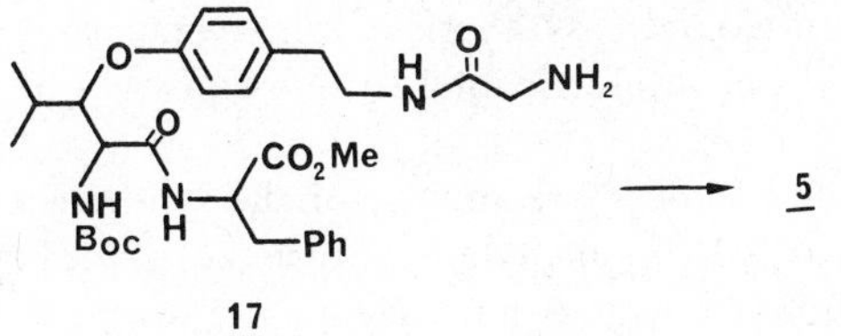

Whereas a high-dilution azide coupling reaction converted **17** to the 17-membered ring system **5,** a similar cyclization reaction using **14** failed to produce **4.** Cyclizations via an acyl azide may result in formation of an undesired

product containing a urea instead of an amide linkage. This product arises via the Curtius rearrangement of the acyl azide to form an isocyanate that can then react with the free amine to afford the urea. Attempted cyclizations using *p*-nitrophenyl ester activation resulted in polymeric products. Therefore, although two different amide bond forming methods were used, the desired 14-membered ring model **(4)** could not be obtained. The failure was attributed to the unusual strain in the 14-membered ansapeptides.

$$\underline{18} \quad \text{a} \quad R_1 = R_2 = (CH_2)_3, \quad R_3 = Me$$

$$\quad \text{b} \quad R_1 = R_2 = (CH_2)_3, \quad R_3 = H$$

$$\underline{19} \quad \text{a} \quad R_1 = H, \quad R_2 = CH_2CHMe_2, \quad R_3 = Me$$

$$\quad \text{b} \quad R_1 = R_3 = Me, \quad R_2 = CH_2CHMe_2$$

$$\quad \text{c} \quad R_1 = R_3 = H, \quad R_2 = CH_2CHMe_2$$

Rapoport and co-workers [43] reported cyclization studies using simplified models such as **18a,b** and **19a–c** for 14-membered cyclopeptide alkaloids. These simplified models did not contain the enamide moiety and the β-hydroxy-α-amino acid moiety. The choice of a saturated model has been followed by most workers in the field. The exclusion of the β-hydroxy-α-amino acid would eliminate diastereomer separations. These models **(18a,b, 19a–c)** were chosen to examine the effect of amide substitution on the course of peptide cyclization. Both amides were alkylated either with methyl groups or by incorporation into a proline residue. Amide resonance is believed to stabilize a planar conformation. If one neglects the contribution of hydrogen bonding, the *trans*-configuration is more favorable than the *cis*. The energy necessary for *cis-trans*-isomerization is approximately 28 kcal/mol [44, 45] for a secondary amide, and 21 kcal/mol [46] for a tertiary amide. Thus, the preference for the *trans*-arrangement diminishes when peptide bonds are alkylated [47]. In linear peptides the presence of a *cis*-amide bond causes the peptide to fold in on itself. If such folding can take place at both peptide ends, it might help cyclization. The influence of amide conformation on the intramolecular reaction between two linear peptide ends has been demonstrated in the synthesis of nine-membered cyclotripeptides such as cyclotrisarcosyl and cyclotriprolyl, which can only be prepared when the amides are tertiary [47].

The preparation of linear precursors for the synthesis of models **18a,b** and **19a–c** began with *p*-hydroxyphenylethylamine **(22)** as the key intermediate. This compound was prepared as shown in Scheme 2. The amine **(22)** was obtained by

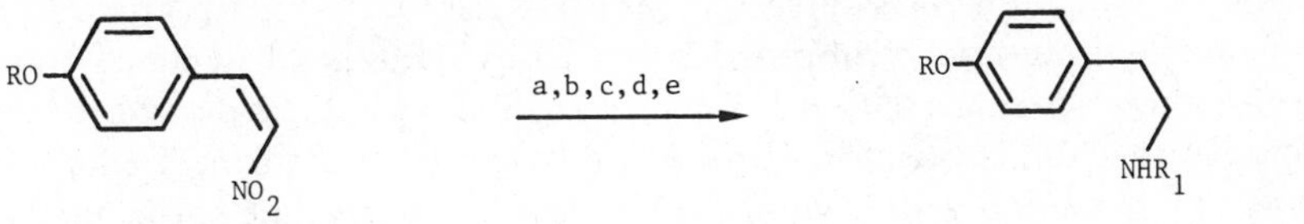

RO—⟨benzene⟩—CH=CH—NO₂ → (a,b,c,d,e) → RO—⟨benzene⟩—CH₂CH₂—NHR₁

20 R = H
21 R = Me

22 R = H R_1 = H
23 R = H R_1 = H·HBr
24 R = H R_1 = CHO
25 R = CH₂Ph R_1 = CHO
26 R = CH₂Ph R_1 = Me
27 R = Me R_1 = Me

[a]Pd/C,H₂,MeCO₂H; [b]HBr; [c]Cl₃CCHO, Et₃N; [d]PhCH₂Cl, K₂CO₃, Me₂CO; [e]LiAlH₄, THF; [f]N-Me morpholine, THF, isobutylchloroformate, N-Boc amino acid; [g]DCC, N-Boc amino acid

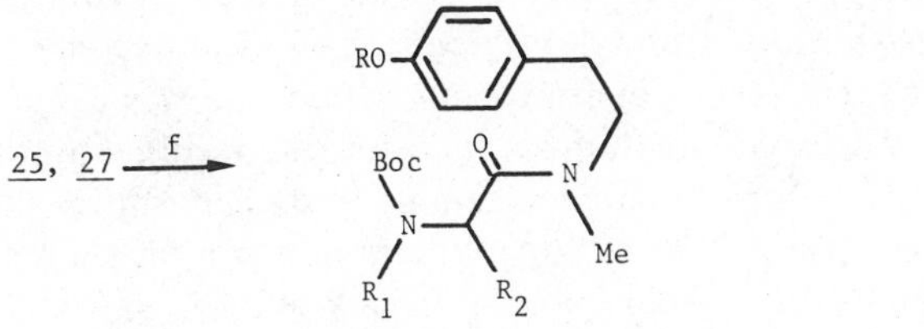

25, 27 —f→

RO—⟨benzene⟩—CH₂CH₂—N(Me)—C(=O)—CH(R₂)—N(Boc)(R₁)

28[g] R = Me, R_1, R_2 = $(CH_2)_3$
29   R = CH₂Ph, R_1 = H, R_2 = CH₂CHMe₂
30   R = CH₂Ph, R_1 = Me, R_2 = CH₂CHMe₂

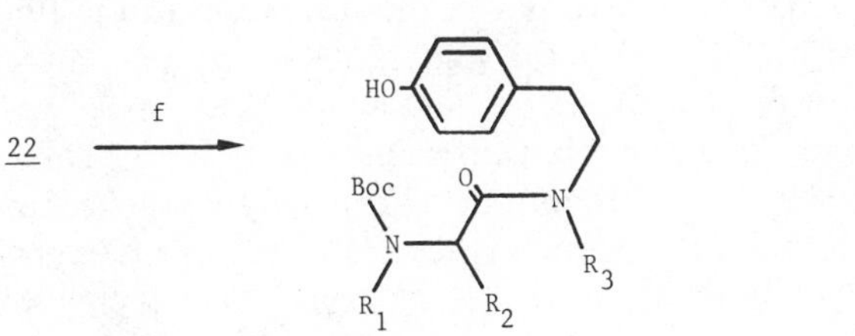

22 —f→

HO—⟨benzene⟩—CH₂CH₂—N(R₃)—C(=O)—CH(R₂)—N(Boc)(R₁)

31a R_1 = R_2 = $(CH_2)_3$, R_3 = Me
31b R_1 = H, R_2 = CH₂CHMe₂, R_3 = Me
31c R_1 = R_3 = Me, R_2 = CH₂CHMe₂
31d R_1 = R_2 = $(CH_2)_3$, R_3 = H
31e R_1 = R_3 = H, R_2 = CH₂CHMe₂

Scheme 2.

the catalytic reduction of the nitrostyrene (20) in acetic acid. Compound 22 was then converted into its hydrobromide salt (23) by refluxing it in concentrated hydrobromic acid. Formylation of 23 with trichloroacetaldehyde and triethylamine afforded 24. Protection of the phenolic group as the benzyl ether (25) followed by reduction with lithium aluminum hydride gave the corresponding N-methyl derivative 26. A similar series of reactions was also carried out on 2-(4'-methoxyphenyl)ethylamine (21) to yield the methoxy analog of 26, compound 27. Amines 26 and 27 were then acylated with N-tert-butoxycarbonyl amino acids using a mixed anhydride procedure (N-methylmorpholine, isobutylchloroformate, THF, −15° C). The yields of peptides 29 and 30 from 27 and 31d and 31e from 22 were greater than 90%. Peptide 28, prepared via dicyclohexylcarbodiimide coupling, was obtained in much lower yields. The N-methyl peptides 29 and 30 were converted in high yields to phenols 31b and 31c, respectively.

Phenols 31a–e were then treated with benzyl propiolate (36) to afford high yields of the corresponding β-aryloxy propenoates 32a–e (Scheme 3). Compounds 32a–e were reduced catalytically and then converted to the corresponding p-nitrophenyl (ONp) esters, 34a–e, using p-nitrophenyl trifluoroacetate. Removal of the N-tert-butoxycarbonyl protecting group was accomplished by dissolving the p-nitrophenyl esters in anhydrous trifluoroacetic acid at 0–5° C. After evaporation of the excess trifluoroacetic acid, the residual amine salts, 35a–e, were dissolved in N,N-dimethylacetamide and then added slowly to pyridine at 90° C. Synthesis of the cyclopeptide monomers on a preparative scale was thus accomplished in yields varying from 9 to 24%. In each case the desired cyclic monomer was accompanied by its respective dimer and could be separated by chromatography on Sephadex LH20. In contrast with the results obtained in the synthesis of cyclotripeptides [47], the yields of cyclopeptide alkaloid models 18a,b and 19a–c were independent of the substitution on the amide group ($-\overset{\overset{\text{O}}{\|}}{\text{C}}-\overset{|}{\text{N}}R_3$) not involved in the formation of the final peptide bond. The yields of cyclopeptides 18a and 18b were similar, although they differed by the substitution pattern of one amide. The same was true for cyclopeptides 19a and 19c, although their yields were lower. Cyclopeptide 19b was obtained in very low yield. The authors concluded that the reactivity of the free amino group in the linear peptide is the major factor in determining the yields of the cyclic monomers. The rate of acylation with tert-butoxycarbonylazide decreased in the series proline > leucine >> N-methylleucine; the yields of cyclopeptides decreased in the same order, that is, with decreasing reactivity of the nucleophile. Although the yields of cyclopeptides were independent of the degree of amide substitution in the linear precursor, the configuration of the cyclic products was dependent on the structure of the amide in the linear peptide. These results are of interest as they represent examples of 14-membered, para-ansa compounds. However, the lack of side chain functionality and different amide substitution patterns made the relevance of models 18a,b and 19a–c to natural cyclopeptide alkaloids very marginal.

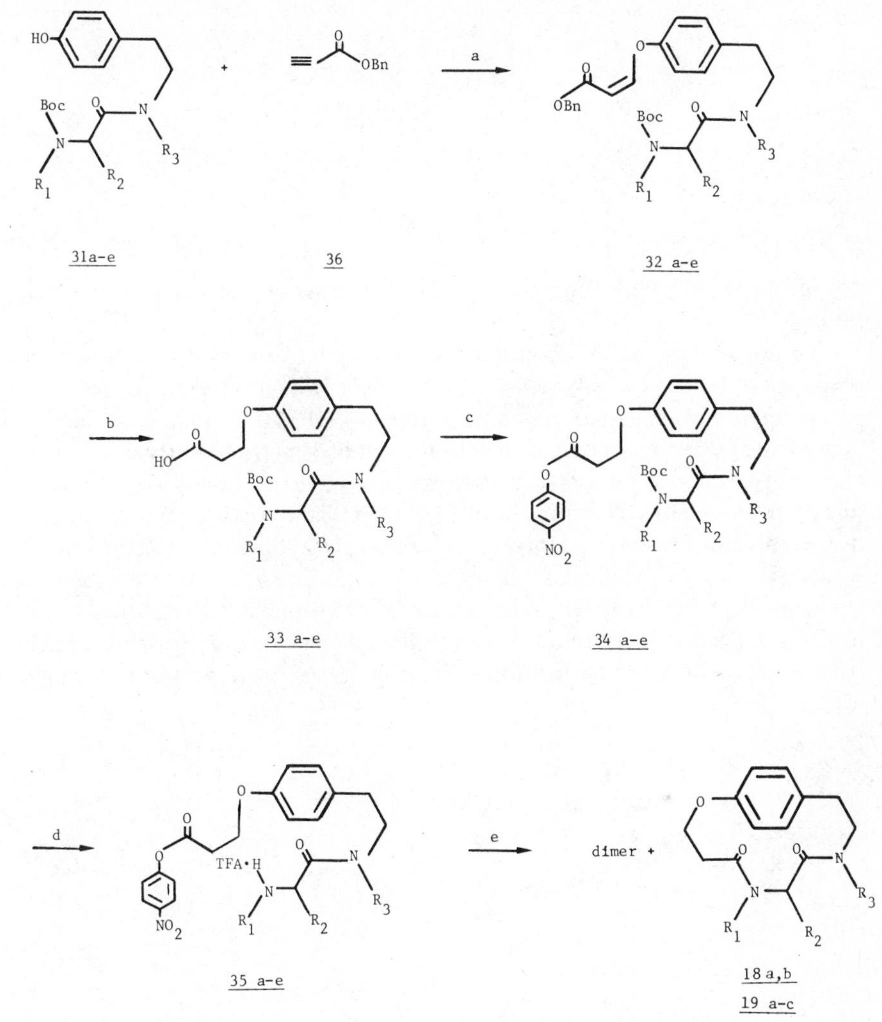

^{a}N-Me-morpholine, THF; ^{b}H$_2$, Pd/C, EtOH; ^{c}CF$_3$CO$_2$Np, py; ^{d}CF$_3$CO$_2$H; ^{e}MeCONMe$_2$, Py

Scheme 3.

The ion binding properties of the synthetic cyclo [3-(4-aminoethyl)phenyl-oxypropanoyl-L-prolyl] **(18b)** and a natural cyclopeptide alkaloid, ceanothine-B, were determined by circular dichroism studies in acetonitrile. Compound **18b** showed selectivity for Mg^{2+} and Ca^{2+} over Li$^+$ and did not interact with Na$^+$ and K$^+$. Ceanothine-B interacted with Mg^{2+} and Ca^{2+} but not with Na$^+$. Cyclic dimers or linear precursors did not exhibit metal complexing [43].

Another investigation on the total synthesis of phencyclopeptines dealt with the introduction of an isopropyl side chain at C(9), a natural product substitution pattern resulting in the synthesis of a new p-phencyclopeptine model (5S,9R)-

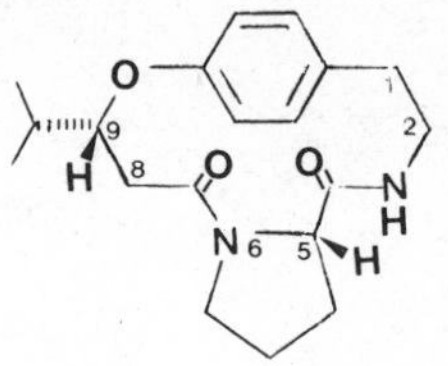

36

9-isopropyl-5,6-trimethylene-8-deamino-1,2-dihydro-*p*-phencyclopeptine (**36**) [48].

The preparation of **36** is outlined in Scheme 4. This synthesis utilized the readily available optically active phenol **37a** which was converted to its potassium salt (**37b**) with potassium hydride in THF at 0° C. Benzyl-4-methyl-2-pentynoate (**38**), prepared from the lithium salt of 3-methyl-1-butyne and benzyl chloroformate, was added to a suspension of **37b** in THF to give isomeric phenoxypentenoates **39a** and **39b** in 37% yield. Catalytic hydrogenation of the mixture afforded racemic phenoxypentanoic acids (**40**) in 92% yield. No attempts were made to separate diastereomers at this stage. The acid mixture was instead converted to the *p*-nitrophenyl active ester **41** with *p*-nitrophenyl trifluoroacetate in pyridine. Subsequent removal of the Boc protecting group with anhydrous trifluoroacetic acid and cyclization as previously described (pyridine, 90° C) [43]

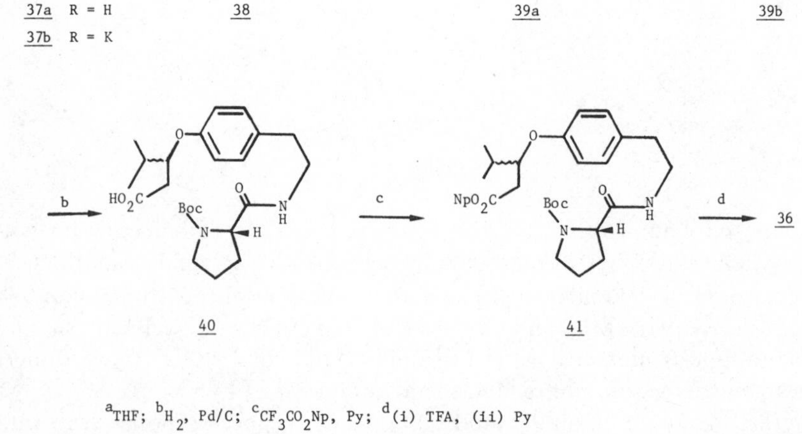

| 37a | R = H | 38 | 39a | 39b |
| 37b | R = K | | | |

^{a}THF; ^{b}H$_2$, Pd/C; ^{c}CF$_3$CO$_2$Np, Py; ^{d}(i) TFA, (ii) Py

Scheme 4.

afforded cyclic monomer **36** in 18% yield after chromatography and sublimation. Asymmetric induction during the cyclization step permitted the isolation of a single isomer with the natural phencyclopeptine stereochemistry at C(9). The ready availability of the starting materials and the stereospecificity of the cyclization makes this approach a promising synthetic route to the phencyclopeptine class of cyclopeptide alkaloids.

6.3.2. Amphibine Class. The first total synthesis of a 14-membered di-hydrocyclopeptide alkaloid, dihydromauritine-A **(42)**, was accomplished from L-proline via an S_N2 displacement and three amide bond forming steps [37]. The nucleophilic displacement generated an aryloxyether from a β-bromopyrroline methyl ester and the thallium salt of Boc-phenylalanyltyramide. Reduction of the pyrroline double bond with dimethylamine-borane resulted in a 60:40 ratio of *trans:cis-β*-substituted proline derivatives. The crucial cyclization step was accomplished by amide bond formation using an active-ester derivative. Macrocyclic ring formation proceeded in similar yields when carried out at 90°C in pyridine or at 25°C using the acylation catalyst 1-hydroxybenzotriazole. Contrary to the results of Rapoport [43, 48], no stereoselectivity was observed in the cyclization step and both proline diastereomers were formed in equal amounts. It should be noted, however, that the cyclization step in this case involved the potentially less reactive phenylalanine and not proline as the nucleophile. The introduction of the dipeptide side chain, *N,N*-dimethylalanyl-valine, was carried out most easily using dicyclohexylcarbodiimide-hydroxy-benzotriazole activation to afford dihydromauritine-A with configuration 8*S*,9*S*,5*S* and its stereoisomer with configuration 8*R*,9*R*,5*R*. Separation of diastereomeric products was only possible at the final-product stage. The synthesis of **42** is shown in Scheme 5.

42

The synthetic scheme began with methyl pyrroline carboxylate **(43a)**, a compound previously described [49]. Compound **43a** was converted to the corresponding allylic bromide **(43b)** which was then condensed with the thallium salt of the *N*-acylated tyramine derivative **44b** to afford an unstable ester **(45a)** which was not purified. Saponification with sodium hydroxide gave salt **45b**. Compound **44b** was best prepared by condensing tyramine with the hydroxysuccinimide ester of *tert*-butoxycarbonyl-L-phenylalanine. Reduction of **45b** with dimethylamine-borane in acetic acid afforded a 60:40 mixture of *trans-* and *cis-*

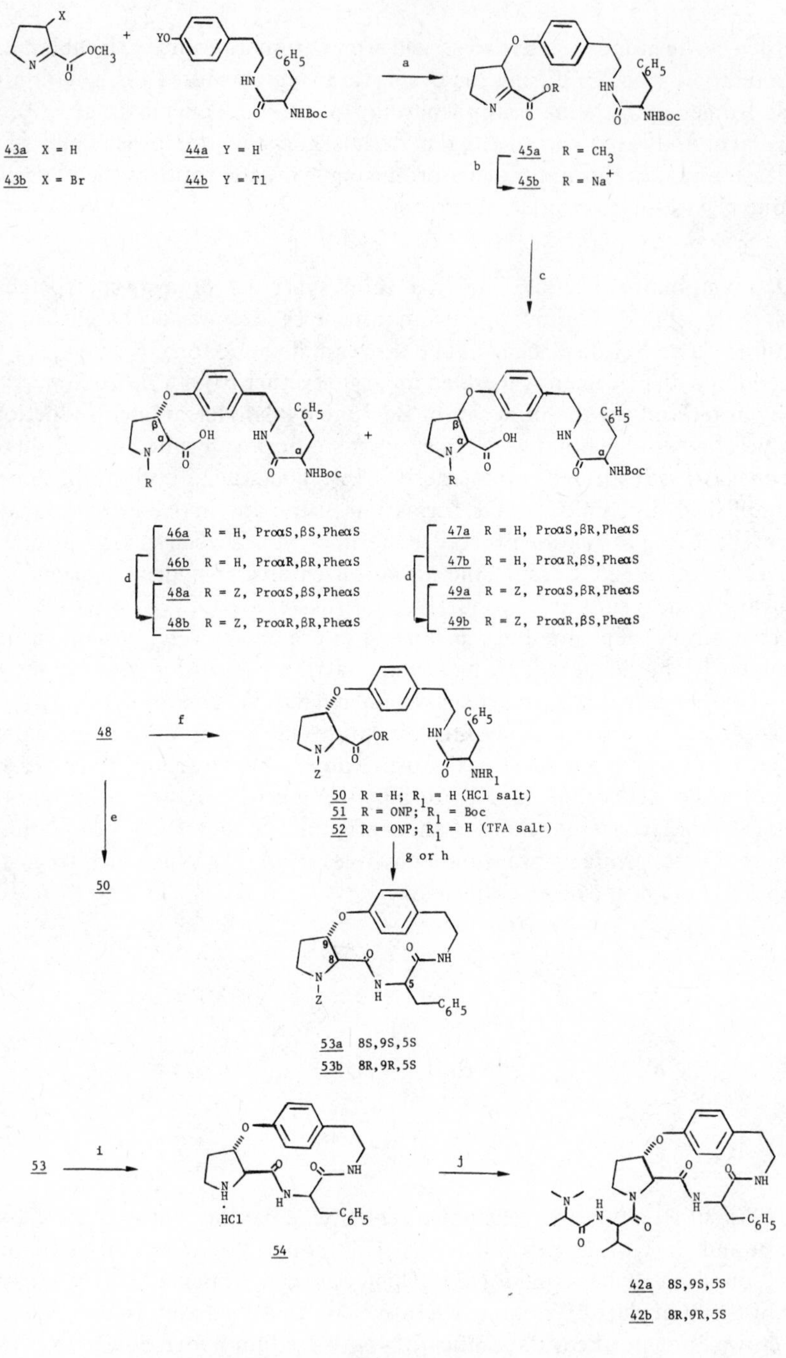

43a X = H
43b X = Br

44a Y = H
44b Y = Tl

45a R = CH₃
45b R = Na⁺

46a R = H, ProαS,βS,PheαS
46b R = H, ProαR,βR,PheαS
48a R = Z, ProαS,βS,PheαS
48b R = Z, ProαR,βR,PheαS

47a R = H, ProαS,βR,PheαS
47b R = H, ProαR,βS,PheαS
49a R = Z, ProαS,βR,PheαS
49b R = Z, ProαR,βS,PheαS

48

50 R = H; R₁ = H (HCl salt)
51 R = ONP; R₁ = Boc
52 R = ONP; R₁ = H (TFA salt)

50

53a 8S,9S,5S
53b 8R,9R,5S

53

54

42a 8S,9S,5S
42b 8R,9R,5S

ᵃDMF; ᵇNaOH/Dioxane; ᶜBH₃·N—; ᵈZ-Cl; ᵉHCl, EtOAc; ᶠ(i) CF₃COO—⟨⟩—NO₂, (ii) TFA; ᵍPy; ʰDMF, HOBT; ⁱH₂, Pd/C, EtOH, HCl; ʲ59, DIPEA, DCC, HOBT, DMF.

Scheme 5.

134

β-aryloxyproline derivatives (**46a,b** and **47a,b**). The overall yield was 52% for three reaction steps starting with the allylic bromide **43b**. The *trans-* and *cis-* isomers were separated by chromatography and converted separately, in almost quantitative yield to their respective *N*-carbobenzoxy derivatives (**48a,b** and **49a,b**), using benzyloxycarbonyloxy-5-norbornene-2,3-dicarboximide. Attempts to separate the *trans-* and *cis-*isomers into their diastereomeric components by various separation techniques were unsuccessful and no determination of their diastereomeric composition could be made at this stage. Therefore, the key intermediate (**48**) was converted into suitable linear precursors for cyclization studies. Selective removal of the *tert-*butoxycarbonyl group in the presence of another acid-labile group, the carbobenzoxy group, was carried out using hydrogen chloride in ethyl acetate at −30°C [50], to afford intermediate **50**. When this compound was subjected to cyclization procedures using carboxyl activation with diphenyl phosphoroazidate or dicyclohexylcarbodiimide and 1-hydroxybenzotriazole in dilute solution, only dimers and polymers were obtained. Compound **48** was then converted into the corresponding *p*-nitrophenyl ester (**51**) by treating it with *p*-nitrophenyl trifluoroacetate in pyridine. After removal of the *tert-*butyloxycarbonyl group with trifluoroacetic acid, the resultant linear precursor (**52**) was subjected to two different cyclization procedures, namely treatment with pyridine under dilute conditions ($c = 0.2$ mM) at 90°C [48], and dimethylformamide at the same concentration, at 25°C, in the presence of 1-hydroxybenzotriazole [50]. Although dimerization still prevailed, some monomeric product (**53**) formed. Isolated yields of monomeric product, as calculated for the three-step conversion starting from **46b,** was 8% in the first case and 10% in the second. These results indicated that elevated temperatures are not essential for cyclization.

The side chain of **42** was synthesized as shown in Scheme 6. L-Valine methyl ester was acylated with carbobenzoxy-L-alanine using the standard mixed anhydride method to afford **55**. The carbobenzoxy blocking group was removed by transfer hydrogenation in the presence of formic acid and 10% Pd/C to give **56**. The free amino group of **56** was then dimethylated using formaldehyde and sodium cyanoborohydride to afford **57** in good yields. The methyl ester (**57**) was then converted to intermediates **58** and **59**, which were suitable for acylation of **54**, either by the azide method or by dicyclohexylcarbodiimide activation. The hydrazide (**58**) was obtained by treating **57** with a 33% hydrazine–methanol solution; the carboxylic acid derivative (**59**) was prepared by treating **57** with 25% aqueous tetrahydrofuran at pH 12–12.9. Hydrazide **58** was then converted to the corresponding azide (**60**) with isoamyl nitrite. In order to carry out acylation studies, the carbobenzoxy group in **53** was removed by hydrogenation and treated with the dipeptide hydrazide employing standard procedures. This acylation attempt, however, afforded only a minimal yield of the desired product. Successful acylation of **54** with the dipeptide was finally accomplished using an excess of the dipeptide acid **59** and activation with dicyclohexylcarbodiimide in the presence of 1-hydroxybenzotriazole and diisopropylethylamine in dimethylformamide at 25°C. A diastereomeric mixture of products was obtained (**42a** and

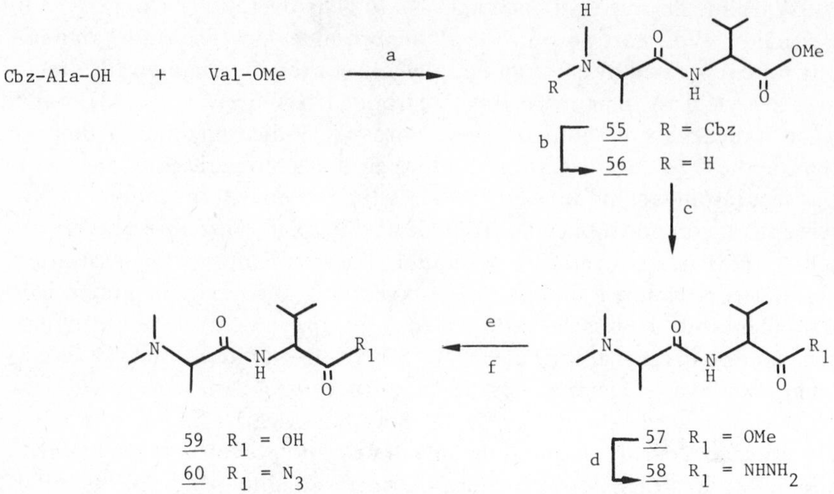

Cbz-Ala-OH   +   Val-OMe $\xrightarrow{a}$

b ⌐ 55   R = Cbz
  └→ 56   R = H

c

59   R_1 = OH
60   R_1 = N_3

$\xleftarrow[f]{e}$

d ⌐ 57 R_1 = OMe
  └ 58 R_1 = NHNH$_2$

^{a}Isobutylchloroformate, NMe morpholine; ^{b}Pd/C, HCOOH; ^{c}CH$_2$O, NaCNBH$_3$; ^{d}NH$_2$NH$_2$;

^{e}H$_2$O-THF, pH 12-12.9; ^{f}i-AmONO

Scheme 6.

42b) which were separable by chromatography. One of the isomers **(42a)** was identical with the product obtained from the catalytic hydrogenation of an authentic sample of mauritine-A.

6.3.3.   Zizyphine Class.   In 1981 Schmidt and co-workers reported a novel ring closure method for the formation of 13- and 14-membered ansa peptides, and illustrated their procedure by synthesizing a 13-membered ring compound **(61)**, which has a 10-membered *meta* bridge, in 80% yield [38, 39]. Ring closure at the unhindered primary amino group, position 2 (Fig. 1), was judged preferable. Under dilute conditions a dioxane solution of the pentafluorophenyl ester **(62)**

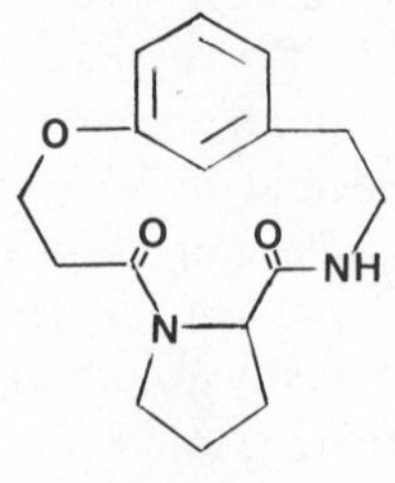

61

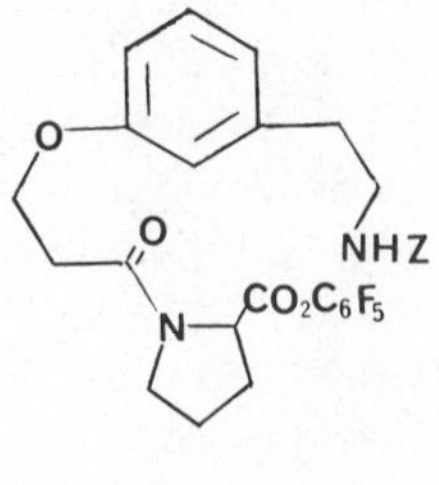

62

was added dropwise to a rapidly stirred suspension of Pd/C in dioxane, at 95° C, into which hydrogen was passed. The solution contained 1 mol of 4-pyrrolidino-pyridine per mole of **62** as catalyst and 2% ethanol (with respect to the solvent). Yields dropped dramatically in the absence of alcohol or when alcohol concentrations exceeded 5%. Lower temperatures (75° C), the use of other bases, or the omission of bases entirely also caused lower yields. This methodology afforded three- to fourfold better yields than the nitrophenyl ester activation method [43]. Other advantages were short reaction times, a facile work-up, and elimination of the need for high-boiling solvents such as dimethylacetamide.

A more detailed account of the syntheses of 13-membered *meta*-ansa and 14-membered *para*-ansa compounds was reported in 1982 [51]. The preparation of **62** and the corresponding 14-membered ring compound (**63**) is shown in Scheme 7.

The allylic bromination of 3-acetoxytoluene with *N*-bromosuccinimide afforded 3-acetoxybenzyl bromide, which was then converted to the corresponding nitrile with tetraethyl ammonium cyanide. A mixture of 3-hydroxy-benzylcyanide, hydroquinone, sodium, and methyl acrylate was refluxed for 24 h to afford **64**. This methyl ester was hydrolyzed to the corresponding acid, 3-[[3-(cyanomethyl)phenyl]oxy]propionic acid, which then was coupled with L-proline benzyl ester in the presence of *N*,*N*-dicyclohexylcarbodiimide. The benzyl ester was removed by hydrogenation and the amine resulting from reduction of the nitrile was acylated with benzyl chloroformate to yield **65**. After reaction with pentafluorophenol to form the pentafluorophenyl ester **62**, cyclization was carried out as previously described.

For the synthesis of the 14-membered ring compound, the key intermediate, **68**, was synthesized from *p*-cresol. *p*-Cresol was treated with hydroquinone, sodium, and methyl acrylate to form methyl 3-[(4-methylphenyl)oxy]propanoate. Allylic bromination of this ester afforded the corresponding bromide, which on treatment with tetraethylammonium cyanide gave the benzyl cyanide (**66**). The ester group in **66** was hydrolyzed and the resultant acid was then coupled with L-proline benzyl ester. Hydrogenation and treatment of the product with benzyl chloroformate afforded **67**. Conversion of **67** to the key pentafluoro-phenyl ester **68** followed by cyclization in the usual manner gave compound **63**. Yields in the cyclization step were 50% for the formation of the 14-membered ring (**63**) and 80% for the 13-membered ring (**61**).

This methodology was applied to the synthesis of the dihydro derivative of a natural product, dihydrozizyphine-G [38, 52]. The synthesis of this compound (**69**) is shown in Scheme 8. Racemic *trans*-3-(*p*-cyanomethylphenoxy)proline (**70**) was obtained by the method described for the preparation of *trans*-3-phenoxyproline [49]. The ester of 3-bromodehydroproline was treated with sodium *p*-cyanomethylphenolate; the product was then saponified and reduced with dimethylamine-borane in acetic acid (25% yield relative to bromodehydro-proline ester). The corresponding Boc-compound (**71**), obtained in 95% yield, was coupled with L-proline benzyl ester. Both diastereomers, **72a** and **72b**, were separated by chromatography. Hydrogenolysis of the ester group followed by

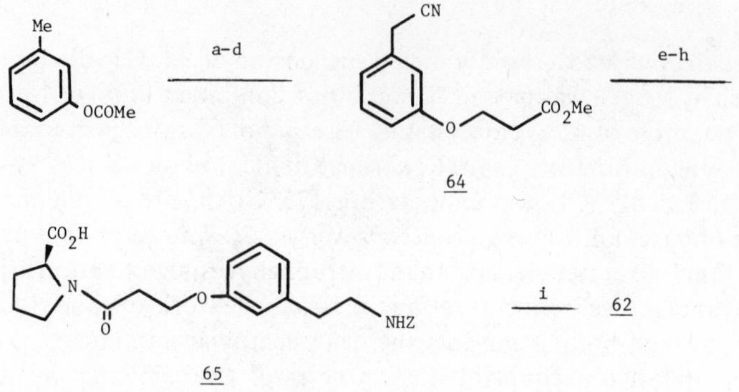

64

e–h

i   62

65

^{a}NBS; ^{b}Et$_4$NCN; ^{c}NH$_3$; ^{d} $\diagup$CO$_2$Me; ^{e}HCl; ^{f}Pro-OBn, DCC; ^{g}H$_2$, Rh; ^{h}Z·Cl; ^{i}C$_6$F$_5$OH, DCC

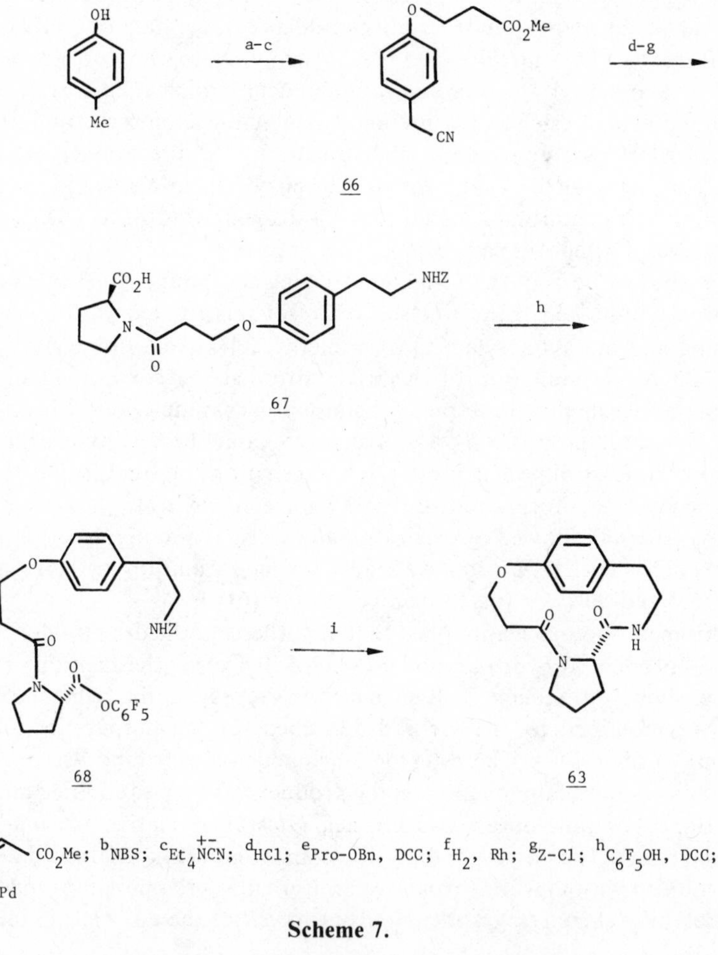

66

67

h

68   63

^{a} $\diagup$CO$_2$Me; ^{b}NBS; ^{c}Et$_4$NCN; ^{d}HCl; ^{e}Pro-OBn, DCC; ^{f}H$_2$, Rh; ^{g}Z-Cl; ^{h}C$_6$F$_5$OH, DCC; ^{i}H$_2$, Pd

Scheme 7.

138

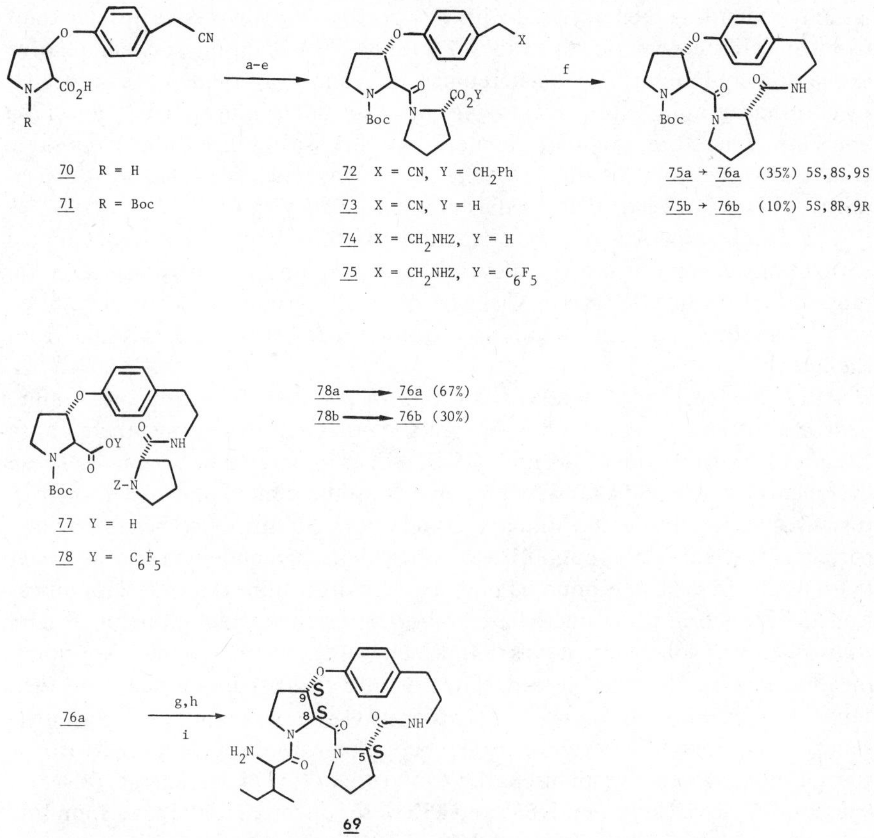

70 R = H

71 R = Boc

72 X = CN, Y = CH$_2$Ph

73 X = CN, Y = H

74 X = CH$_2$NHZ, Y = H

75 X = CH$_2$NHZ, Y = C$_6$F$_5$

75a → 76a (35%) 5S,8S,9S

75b → 76b (10%) 5S,8R,9R

78a ——→ 76a (67%)

78b ——→ 76b (30%)

77 Y = H

78 Y = C$_6$F$_5$

76a —g,h→ 69
      i

69

^{a}Pro-OBn; ^{b}H$_2$, Pd/C, dioxane; ^{c}H$_2$, Rh/Al$_2$O$_3$, 10% NH$_3$ in EtOH; ^{d}Z-Cl; ^{e}C$_6$F$_5$OH/DCC; ^{f}H$_2$, Pd/C; ^{g}TFA; ^{h}Boc-Ile; ^{i}TFA, m-C$_6$H$_4$(OMe)$_2$

Scheme 8.

catalytic hydrogenation of the nitrile group and acylation with benzyl chloroformate afforded **74a** and **74b,** respectively (88% and 84% yields relative to **72a** and **72b**). Compounds **74a** and **74b** were converted to the corresponding active esters (88% and 83% yields) using pentafluorophenol and dicyclohexylcarbodiimide. The diastereomeric series designated as **a** and **b** could not be assigned until **69** had been synthesized. Using the previously described cyclization method [39, 51], cyclopeptide **76a** was obtained in 35% yield from the appropriate diastereomer (**75a**). Cyclopeptide (**76b**) was also prepared in 10% yield. Cyclopeptides **76a** (5S,8S,9S) and **76b** (5S,8R,9R) could be sublimed without decomposition at 120°C at very low pressure.

Ring closure at position 1 (Fig. 1) was found to be more favorable and resulted in the formation of **76a** in 67% yield. For this purpose, Boc-*trans*-3-(*p*-cyanomethylphenoxy)proline was hydrogenated to give the *p*-aminoethyl compound, which was then acylated with Z-L-proline-N-hydroxysuccinimide ester to afford diastereomeric dipeptides **77a** and **77b** (quantitative yield relative to the Boc-compound). The diastereomers (**77a** and **77b**) could not be separated by chromatography, either at this stage or as the pentafluorophenyl esters (**78a** and **78b**). Therefore, the diastereomeric mixture was cyclized under the same conditions used for **75a** and **75b**. The desired cyclopeptide (**76a**) was formed more easily (67% yield) than its diastereomer (30% yield).

The diastereomers, which were separable by chromatography, were coupled with Boc-isoleucine using the dicyclohexylcarbodiimide procedure, after removal of the proline Boc group. Cleavage of the Boc group on isoleucine resulted in the formation of dihydrozizyphine-G from the **a** series (**69**) and its isomer from the **b** series.

After the successful synthesis of the 14-membered ansa peptide dihydrozizyphine-G, Schmidt and co-workers synthesized two 13-membered ansa peptides dihydrozizyphines-A and -B [53, 54], as illustrated in Scheme 9. These cyclopeptides (**79a** and **80a**) were prepared from substituted phenol **81**, which in turn was prepared by conventional methods from 2,5-dihydroxybenzoic acid in 65% overall yield. Compound **81** was treated with bromodehydroproline ester (**43b**) to afford the corresponding phenoxydehydroproline ester **82**. Saponification and reduction with dimethylamine-borane resulted in a mixture of *cis*- and *trans*-phenoxyproline derivatives (**83**), which were converted to the corresponding Boc esters (**84**) in 90% yield. The *cis*- and *trans*- compounds were then separated by chromatography. The *trans*-methyl ester was saponified, the nitrile group reduced, and the resulting product coupled with Z-L-proline-N-hydroxysuccinimide ester to give an inseparable mixture of two diastereomers (**85a** and **85b**) in 94% yield. Compounds **85a** and **85b** were converted to the corresponding pentafluorophenyl esters (**86a** and **86b**) in 70% yield. Cyclization was accomplished by the standard procedure and the desired isomer, **79a**, $(3S,18S,13S)$ was isolated in 98% yield. Compound **80a** $(3R,18R,13S)$ was obtained in 65% yield. Removal of the Boc group with trifluoroacetic acid and coupling of the resultant product with Boc-isoleucine using the dicyclohexylcarbodiimide method gave a 60% yield of product. Deblocking of the Boc group followed by reaction with activated N,N-dimethylisoleucine gave dihydrozizyphine-A (**79a**) in 60% yield. Coupling of **87a**, after removal of the Boc group, with Boc-N-methylisoleucylisoleucine using standard procedures, and subsequent removal of the second Boc group afforded dihydrozizyphine-B (**80a**) in 60% yield.

Finally, in 1983, the first synthesis of a 13-membered cyclopeptide alkaloid, zizyphine-A (**88**), was reported [55]. This compound contains a 10-membered *meta*-ansa bridge. Most investigators have recognized that the practical approaches to the synthesis of cyclopeptide alkaloids must involve the introduction of the double bond after cyclization. Schmidt and co-workers used this strategy and investigated different ways of constructing an ethanolamide moiety. The best method was a selenoxide elimination. The synthesis of zizyphine-A (**88**) using this route is shown in Scheme 10.

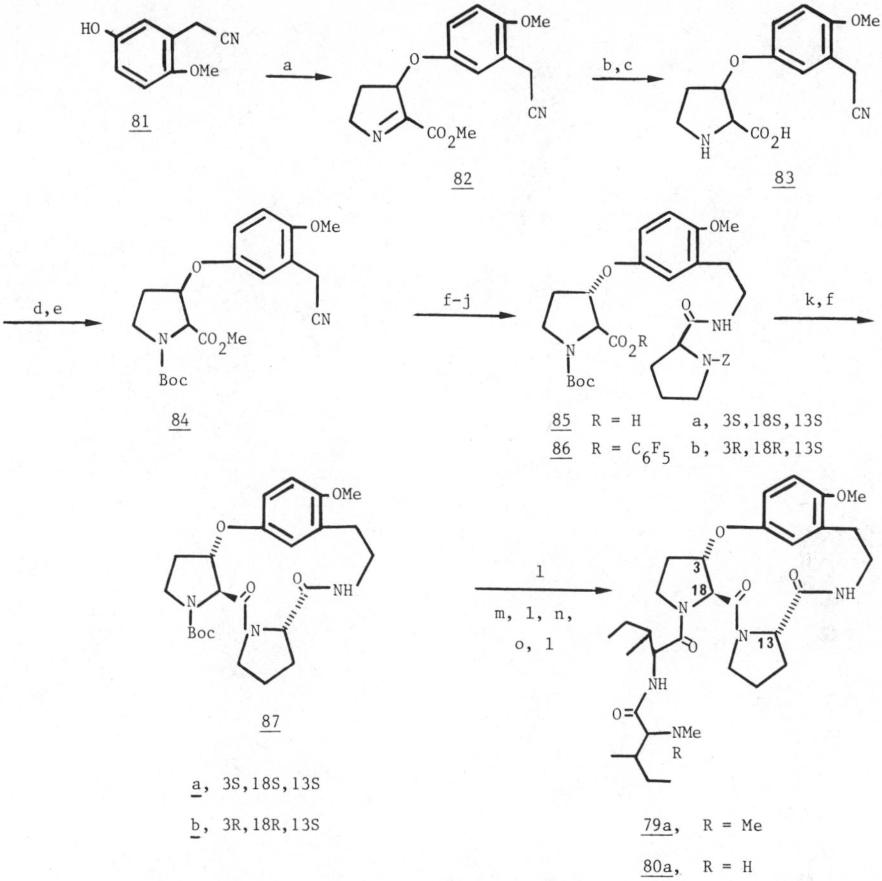

$$81 \xrightarrow{a} 82 \xrightarrow{b,c} 83$$

$$\xrightarrow{d,e} 84 \xrightarrow{f-j} 85 \quad R = H \quad a, 3S,18S,13S \quad / \quad 86 \quad R = C_6F_5 \quad b, 3R,18R,13S \xrightarrow{k,f}$$

87

a, 3S,18S,13S

b, 3R,18R,13S

$$\xrightarrow[\substack{m, l, n, \\ o, l}]{l} \quad 79a, \ R = Me \qquad 80a, \ R = H$$

[a] NaOMe, 43b, DMF; [b] H_2O, OH^-; [c] $Me_2NH \cdot BH_3$, AcOH; [d] $(t\text{-}BuO_2C)_2O$; [e] CH_2N_2, Et_2O; [f] chromatography; [g] OH^-, H_2O; [h] H_2, Rh/Al_2O_3; [i] Z-Pro-OSu; [j] C_6F_5OH, DCC; [k] H_2, Pd,dioxane; [l] TFA; [m] Boc-Ile, DCC; [n] activated Me_2Ile; [o] Boc-Me-Ile-Ile/DCC.

Scheme 9.

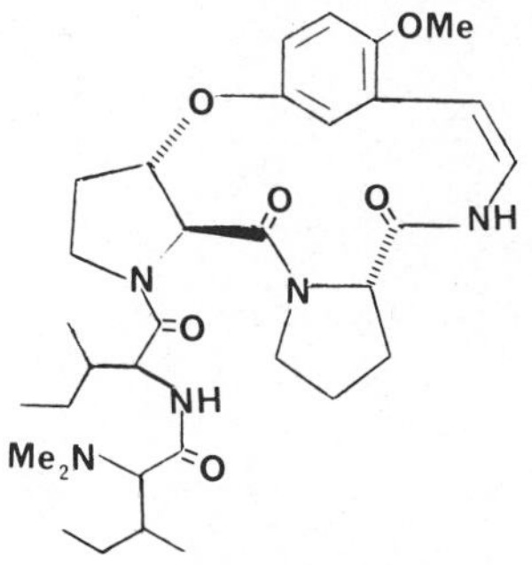

88

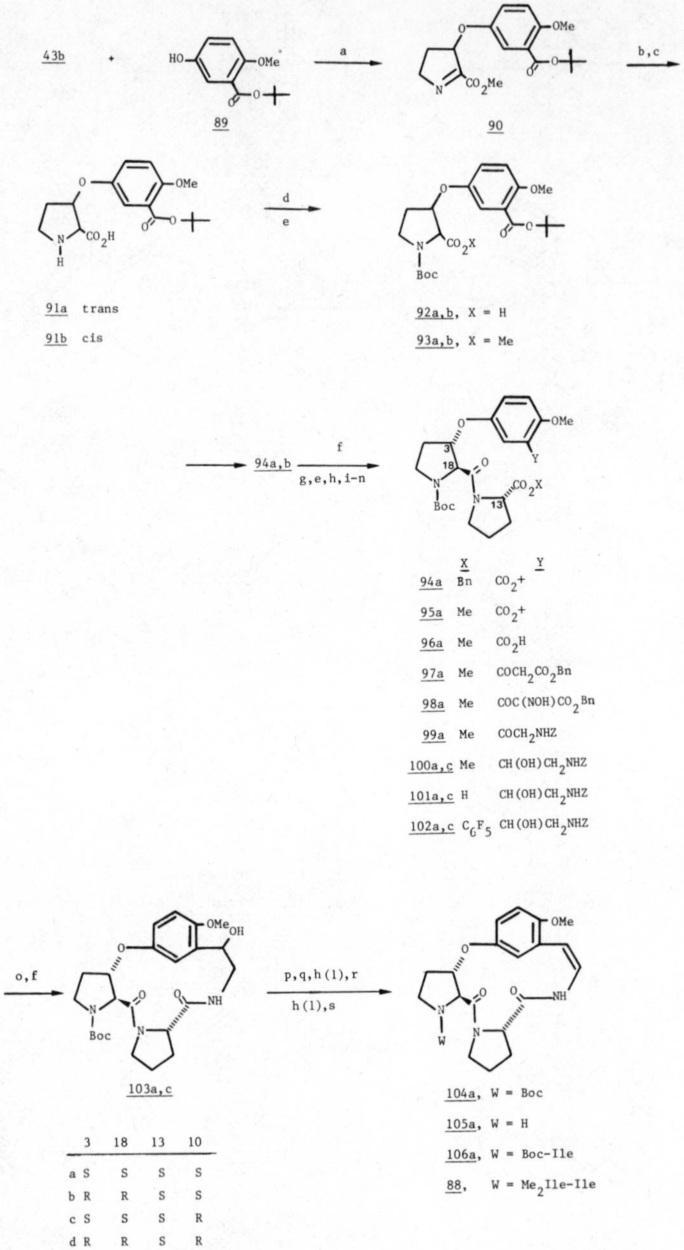

	X	Y
94a	Bn	CO$_2$+
95a	Me	CO$_2$+
96a	Me	CO$_2$H
97a	Me	COCH$_2$CO$_2$Bn
98a	Me	COC(NOH)CO$_2$Bn
99a	Me	COCH$_2$NHZ
100a,c	Me	CH(OH)CH$_2$NHZ
101a,c	H	CH(OH)CH$_2$NHZ
102a,c	C$_6$F$_5$	CH(OH)CH$_2$NHZ

104a,	W = Boc
105a,	W = H
106a,	W = Boc-Ile
88,	W = Me$_2$Ile-Ile

	3	18	13	10
a	S	S	S	S
b	R	R	S	S
c	S	S	S	R
d	R	R	S	R

[a]NaOMe, DMF; [b]θOH, dioxane, water; [c]Me$_2$NH·BH$_3$/AcOH; [d](t-BuO$_2$C)$_2$O; [e]CH$_2$N$_2$, Et$_2$O; [f]chromatography; [g]H$_2$, Pd/C, EtOH; [h](1) TFA, m-C$_6$H$_4$(OMe)$_2$ (2) (t-BuO$_2$C)$_2$O; [i]CDI, MgBzl malonate; [j]NaNO$_2$, HOAC; [k](1) H$_2$, Pd/C, HOAC, (2) Z-Cl; [l]NaCNBH$_3$; [m]LiOH, dioxane; [n]C$_6$F$_5$OH, DCC; [o]Pd/C, 4-pyrrolidino pyridine, dioxane; [p](n-Bu)$_3$P, P-NO$_2$-C$_6$F$_4$-SeCN; [q]H$_2$O$_2$, Py; [r]Boc-Ile, DCC; [s]activated Me$_2$Ile, DMAP.

Scheme 10.

142

Phenol **89** was prepared by conventional methods from 2,5-dihydroxybenzoic acid. Previously described bromodehydroproline **43b** was treated with **89** according to standard procedures to give **90**, which was subsequently saponified and reduced with dimethylamine-borane. A mixture of *cis*- and *trans*-(aryloxy) prolines (**91b** and **91a**) resulted; crystallization afforded a pure sample of the racemic *trans*-compound (**91a**). Attempts to epimerize **91b** to **91a** with bases or acetic anhydride were not successful. Therefore, the isomers were converted into a mixture of *cis*- and *trans*-Boc methyl esters (**93a** and **93b**). In the synthesis of dihydrozizyphine-G (**69**), a rigid 10-membered *para*-ansapeptide, ring closure at the 1-position proceeded more readily than at the 2-position. For the saturated *meta*-ansa peptides dihydrozizyphines-A and -B (**79a** and **80a**), cyclization proceeded in equal and excellent yields at both positions. In the case of zizyphine-A (**88**) ring closure at the 2-position afforded chromatographically separable diastereomeric dipeptides **94a** and **94b** (prepared by treatment of mixture **92a,b** and benzyl-L-prolinate with dicyclohexylcarbodiimide). Based on previous results [38, 52], the compound with the higher R_f value was suspected of being the 3*S*,18*S*,13*S* isomer **94a**. Hydrogenolysis of the benzyl ester followed by reaction with diazomethane gave methyl ester **95a**. Cleavage of the *tert*-butyl ester afforded the carboxylic acid **96a**, which was further transformed in high yields (via **97a** → **99a**) to an amino ketone unit. These conversions involved the reaction of a carboxylic acid imidazolide with the magnesium salt of benzyl malonate [56] to give a benzyl aroylacetate (**97a**). Nitrosation of **97a** gave the corresponding oxime (**98a**). Catalytic hydrogenation reduced the oximino group and cleaved the benzyl ester. Subsequent decarboxylation gave the amino ketone. The amine was then protected as the carbobenzoxy derivative. Sodium cyanoborohydride reduced **99a** nonstereoselectively to diastereomers **100a** and **100c**. Saponification of the methyl esters (**101a** and **101c**) followed by formation of a mixture of the two diastereomeric pentafluorophenyl esters (**102a** and **102c**) using dicyclohexylcarbodiimide-mediated condensation yielded the linear precursors for the cyclization step. Catalytic hydrogenation of **102a** and **102c** under standard conditions afforded a mixture of cyclic alcohols (**103a** and **103c**) in 80% yield. Chromatographic separation of these alcohols gave a 60% yield of each isomer.

The presence of a hydroxyl group at the 10-position increased the number of diastereomers and led to mixtures that were difficult to purify. This disadvantage was counterbalanced by the high yields obtained in the cyclization step. Ring closure using the ketone (**99a**, $X = C_6F_5$) proceeded in poor yield and led to partial reduction of the ketone to a hydroxyl group. The hydroxyl required for the elimination step was therefore introduced before cyclization.

Transformation of diastereomeric alcohols **103a** and **103c** into the selenides and oxidative elimination was carried out with both isomers separately, the yield of olefin **104a** being identical (65%). After cleavage of the Boc group with trifluoroacetic acid, the amine **105a** was treated with *tert*-butoxycarbonylisoleucine hydroxysuccinimide ester, to give **106a**. Removal of the protecting group with trifluoroacetic acid and treatment of the resulting amine with dimethylisoleucine pentafluorophenyl ester gave zizyphine-A (88) in 38% yield from 104a.

The ring closure at position 1 proceeded in 80% yield. Since separation of the diastereomers could only be achieved after ring closure, olefin 104a had to be separated from diastereomer 104b, making this sequence less desirable.

6.3.4. Mucronine Class.

The synthesis of a 15-membered ring cyclopeptide alkaloid, mucronine-B (107), was also reported by Schmidt [57]. Mucronine-B contains a benzene ring with a 12-membered handle over the *meta*-position. In contrast to the 14- and 13-membered peptide alkaloids, which are derived biogenetically from tyrosine and contain a β-phenoxy amino acid unit in the ring, the handle in compound 107 is not constructed with an ether linkage but rather involves a C—C bond. The biogenesis of these compounds presumably originates from *m*-phenylenediamine.

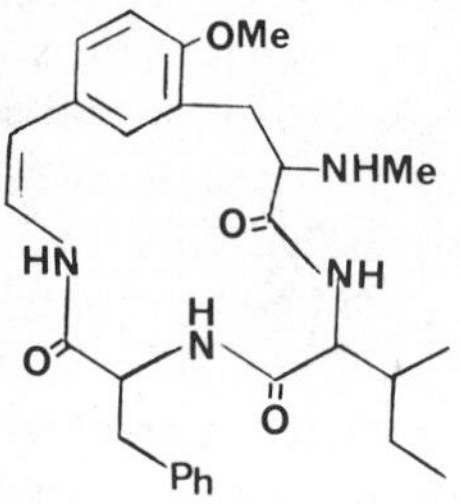

107

The synthesis of mucronine-B (107) is shown in Scheme 11. The dehydro-amino acid 108 was used as the starting material for this approach [58]. Compound 108 was prepared by a novel route starting with the corresponding aldehyde, 2-methoxy-5-benzoxycarbonylbenzaldehyde. Ethyl 2-benzyloxycar-bonylamino-2-ethoxyacetate was successively treated with phosphorus tri-chloride and triethyl phosphite to afford ethyl 2-benzyloxycarbonylamino-2-diethyloxyphosphorylacetate in 80% yield. This phosphoryl ester was metalated using sodium hydride in tetrahydrofuran and converted in 85% yield to 108 by condensation with 2-methoxy-5-benzoxycarbonylbenzaldehyde. N-Methylation of 108 with methyl iodide, followed by catalytic reduction of the double bond, removal of the carbobenzoxy group, and introduction of the Boc group by conventional methods afforded 109. The carboxyl group was converted into the amino ethanol side chain using a previously described procedure [56, 55] (109 → 110 →111). Hydrolysis of 111 gave the amino acid that was coupled with Ile-Phe-OMe using dicyclohexylcarbodiimide to yield tripeptide 113. The latter compound was converted to its pentafluorophenyl ester (114), the key precursor for ring closure. Cyclization was carried out as already indicated. The high yields (85%) of cyclic products (115a and 115b) further supported the superiority of this cyclization methodology compared to conventional approaches for the synthesis of 13- to 15-membered cyclopeptides. The acetate mixture (116a and 116b) was purified by treatment with trifluoroacetic acid–selenophenol to afford a mixture

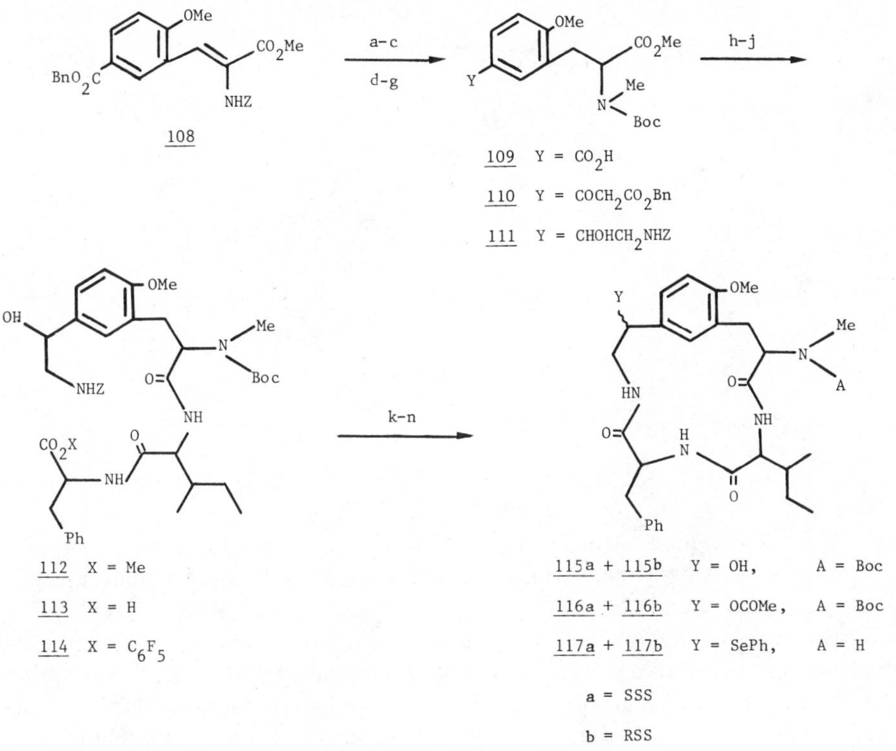

→ 107a + 107b → 107a

108 → 111 50% 114 → 116a + 116b 60%

111 → 114 65% 116a + 116b → 107a + 107b 40%

[a]MeI, K_2CO_3; [b]H_2/Pd, MeOH; [c]$(t\text{-}BuO_2C)_2O$; [d]CDI, MgBzl malonate; [e]$NaNO_2$, HOAc; [f]H_2/Pd, HOAC; [g]Z-Cl; [h]H_2O, $\bar{O}H$; [i]Ile-Phe-OMe, DCC; [j]C_6F_5OH, DCC; [k]H_2/Pd, dioxane; [l]Ac_2O, Py; [m]TFA, PhSeH; [n]$NaIO_4$.

Scheme 11.

of diastereomeric phenyl selenides (**117a** and **117b**). Oxidative elimination with sodium periodate gave exclusively the Z-olefin and yielded synthetic mucronine (**107a**) as well as its diastereomer, isomucronine (**107b**), which were separated by chromatography.

6.4. Cyclization at Site 3. Integerrine Class.

The largest class of cyclopeptide alkaloids are those containing 14-membered rings, the most synthetically challenging of the macrocycles. Within this class the integerrine types illustrate well the inherent difficulties encountered.

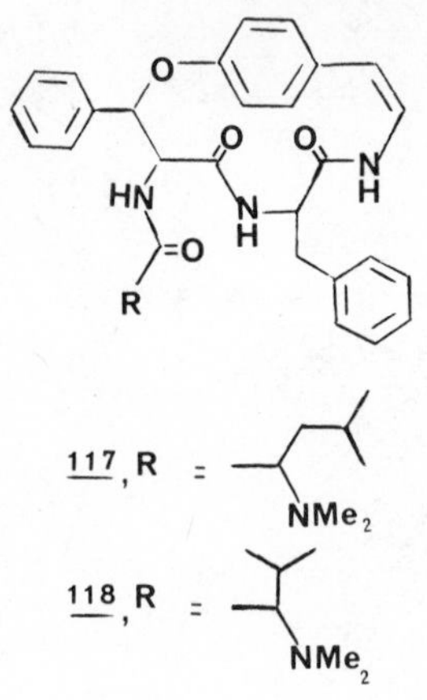

117, R = [structure]
NMe₂

118, R = [structure]
NMe₂

Crenatine-A (117) and integeressine (118) are representative examples of the integerrine type.

The styrylamide unit in 117 is believed to be derived biogenetically from a decarboxylated dehydrotyrosine residue. Once formed, this enamide unit requires careful manipulation [59]. Investigations carried out in Lawton's laboratory have focused on the cyclization of linear precursors at position 3. Formation of the phenolic ether linkage was attempted by both oxidative coupling and nucleophilic displacement reactions on appropriate compounds. The strategy employed by this research group was intended to mimic a possible biosynthetic pathway. Although little is known about the biogenesis of cyclopeptide alkaloids, the presence of different types of compounds in any one plant suggests that the enzyme responsible for their formation cannot be very specific. It has been proposed that the ether oxygen bond forms last although no evidence is available for this pathway. A quinone methide, 119, composed of a dehydrophenylalanylphenylalanine coupled to an oxidized tyrosine unit, has been

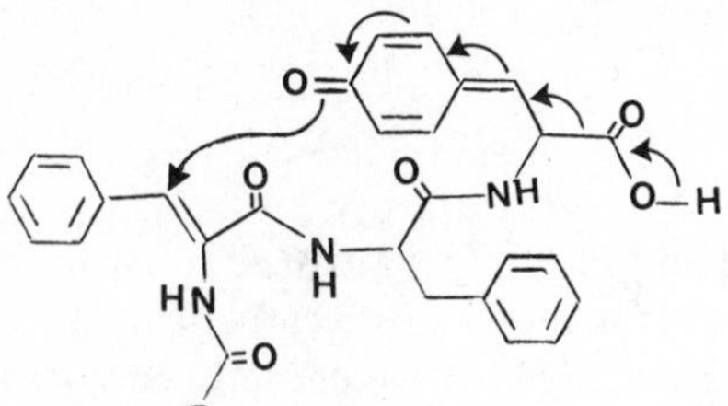

119 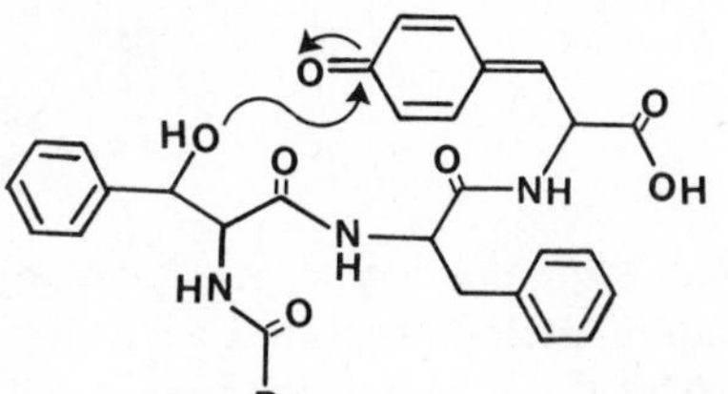   120

suggested as a precursor. Decarboxylation and subsequent Michael addition would form the expected macrocycle]59]. Alternatively, a phenylserylphenylalanyltyrosine derivative (120) could also undergo cyclization via attack of the β-hydroxyl oxygen to the quinone group, followed by decarboxylation and subsequent expulsion of the phenolic hydroxyl group to afford the macrocycle. From a synthetic view-point, neither strategy appears easy to achieve. Lawton and Stack [60] chose to approach the synthesis of **117** along the lines of the proposed biogenesis. Towards this end, several tripeptide precursors were constructed containing both a dehydrophenylalanine and a tyrosine residue. Synthesis of two of these precursors (**121** and **122**) are shown in Scheme 12.

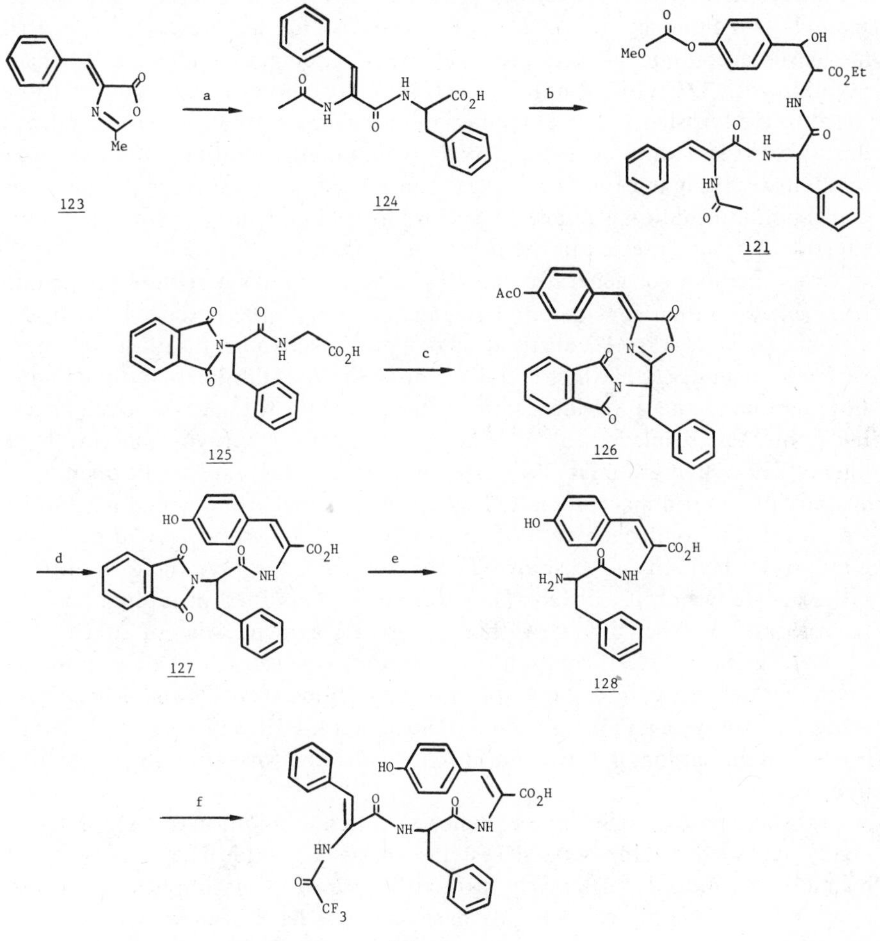

^{a}dl-PheOH, NaOH; ^{b}MeO—C(O)—O—⟨⟩—CH(OH)—CO₂Et, N-ethyl-5-phenylisoxalium-3'-sulfonate (Woodward reagent K), Et₃N; ^{c}Ac₂O, NaOAC, p-AcO-C₆H₄CHO; ^{d}H⁺, H₂O; ^{e}NH₂NH₂, HOAc; ^{f}Ph-CH=C—C(O)—N=C-O with CF₃, CH₃-C(=O)-N((CH₃)₃Si)₂.

Scheme 12.

Compound **121** was obtained from the opening of an appropriate azlactone (**123**) by *dl*-phenylalanine followed by coupling of the resulting dipeptide (**124**) with a tyrosine derivative using Woodward's reagent K.

The synthesis of **122** proceeded via *dl*-phenylalaninedehydrotyrosine (**128**). Treatment of *N*-phthaloyl-*dl*-phenylalanylglycine (**125**) with acetic anhydride and sodium acetate in the presence of *p*-acetoxybenzaldehyde gave the azlactone **126** in 54% yield. The azlactone was hydrolyzed with acid to dipeptide **127** in high yield. Hydrazinolysis of **127** afforded **128**. Subsequent opening of the trifluoromethyl oxazolone by this free dipeptide (**128**) in the presence of *bis*(trimethylsily)acetamide yielded precursor **122**. Attempts to effect a Michael addition with **121** resulted only in a retroaldol reaction. It was hoped that the unsaturated tyrosine **122** would eliminate this problem, but the compound was unstable under acidic and phenolic oxidative coupling conditions, all cyclization attempts resulting in peptide cleavage. These attempts support the premise that C—O bond formation is not easily accomplished in the presence of the various functional groups present in the peptide precursors.

Other methods for generating an alkyl aryl ether linkage within the peptide fragment were then considered. Three major retrosynthetic schemes were envisioned (Scheme 13): (1) construction of a highly activated unsaturated system within the peptide chain which could promote Michael addition of the phenolate. Elimination of the alkoxide group would generate the macrocycle **129**; (2) intramolecular displacement of a halogenated phenylpyruvoyl peptide by a phenol followed by a similar alkoxide elimination; and (3) a solvolytic opening of an epoxide to form macrocycle **129** whose α-hydroxyl group would need to be transformed to a nitrogen-containing function. Although the formation of the alkyl aryl ether linkage is shown as the last step, the investigators did not necessarily envision it as such. They were more interested in finding ways to incorporate the ether linkage within a peptide fragment. Also of concern, of course, was macrolide formation. In addition to cyclization at the 3-position (path 2 in Scheme 13), they investigated cyclizations at the 2- and 4-positions, using a model system (**130**) as shown in Scheme 14. Pathway 4 was designed to involve an amination; pathway 5 utilized C—C bond formation in the cyclization step.

Pathway 1 shown in Scheme 13 called for Michael addition of phenolates to highly activated double-bonded systems within a peptide chain. A Michael reaction was judged particularly attractive because of its implication in the biogenesis of natural products. Michael reactions have been shown to occur under physiological or "cell possible" conditions [61]. The authors thought that the Michael acceptor should be highly activated to provide a driving force for addition of a phenolate. A number of such acceptors could presumably be prepared by condensation of amino acids and esters with aromatic aldehydes, particularly 5-nitrosalicylaldehyde. Although several precursors for such Michael acceptors were prepared, attempts to isolate the acceptors themselves met with failure because of the instability of these products in acid. Generation of the acceptor in the presence of trapping agents such as benzylthiol or *p*-methoxy-

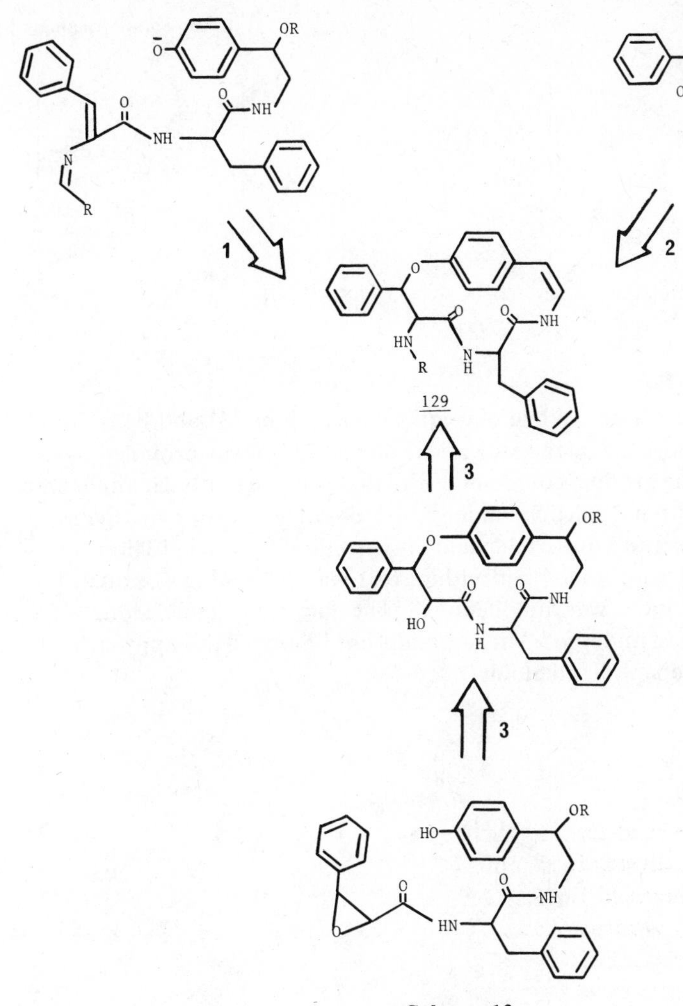

Scheme 13.

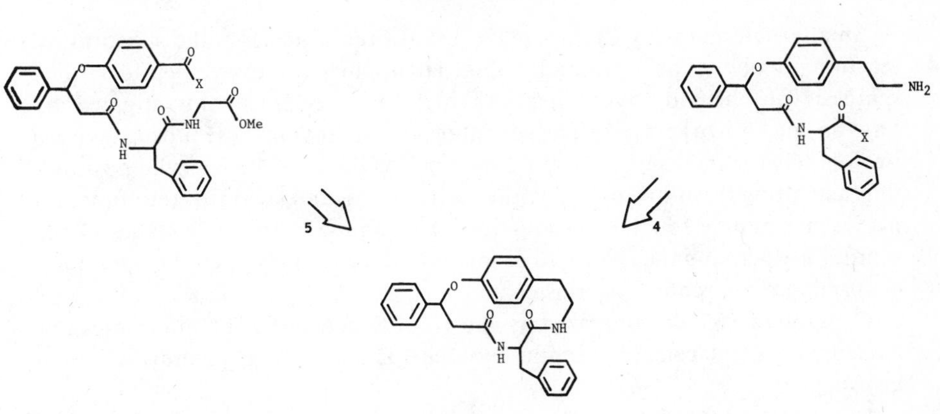

Scheme 14.

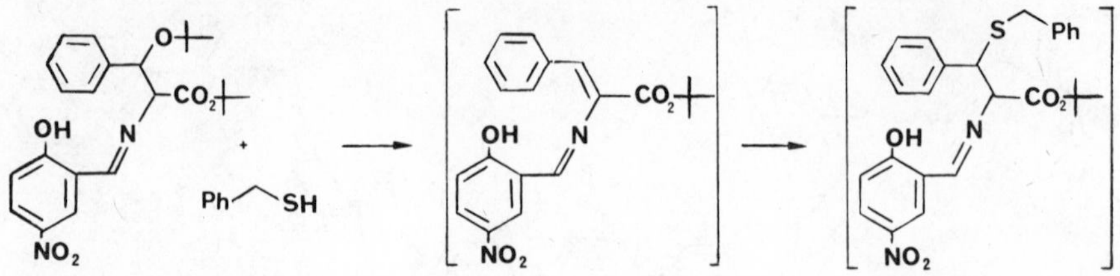

131   132   133

phenol proved unsuccessful. When one equivalent each of **131** and benzylthiol was warmed in pyridine, evidence for a Michael addition was provided by ^{1}H NMR studies, but the product could not be isolated. Concurrently, Schmidt and co-workers [62, 63] reported the condensation of dehydrovaline methyl ester with salicylaldehyde and 4-nitrosalicylaldehyde to give **134a** and **134b**, respectively. The Michael addition of benzylthiol to **134a** and **134b** was noted, but isolation of the adducts was not detailed. The fact that thiols alone were nucleophilic enough to undergo a Michael addition eliminated this approach as a viable route to cyclopeptide alkaloids.

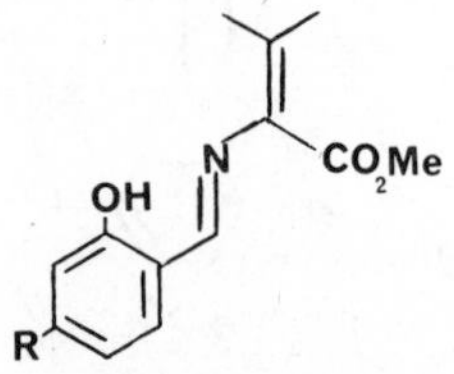

134 a , R = H
134 b , R = NO$_2$

An alternate pathway (2 in Scheme 13) utilized a nucleophilic substitution reaction involving phenols and halogenated phenylpyruvoyl peptides. The synthesis of phenylpyruvoyl peptides had been previously investigated by Lawton and co-workers [64]. The advantages of employing these substrates were twofold: (1) the intermediates could be functionalized at the benzylic position, thus activating them toward substitution reactions, and (2) the α-keto function provided a handle for the introduction of an amine in the later stages of the synthesis. Retaining the α-keto function until the end eliminated the possibility of forming diastereomers at this center.

Compound **135** was prepared as shown in Scheme 15. The intermolecular nucleophilic displacement reaction between **135** and various phenols was then investigated.

Compound **(136),** the precursor for the synthesis of **135,** was prepared via the method described by Stack [60] from α-pivaloxycinnamic acid and L-phenyla-

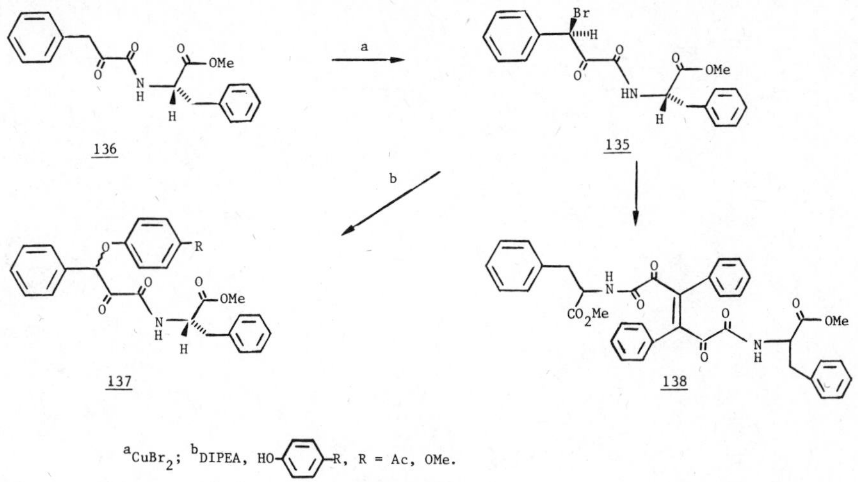

^{a}CuBr$_2$; ^{b}DIPEA, HO—⟨ ⟩—R, R = Ac, OMe.

Scheme 15.

lanine methyl ester hydrochloride. Treatment of **136** with two equivalents of cupric bromide afforded a 40% yield of bromoketone **135** as an equal mixture of diastereomers racemic at the benzylic center. Crystallization afforded pure isomers. The *R*-isomer was treated with one equivalent of *N, N*-diisopropylethylamine and the appropriate phenol in dry acetonitrile. Compound **137** formed in 10–50% yields, but a pure ether product could not be isolated. Further investigation of this reaction showed that base removed the highly acidic benzylic proton and facilitated the condensation of **135** with its anion. This condensation occurred as fast as the reaction of **135** with phenols. Acidic conditions did not eliminate these complications. Replacement of bromine with chlorine, a harder leaving group, also failed to prevent undesired side reactions. When **135** was treated with benzylthiol, an isolable thiol ether was produced (38% yield), but halogen transfer was a major side reaction and could not be excluded despite a change in the nucleophile. The instability of **135** to bases was also a problem; therefore, the displacement reaction was examined under neutral conditions. When compound **135** was heated with the thallium salts of four representative phenols, the expected ethers **(139a–d)** formed, although in low yields (8–34%).

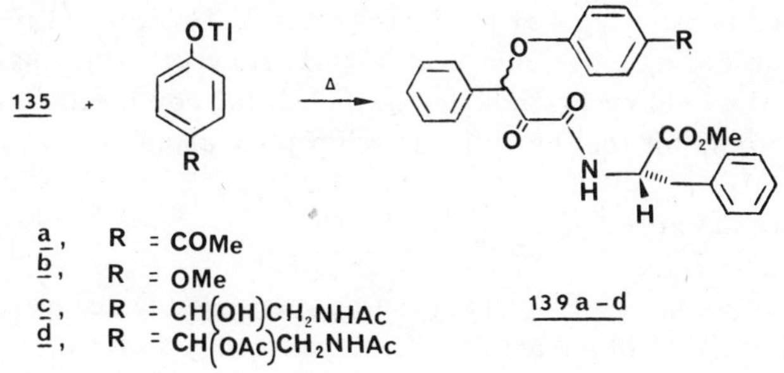

a, R = COMe
b, R = OMe
c, R = CH(OH)CH$_2$NHAc
d, R = CH(OAc)CH$_2$NHAc

139 a–d

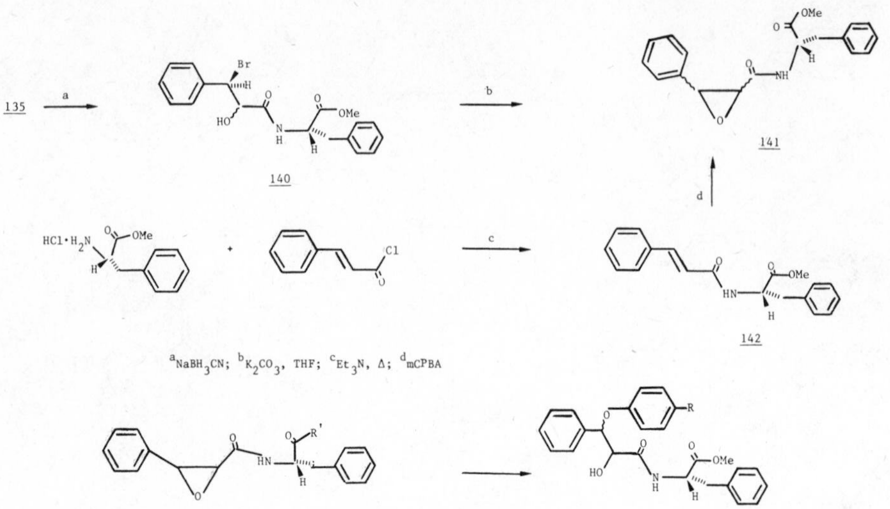

$135 \xrightarrow{a}$ **140**

141

HCl·H$_2$N ... + ... Cl $\xrightarrow{c}$ **142**

d

^{a}NaBH$_3$CN; ^{b}K$_2$CO$_3$, THF; ^{c}Et$_3$N, Δ; ^{d}mCPBA

141 R' = OMe

143 R' = CONHCH$_2$CO$_2$Me

			Yield
144a	R' = OMe ;	R = Me	38%
144b	R' = OMe ;	R = CHO	20%
144c	R' = OMe ;	R = CH$_2$CO$_2$Me	40%
144d	R' = CONHCH$_2$CO$_2$Me;	R = CHO	20%
144e	R' = CONHCH$_2$CO$_2$Me;	R = CO$_2$Me	70%

Scheme 16.

The most synthetically useful approach to cyclization at site 3 involved a regioselective solvolytic opening of epoxy-peptides with various phenols. This procedure produced the alkyl aryl ether linkage in moderate-to-good yields via a simple procedure. Reduction of **135** with sodium cyanoborohydride afforded *erythro-* and *threo-*bromohydrins (**140**, Scheme 16) in 40:60 ratio. The isomers could be separated by HPLC. Treatment of the respective bromohydrins with potassium carbonate gave a quantitative yield of the corresponding epoxides as diastereomeric mixtures. A simpler approach utilizing epoxidation of cinna-moyl-L-phenylalanine methyl ester (**142**) was also devised. The stereochemistry of these epoxides has been the subject of much study [65]. Eventually two epox-ides were selected; fusion of these compounds (**141** and **143**) with various phenols afforded the desired ethers in moderate-to-good yields. The more nucleophilic phenols gave higher yields of products (**144a–e**). This proved to be the best method of producing a phenolic ether linkage. However, attempts to effect the reaction in the final cyclization step all failed, thereby precluding use of the biomimetic route for the synthesis of cyclopeptide alkaloids.

6.5. Cyclization at Site 4

Generation of a final C—C bond at position 4 was investigated using precursors such as **145** and the aldol reaction [65]. It was readily apparent, however, that peptide fragments possess several protons of very comparable acidity and that

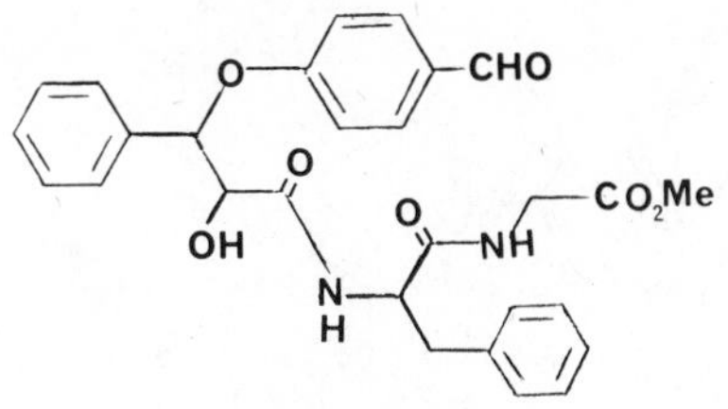

145

abstraction of the desired proton to generate the necessary carbanion could not be done selectively. Furthermore, the base instability of a precursor such as **145** resulted in facile β-elimination of phenolate via a retroaldol. These results prompted the investigation of acylation-type reactions with precursors such as **146**. Although these diacid compounds were stable, the manipulation of their various isomers was not predictable. A large variety of acylation-type reactions were examined using various functional derivatives of **146,** such as aldehydes, esters, diacid chlorides, mixed anhydrides, and the diacid itself, but no cyclization was observed even under conditions as drastic as those of the Thorpe reaction.

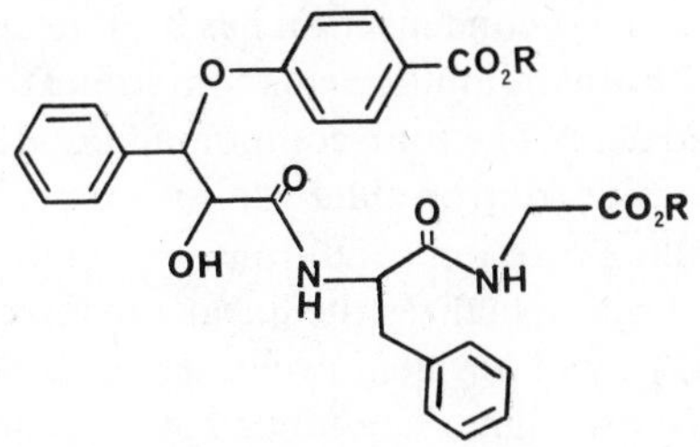

146 R = H
147 R = Me

The last approach in this series utilized diester **147,** an appropriate precursor for a Dieckmann condensation. Four diastereomeric forms of compound **147** had been previously prepared in good yields [65]. Treatment of each diastereomer with excess *bis*(trimethylsilyl)acetamide gave the corresponding *O,N,N*-silylated diester, **148.** Silylation of both amide nitrogens removed the possibility of amide proton abstraction and enhanced the possibility of deprotonating the glycine methylene protons. Silylation also offered a structural advantage. Although there is an equilibrium between *N*- and *O*-silylated tautomers, silylation of amides temporarily converts them to tertiary amides, thereby promoting a *cis*-conformation that should bring the ends of the chain closer together and enhance the probability of cyclization. The choice of base used in the Dieckmann condensation was governed by the propensity of cyclopeptide macrolides to complex with cations such as Mg^{2+}, Ca^{2+}, and Li^+. A base such as lithium di-

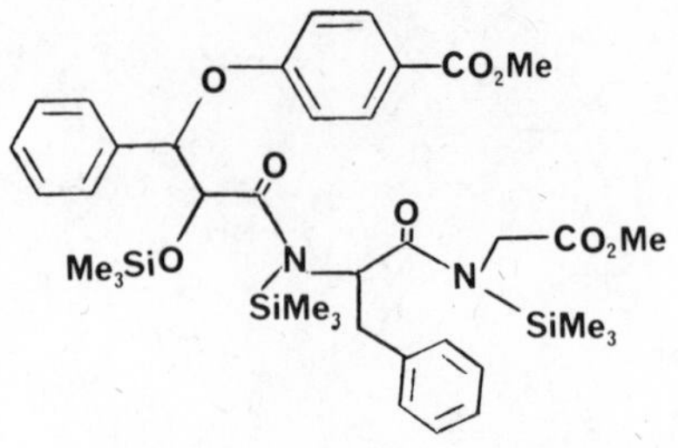

148

isopropylamide was employed with the expectation that it would induce a template effect and promote cyclization. Only epimerization, however, was observed. Other attempts yielded starting material and polymers. Although evidence was obtained for the formation of the appropriate anion, condensation of this anion did not occur under a variety of conditions.

6.6. Four-Component Condensation. A novel approach to the synthesis of cyclopeptide alkaloids.

Recently, the four-component condensation has been proposed [66] as a way of introducing, in one step, both the amino acids containing the basic side chain and the ring bound α-amino acid. The four-component condensation generates an N-acylated amino acid amide from an aldehyde, an amine, a carboxylic acid, and an isonitrile. By controlling reaction conditions, a high degree of stereoselectivity has been obtained. If one visualizes the linear precursor 149 (Scheme 17) of dihydromauritine-A (42) as an acylated cyclic secondary amino acid amide, a retrosynthetic analysis shows that compounds 150, 151, and 152 are needed for carrying out the four-component condensation. Compound 150, an (aryloxy)pyrroline, reprsents an intramolecular condensation product of the amine and aldehyde components. This key intermediate (150) can then react with 151, the isonitrile of the desired amino acid, L-phenylalanine in this case, and a carboxylic acid, the appropriate dipeptide side chain of the cyclopeptide alkaloid, N,N-dimethyl-L-alanyl-L-valine (152), to afford the desired acylated β-(aryloxy)prolyl peptide (149). This strategy would generate the N-acyl bond, the bond to the α-carbon of proline, the prolyl amide, and the trans-stereochemistry of the proline derivative all in one reaction step. Such a convergent approach, coupled with Schmidt's cyclization and the subsequent introduction of unsaturation in the tyramine side chain, could prove to be a most efficient and viable route to cyclopeptide alkaloids.

The success of the four-component condensation strategy is dependent on both the availability of the key intermediate (150) and the feasibility of preparing cyclic secondary amino acid peptides by this method, a feature without literature precedent. The condensation of an unsymmetrically substituted pyrroline

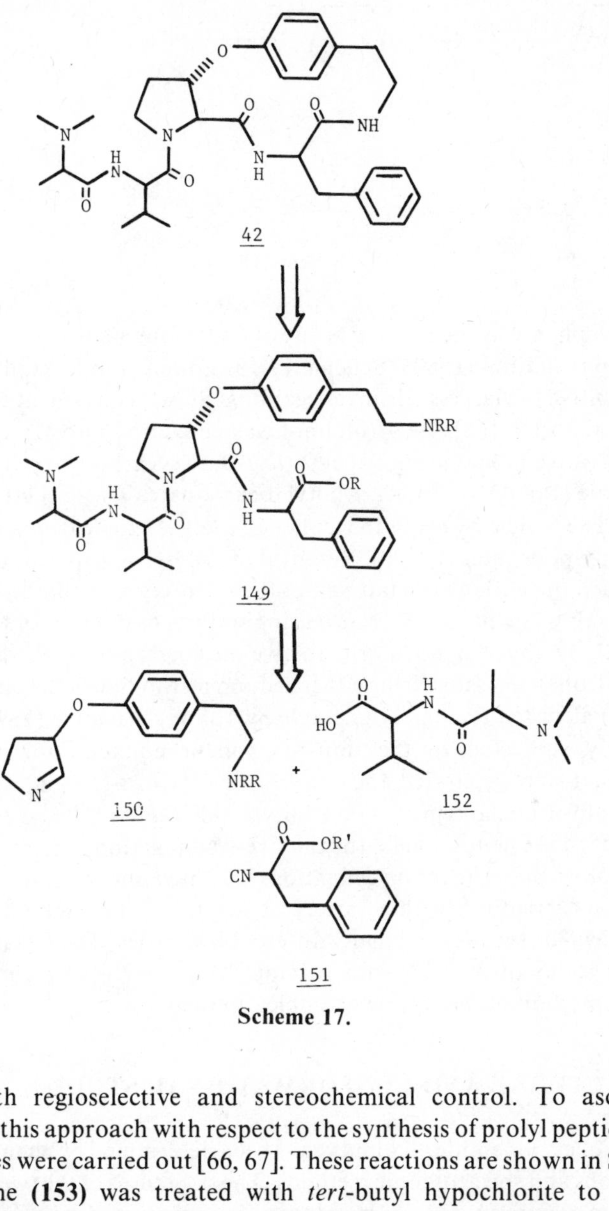

Scheme 17.

requires both regioselective and stereochemical control. To ascertain the feasibility of this approach with respect to the synthesis of prolyl peptides, several model studies were carried out [66, 67]. These reactions are shown in Scheme 18.

Pyrrolidine (153) was treated with *tert*-butyl hypochlorite to afford the corresponding *N*-chloro derivative which was dehydrohalogenated *in situ* using freshly prepared sodium methoxide to yield the pyrroline 154. Because of its instability, compound 154 was trapped with *tert*-butylisonitrile in the presence of Boc-L-valine to afford the *tert*-butylamide of Boc-valylproline (155) in 56% yield as a diastereomeric mixture with DL-stereochemistry at proline.

The synthesis of intermediate 150 and the feasibility of using (aryloxy)pyrrolidines in the four-component condensation was first examined in a model

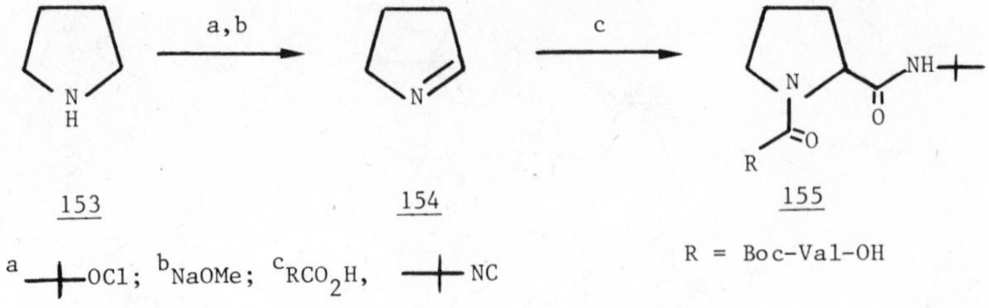

Scheme 18.

system with phenol as the aryloxy group. The starting material for the synthesis of **150**, 3-pyrrolidinol **(156b)** (Scheme 19), although commercially available, is cost prohibitive for large-scale investigations. Short, convenient routes to **156b** have been reported [68]. 3-Pyrrolidinol was converted into the corresponding N-Boc derivative **(156a)** using either 2-(*tert*-butoxycarbonyloxyimino)-2-phenylacetonitrile (Boc-ON) or di-*tert*-butyl dicarbonate. The aryl ether **(158a)** was prepared in 82% yield by using **156a,** phenol, diethyl azodicarboxylate (DEAD), and triphenylphosphine [69, 70]. Removal of the Boc group with trifluoroacetic acid afforded the trifluoroacetate salt of the phenoxy pyrrolidine **(158b)** in 99% yield. The salt was neutralized *in situ* using sodium carbonate and subsequently treated with *tert*-butyl hypochlorite to give the corresponding N-chloro derivative. Dehydrohalogenation using either sodium methoxide or diazabicycloundecene (DBU) afforded the highly unstable pyrroline derivative **(159a)** which was immediately subjected to the four-component condensation to yield two products that were separated and identified as the α,β-*cis*- and -*trans*- isomers of benzamido-β-phenoxyproline *tert*-butylamide **(160a)**. The *cis:trans*-isomeric ratio was 55:45. The total yield for the five-step reaction scheme, starting with **158a,** was 58%. Subsequent investigations [67] have shown that these reactions may also be carried out with a variety of substituents on the phenoxyl group **(158a–h, 159a–d, 160a–e)** and many different isonitriles. The Ugi products were isolated in yields of 30–60% with varying ratios of *cis*- to *trans*-isomers, the *cis*-species predominating in the examples studied.

7. STRUCTURAL AND CONFORMATIONAL STUDIES

Structures of cyclopeptide alkaloids have been determined by chemical degradation reactions and spectroscopic methods. The structures of the two cyclopeptide alkaloids mauritine-A [71, 72] and the frangulanine derivative, N,N,N-trimethylfrangulanine methiodide [73, 74] have been confirmed by X-ray analyses.

7.1. Chemical Degradation

Methods used for the chemical degradation of cyclopeptide alkaloids into their components have been summarized in detail by Tschesche [7]. Acid hydrolysis is the most commonly used method for amino acid determinations. Hydrolysis is

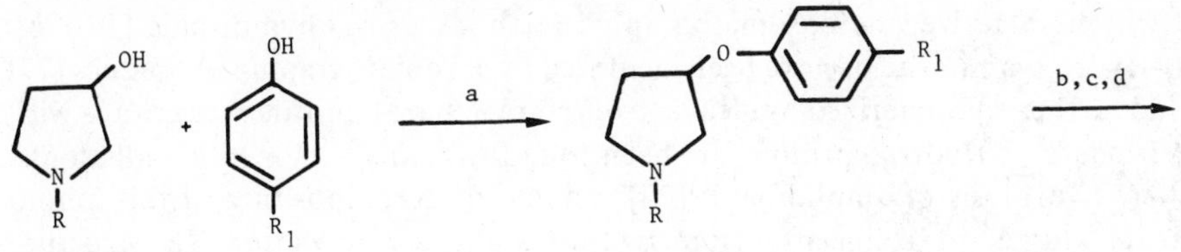

156 + 157 →(a) 158 →(b,c,d)

a R = Boc, R_1 = H

b R = R_1 = H

c R = H, R_1 = Me

d R = H, R_1 = CN

e R = H, R_1 = CH_2CN

f R = H, R_1 = CHO

g R = H, R_1 = $CHOMeCH_2N$-Phth

h R = H, R_1 = CH_2CH_2N-Phth

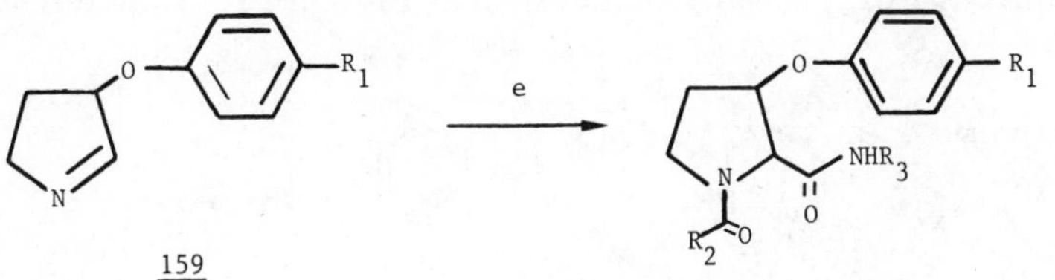

159 →(e) 160 (cis and trans)

a R_1 = H

b R_1 = Me

c R_1 = CN

d R_1 = CH_2CH_2N-Phth

a R_1 = H, R_2 = Ph, R_3 = ⊣

b R_1 = CH_2CH_2N-Phth, R_2 = Ph, R_3 = ⊣

c R_1 = H, R_2 = Ph, R_3 = $CHPh_2$

d R_1 = Me, R_2 = Ph, R_3 = $CHPh_2$

e R_1 = CN, R_2 = Ph, R_3 = $\underset{CH_2Ph}{CHCO_2Me}$

[a] DEAD, Ph_3P; [b] TFA; [c] Na_2CO_3, ⊣OCl; [d] DBU or NaOMe; [e] R_2COOH, R_3NC

Scheme 19.

often carried out after reduction or ozonolysis of the styrylamine functionality. Alkaline hydrolysis has been used to ascertain the tryptophan content in these alkaloids [75] and the substituents on the aromatic ring of the aryl ether moiety. Direct acidic hydrolysis, followed by identification of amino acids via paper chromatography, has been used by Tschesche for the structure determination of scutianines [17]. Amino acid components have also been determined quantitatively as their DNP derivatives [17, 32]. The *N,N,*-dimethylated amino acids have been characterized by chromatographic methods using thymol blue [17, 76]. β-Hydroxyamino acids have been identified by chromatographic techniques [77] and further characterized by cleavage of their vicinal amino alcohol groups with periodate. 3-Hydroxyproline has been found in many cyclopeptide alkaloids. Acid hydrolysis of amphibine-E [75] resulted in decomposition of this amino acid giving Δ^1-pyrroline after isomerization and decarboxylation. The structure of 3-hydroxyproline has been determined by exhaustive ozonolysis of the aryl ether to an oxalic acid monoester followed by mild hydrolysis. Final comparative studies with synthetic cis- and trans-3-hydroxyprolines confirmed the latter as a structural component of many cyclopeptide alkaloids such as amphibine-E [75] and -D [26] and mauritine-A and -B [26]. Acid hydrolysis of the intact cyclopeptide alkaloids usually results in decomposition of the hydroxy styrylamine portion of the molecule. Hydrogenation [17] or ozonolysis of the enamide double bond is generally carried out before hydrolysis in order to obtain the substitution pattern of the aryl ether moiety. A better analysis of the polysubstituted aromatic chromophores is obtained after alkaline hydrolysis of the alkaloids. Characteristic R_f values [78] and, after pretreatment with diazosulfanilic acid, distinctive color reactions [79] were obtained for ortho- and meta-substituted hydroxyphenethylamines such as 161 and 162. More recently

161   162

Tschesche [80] developed a gas chromatographic method for the identification of tyramine hydrolysis products obtained from cyclopeptide alkaloids. After conversion to $O, N, N,$-tris-(trimethylsilyl) derivatives, the three isomeric tyramines and five isomeric methoxytyramines were separated completely on capillary columns. Trimethylsilated amino acids did not interfere because of their different retention times. The method was used to confirm the structures of the chromophores in integeressine (14-membered ring) and zizyphine-A (13-membered ring).

In order to obtain information on the peptide sequence, the alkaloid was treated with 6 N sulfuric acid; this resulted in partial hydrolysis of the peptide linkages. In scutianine [17] the exocyclic dipeptide dimethylphenylalanylproline was cleaved selectively, keeping intact the 14-membered ring which then yielded tyramine, phenylalanine, and β-hydroxyleucine upon further hydrolysis. In a selected number of cyclopeptide alkaloids the terminal side chain amino acid is monomethylated rather than dimethylated as in the majority of natural products. In those cases characterization is usually carried out after reductive alkylation to generate the dimethylated product, followed by hydrolysis to give standard amino acids [24b].

7.2. Stereochemistry

The stereochemistry of the hydroxyamino acids in cyclopeptide alkaloids has been determined by paper [17] or thin layer chromatography [77] after total hydrolysis of the natural products. Marchand [81] found that NMR spectroscopy did not show any significant differences between the α,β-coupling constants when the *threo-* and *erythro*-hydroxyamino acids were examined directly.

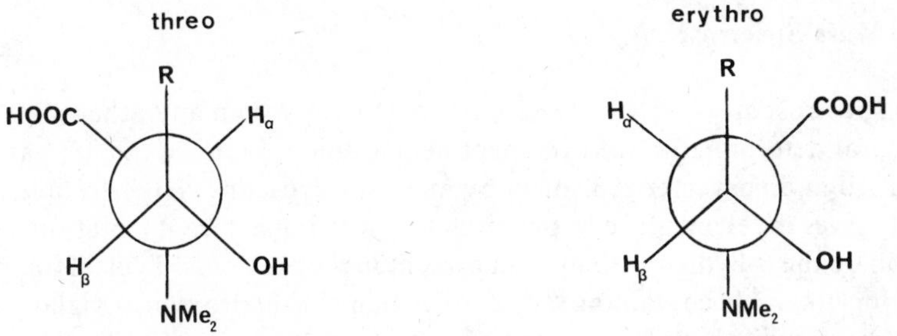

R = Me₂CH, Me, Ph

However, after reductive methylation to give the N,N-dimethylamino acids, the increased steric bulk resulted in a rotameric preference in which the side chain R is *trans* to the NMe₂ group. The enhanced rotamer population in the dimethylamino acids resulted in significantly different α,β-coupling constants for amino acids with *threo* ($J_{\alpha,\beta} = 8.5$–10 Hz) and *erythro* ($J_{\alpha,\beta} = 2.5$–4 Hz) configurations. β-Hydroxyleucine has been found only as the *erythro* form in cyclopeptide alkaloids.

Amino acids in cyclopeptide alkaloids generally occur in the L- form [7]. Exceptions are lasiodine-A [82] and scutianine-G [24a], in which the ring amino acid is D-phenylserine, and scutianine-E, which contains D-*threo*-β-phenylserine and D-*erythro*-β-hydroxyleucine [23]. Configurational assignments of component amino acids have been made on the basis of optical rotation of isolated peptides or free amino acids [5]. Enzymatic digestion with L- and/or D-amino acid oxidase, after acid hydrolysis of the cyclopeptide alkaloid, has been used successfully in many cases. Recently [77] pubescine-A was shown to be isomeric with melonovine-A and was found to be one of the few alkaloids to contain an amino acid of D- configuration. This is the first time D-leucine has been identified in a cyclopeptide alkaloid. The ether-linked amino acid was found to be L-*erythro*-β-hydroxyleucine.

Chemical conversion of amino acids into diastereomeric derivatives and subsequent identification by GLC has been used successfully to determine the chirality of N,N-dimethylamino acids [21]. In this manner, dihydroscutianine-A was hydrolyzed to the component amino acids proline, phenylalanine, and N,N-dimethylphenylalanine. Selective conversion of proline and phenylalanine to their methyl esters, followed by coupling with trifluoroacetyl proline acid chloride and examination of the products by GLC, revealed that the two amino acids had the L configuration. Coupling of N,N-dimethylphenylalanine to L-leucine methyl ester via the mixed anhydride method followed by GLC analysis showed N,N-dimethylphenylalanine to be of L configuration. Configurational assignments required 10 mg of cyclopeptide alkaloid.

7.3. Mass Spectroscopy

Mass spectroscopy has been used more extensively than any other method for structural determination of cyclopeptide alkaloids. Many alkaloids have been identified and characterized solely by mass spectroscopy. High-resolution MS readily gives the elemental composition. Electron impact fragmentation patterns depend on the β-hydroxyamino acid present in the alkaloids. Typical fragmentations for alkaloids containing β-hydroxyleucine, β-hydroxyphenylalanine, and β-hydroxyproline have been described in detail by Tschesche [7]. The fragmentation patterns compliment degradative studies by allowing the deduction of structural assembly of the alkaloids. Until recently MS did not allow differentiation between leucine and isoleucine. High-resolution NMR or amino acid analysis was needed to identify the two isomeric amino acids. In 1974, however, McLafferty [83] reported a novel application of metastable ion and collisional activation spectra to peptides containing leucine and isoleucine. Using this new technique, unequivocal differentiation between N,N-dimethylated leucine, isoleucine, and norleucine could be made [24c]. Analyses of MIKES (metastable ion kinetic energy spectra) of scutianine-C and -H were consistent with the presence of N,N-dimethylisoleucine in both alkaloids [24c].

Recently field desorption mass spectroscopy has been used to examine crude cyclopeptide alkaloid extracts. Lagarias [2b] has identified five components in the extract of *C. sanguineus* with molecular weights of 504, 520, 534, 559, and 573. Isolation and structure determination was carried out by HPLC, amino acid analysis, electron impact MS, and NMR. The fraction with molecular weight 534 was found to be a two-component mixture containing alkaloids with leucine and isoleucine residues.

7.4. Infrared Spectroscopy

The IR spectra of cyclopeptide alkaloids show expected bands for NH (3285–3400), N-methyl (2780–2790), phenol ether (1230–1240), styryl double bond (1623), and amide functionality (1680–1690 and 1630–1655 cm^{-1}). The methoxy group present in some of the 13- and 15-membered ring systems is characterized by a band at 2830 cm^{-1}. Alkaloids with ring-bound phenylserine show a characteristic OH band at 3600 cm^{-1} [24c].

7.5. Ultraviolet Spectroscopy

The UV spectra of 14-membered cyclopeptide alkaloids generally do not exhibit the characteristic peak indicative of the conjugated styrylamine chromophore [7, 4]. This observation was attributed to strain in the ring system which precluded coplanarity of the aromatic ring and the enamido chromophore, thereby preventing π-orbital overlap. The lack of planarity in the conjugated system has been confirmed by X-ray studies [71–74]. Typical UV spectra of the 14-membered alkaloids, as exemplified by scutianines [17] and amphibines [32, 75], show aromatic end absorption with a shoulder at 252 nm (log ϵ = 3.78) and 277 nm (log ϵ = 3.22). After comparisons with the dihydro derivative, the shoulder at 252 nm was attributed to absorption of the enamide and the 277-nm band was assigned to the phenol ether chromophore. No evidence for a styrylamide absorption at 272 nm (log ϵ = 4.35) has been observed in 14-membered alkaloids. The tryptophan moiety in alkaloids is recognized readily by UV absorptions at 270, 282, and 290 nm [32].

The UV spectra of the 13-membered ring alkaloids, which often contain a methoxy substituent in the styryl portion, show absorption bands at 268 nm (log ϵ = 4.06), 321 nm (log ϵ = 3.8), and 210 nm (log ϵ = 3.76), typical for 2,5-dialkoxy styrylamine derivatives. The aromatic hydroxyl substituent in the 13-membered zizyphine-F is easily identified by a characteristic bathochromic shift from 322 nm (log ϵ = 3.8) to 355 nm (log ϵ = 3.8) [84]. The UV spectra of 13-membered ring alkaloids provide confirmatory evidence that considerably less ring strain exists than in the 14-membered ring alkaloids and therefore π-orbital overlap of the styrylamine system is possible. Dihydro derivatives show absorption maxima at 290 nm.

The less strained 15-membered ring bases (mucronines) produce characteristic maxima, resulting from the styrylamine portion, at 273 nm, very similar to N-styrylamides. The dihydro derivatives show maxima at 227, 276, and 283 nm, typical of p-alkyl phenol ethers.

7.6. Nuclear Magnetic Resonance Spectroscopy

Detailed NMR spectral studies have been reported for ceanothine-B, pandamine, americine, frangulanine, and the discarines-A and -B, all members of the hydroxyleucine-containing alkaloids [7]. Aralionine-A, containing hydroxyphenylalanine, and amphibines, which contain hydroxyproline have also been investigated [7, 24c]. Additional studies using ^{1}H and ^{13}C NMR have been used to study conformational aspects of the 14-membered ring alkaloids discarine-B [22] and frangulanine [22, 85–88]. The NH groups of discarine-B show both solvent- and temperature-dependent chemical shifts. Compared to CDCl$_3$, d_6-DMSO strongly deshields the NH protons. The tryptophan amide unit shows the weakest and the indole NH the strongest deshielding effect. This fact suggests that the amide moiety is most protected from solvent, possibly because of hy-

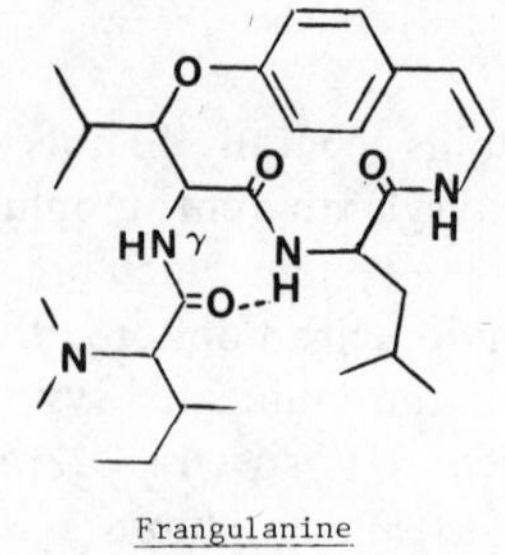

Frangulanine

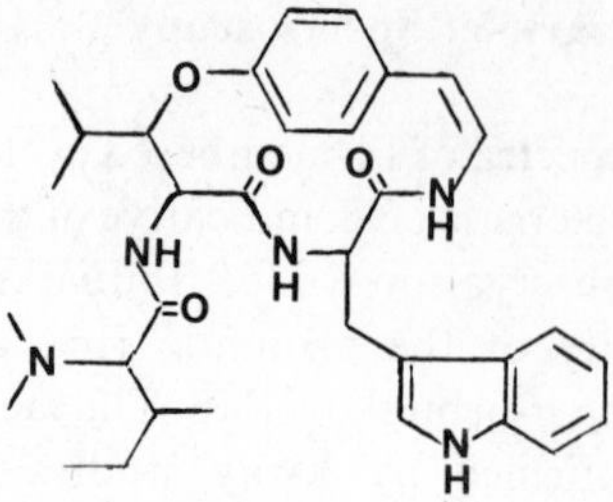

Discarine-B

drogen bonding with the carbonyl group of the isoleucine unit. Deuterium exchange rate studies in d_6-DMSO-CDCl$_3$, however, show that both the tryptophan amide and indole NH protons have very slow rates of exchange. Addition of trifluoroacetic acid to a d_6-DMSO solution of discarine-B also points to the tryptophan amide NH as the most protected proton. The acid causes deshielding of the styrylamide and the hydroxyleucine NH protons by 0.23 and 0.45 ppm, respectively, while leaving the amide of tryptophan nearly unchanged ($\Delta\delta = -0.05$ ppm). The almost complete immunity of the coupling constants of the hydrogens on vicinal carbon sites in the 14-membered ring of discarine-B to solvent and temperature changes suggests minimal conformational change. Conformational rigidity is also substantiated by the apparent solvent–solute interaction between one face of the peptide frame and polar solvents. A dramatic shielding of the olefinic α-H, H$_4$ and one of the methylene protons of tryptophan was observed upon addition of d_6-DMSO to a CDCl$_3$ solution of discarine-B. Partial stacking of the tryptophan and isoleucine side chains was proposed from chemical shifts of the isoleucine methyl groups [22].

Haslinger [85–88] used ^{1}H NMR, ^{13}C NMR, NOE measurements, solvent titration, and variable-temperature experiments to conclude that frangulanine exists in a γ-turn conformation in CDCl$_3$, with the NH of leucine providing the hydrogen bonding proton in the turn. Contrary to discarine-B, frangulanine undergoes a conformational change in d_6-DMSO, which is evidenced by the disappearance of the γ-turn and by the large solvent dependency of chemical shifts and vicinal coupling constants.

A combined NMR spectroscopic and X-ray crystallographic approach to the analysis of the conformations of the two synthetic model structures **163a** and **163b** of 14-membered dihydrocyclopeptide alkaloids has been carried out [89].

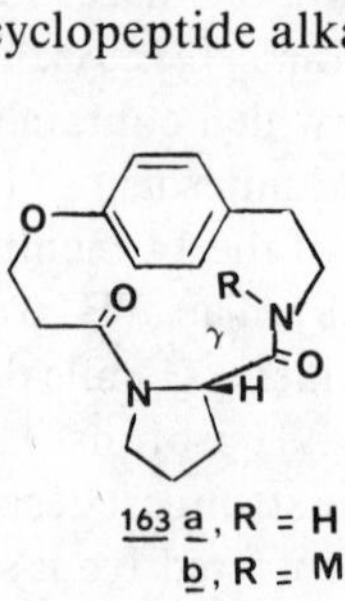

163 **a**, R = H
    b, R = Me

Total spectral assignments were accomplished using two-dimensional homo-
nuclear J-spectral analysis, decoupling experiments, computer-generated spectral simulations, steady-state ^{1}H NOE measurements at 270 and 360 MHz, and
^{13}C NMR analyses. The solution conformations of **163** are similar and are in
good agreement with the X-ray structure of **163a**. Both structures have amides
with *trans*-geometry. Compound **163a** adopts a conformation that accommo-
dates an internal hydrogen bond to form a γ-turn.

The NMR spectra of the amphibine-type alkaloids, which contain hydroxyproline as the ether-linked amino acid, offer evidence distinguishing between 3-
and 4-hydroxyprolines. The coupling constant of 3-hydroxyproline, $J_{\alpha,\beta} = \sim 5$
Hz, is characteristic of a *cis*-stereochemistry. Chemical degradation, however,
indicated the *trans*-stereochemistry. X-ray analysis of mauritine-A [71, 72] confirmed the *trans*-stereochemistry of the 3-substituted proline. The inclusion of
this amino acid in the strained 14-membered ring system involves an alteration of
its normal conformation ($J_{\alpha,\beta} = \sim 1$ Hz). The unstrained conformation is seen
again after oxidative opening of the ring at the double bond. Reduction of the
double bond to give dihydrocyclopeptides has no effect on the strained conformation of proline. As in the case of mauritine-A, the dihydro derivative exhibits a
coupling constant of 6 Hz for the α,β-protons of proline [37a]. However, ring-
opened derivatives such as **48,** which were obtained as intermediates during the
synthesis of dihydromauritine-A [37], showed the typical unstrained α,β-cou-
pling constant of $\leqslant 1$ Hz. Isomer **49** showed a coupling constant of 6 Hz, which is
typical of *cis*-stereochemistry. In the synthetic model studies using the Ugi four-
component condensation to convert β-(aryloxy)pyrrolines to acylated β-(aryloxy)proline amides, as depicted in Scheme 19, the singlet character of the
proline α-proton in one of the reaction products **(160)** was used to assign the
trans-stereochemistry [66].

Several detailed ^{13}C NMR studies of cyclopeptide alkaloids have appeared
recently [24c, 90, 91]. The 14-membered ring alkaloids show distinctive NMR
features that indicate conformational rigidity and macrocyclic ring strain. The
β-aryloxyleucine side chain residue in discarine-B reveals nonequivalence of the
δ-methyl resonances ($\Delta\delta = 5.7$ ppm). Restricted motion of the aromatic ring
results in the four aromatic methines being in dissimilar environments; they
therefore exhibit different chemical shifts. The two styrene carbons are almost
equivalent, consistent with a nonconjugated double-bond system. In contrast,

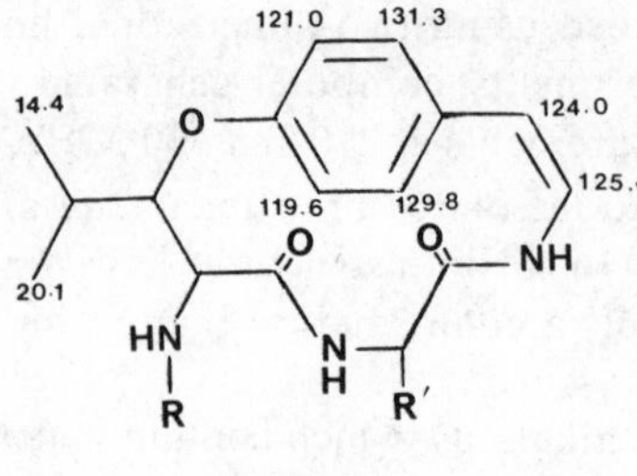

Discarine-B

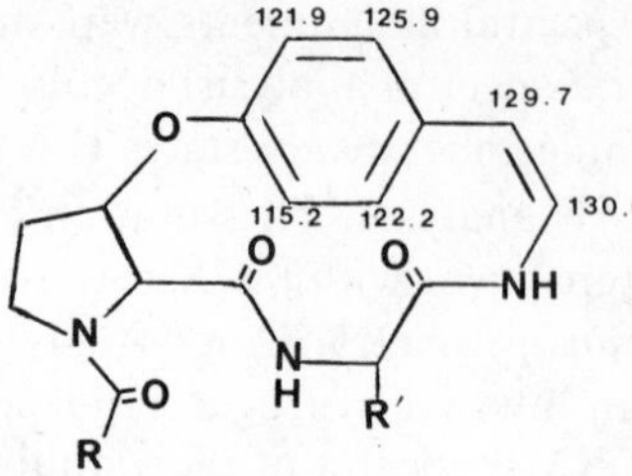

Amphibine-D and -E

acyclic derivatives that are strain free exhibit greatly different chemical shifts, 111.7 and 129.2 ppm, respectively, for the analogous carbons.

Alkaloids with 13- and 15-membered macrocycles show chemical shifts similar to the unstrained noncyclic derivatives. These values are consistent with conjugation of the styrylamide π-electron system.

7.7. Circular Dichroism

Circular dichroism (CD) spectra of 14-membered ring alkaloids reveal a weak positive band at 285 nm and a strong negative one at 237 nm. Thirteen-membered rings show strong negative maxima at 324, 276, 254, and 218 nm and a weak positive band at 232 nm [7]. More recently CD studies of ceanothine-B, in the presence of various metal cations, have been carried out [43]. Both ceanothine-B and a synthetic model of the 14-membered macrocyclic system bind Li^+, Ca^{2+}, and Mg^{2+} [43, 89].

7.8. Crystal Structure

Crystal structures of cyclopeptide alkaloids have been elucidated by X-ray diffraction analyses. Mauritine-A [71, 72] and the *N, N, N*-trimethylated derivative of frangulanine [73, 74] were so examined, both being representatives of 14-membered ring alkaloids. X-ray analysis has also been carried out on a simplified synthetic model of the 14-membered dihydro macrocycle [89]. In both frangulanine and mauritine-A all of the amino acids were found to be of L- configuration, with the amide bonds having a *trans*-geometry. Ring strain in the 14-membered macrocycles, which had been postulated from UV, ^{1}H, and ^{13}C NMR studies, was clearly evident in the crystals. For example, in mauritine-A the benzene ring in the central ring system is slightly bent and the attached atoms are considerably out of the benzene plane. A pronounced deviation from coplanarity is apparent in the styrylamide system, preventing π-orbital overlap. The *trans*-stereochemistry of β-hydroxyproline was clearly established, confirming degradative studies and clarifying assignment ambiguities that had arisen from NMR studies. In frangulanine the *erythro*-stereochemistry of hydroxyleucine was confirmed.

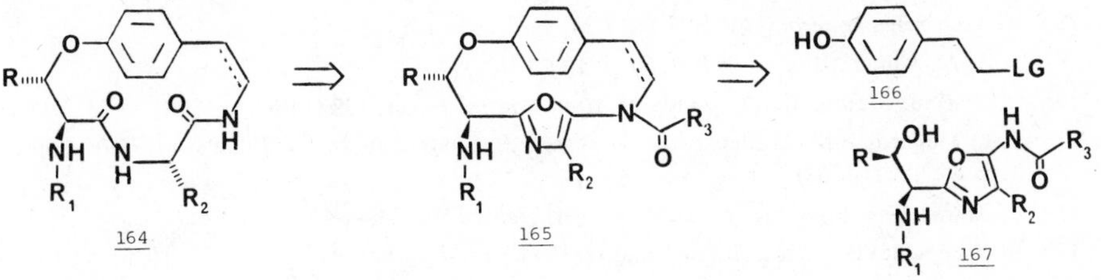

Scheme 20.

ADDENDUM

Dipeptide Equivalents: A Different Approach to Cyclopeptide Alkaloids. A different approach to cyclopeptide alkaloids has been recently proposed by Lipshutz and co-workers [92]. The strategy envisioned by these authors is based on the use of heterocycles, such as imidazoles or oxazoles as masked diamide-dipeptide equivalents. To this end, several 2,4(5)-disubstituted imidazoles [93] and 5-(acylamino)oxazoles [94] have been prepared as intermediates for the synthesis of heterocyclophanes, potential precursors of cyclopeptide alkaloids (Scheme 20). Unmasking of these heteroaromatic moieties to their dipeptide equivalents has been demonstrated. Although carbon and nitrogen alkylation chemistry was examined as model studies for subsequent elaboration to specific heterocyclophanes, the potential value of this concept as applied to the synthesis of cyclopeptide alkaloids still remains to be exploited.

ACKNOWLEDGMENTS

We wish to express our sincere thanks to Dr. Kris Bhat for proofreading the entire manuscript and making many useful suggestions. We also thank Ms. P. Lovelace for typing the manuscript. Support by the National Science Foundation is gratefully acknowledged.

REFERENCES

1. M. Païs, X. Monseur, X. Lusinchi, and R. Goutarel, *Bull. Soc. Chim. France*, 817 (1964).

2. (a) J. C. Lagarias, D. Goff, F. K. Klein, and H. Rapoport, *J. Nat. Prod.* **42**, 220 (1979); (b) J. C. Lagarias, D. Goff, and H. Rapoport, *J. Nat. Prod.* **42**, 663 (1979).

3. M. Païs, F.-X. Jarreau, X. Lusinchi and R. Goutarel, *Ann. Chim. (Paris)*, 83 (1966).

4. E. W. Warnhoff, *Fortschr. Chem. Org. Naturst.* **28**, 162 (1970).

5. M. Païs and F.-X. Jarreau, in *Chemistry of Amino Acids*, "Peptides and Proteins," Vol. 1, B. Winstein, Ed., Dekker, New York, 1971, Chap. 5.

6. Y. Ogihara, *Ann. Rep. Fac. Pharm. Sci. Nagoya City Univ.* **22**, 1 (1974).

7. R. Tschesche and E. U. Kaussmann, in *The Alkaloids*, Vol. 15, R. H. F. Manske, Ed., Academic, New York, 1975, Chap. 4.

8.   R. Tschesche, *Heterocycles* **4**, 107 (1976).

9.   J. H. M. Clinch, *Am. J. Pharm.* **56**, 131 (1884).

10. M. Païs, J. Mainil, and R. Goutarel, *Ann. Pharm. Fr.* **21**, 139 (1963).

11. E. L. Ménard, J. M. Müller, A. F. Thomas, S. S. Bhatnagar, and N. J. Dastoor, *Helv. Chim. Acta* **46**, 1801 (1963).

12. E. Zbiral, E. I. Ménard, and J. M. Müller, *Helv. Chim. Acta* **48**, 404 (1965).

13. R. Tschesche, E. U. Kauffmann, and G. Eckhardt, *Tetrahedron Lett.*, 2577 (1973).

14. R. Wasicky, M. Wasicky, and R. Joachimovits, *Plants Med. (Stuttgart)* **12**, 13 (1964).

15. E. W. Warnhoff, S. K. Pradhan, and J. C. N. Ma, *Can. J. Chem.* **43**, 2594 (1965).

16. E. W. Warnhoff, J. C. N. Ma, and P. Reynolds-Warnhoff, *J. Am. Chem. Soc.* **87**, 4198 (1965).

17. R. Tschesche, R. Wetters, and H. W. Fehlhaber, *Ber.* **100**, 323 (1967).

18. M. Païs, J. Marchand, F.-X. Jarreau, and R. Goutarel, *Bull. Soc. Chim. France,* 1145 (1968).

19. R. Tschesche, E. Ammermann, and H.-W. Fehlhaber, *Tetrahedron Lett.*, 4405 (1971).

20. M. González Sierra, O. A. Mascaretti, F. J. Diaz, E. A. Ruveda, C.-J. Chang, E. W. Hagaman, and E. Wenkert, *J. Chem. Soc. Chem. Commun.*, 915 (1972).

21. M. González Sierra, O. A. Mascaretti, V. M. Merkuza, E. L. Tosti, E. A. Ruveda, and C.-J. Chang, *Phytochemistry* **13**, 2865 (1974).

22. V. M. Merkuza, M. González Sierra, O. A. Mascaretti, E. A. Ruveda, C.-J. Chang, and E. Wenkert, *Phytochemistry* **13**, 1273 (1974).

23. R. Tschesche and R. Ammermann, *Ber.* **107**, 2274 (1974).

24. (a) R. Tschesche and D. Hillebrand, *Phytochemistry* **16**, 1817 (1977); (b) R. Tschesche, D. Hillebrand, H. Wilhelm, E. Ammermann and G. Eckhardt, *Phytochemistry* **16**, 1025 (1977); (c) A. F. Morel, R. V. F. Bravo, F. A. M. Reis, and E. A. Ruveda, *Phytochemistry* **18**, 473 (1979).

25. (a) R. Tschesche, A. H. Shah, and G. Eckhardt, *Phytochemistry* **18**, 702 (1979). (b) A. H. Shah, V. B. Pandey, G. Eckhardt, and R. Tschesche, *Phytochemistry* **23**, 931 (1984). (c) A. H. Shah, private communication, 1983.

26. R. Tschesche, H. Wilhelm, and H.-W. Fehlhaber, *Tetrahedron Lett.* **26**, 2609 (1972).

27. B. K. Cassels, G. Eckhardt, E. V. Kaussman, and R. Tschesche, *Tetrahedron* **30**, 2461 (1974).

28. (a) R. Tschesche, I. Khobhar, H. Wilhelm, and G. Eckhardt, *Phytochemistry* **15**, 541 (1976); (b) V. B. Pandey, J. P. Singh, K. K. Seth, A. H. Shah, and G. Eckhardt, *Erster Gesamt Kongress der Pharmazeutische Wissenschaften.* München (W. Germany) 17–20 April 1983, p. 121.

29. G. J. Kapadia, Y. N. Shukla, J. F. Morton, and H. A. Lloyd, *Phytochemistry* **16**, 1431 (1977).

30. M. I. D. Chughtai, I. Khokhar, A. Ahmad, U. Ghani, and M. Anwar, *Pak. J. Sci. Res.* **31**, 79 (1979).

31. M. I. D. Chughtai, I. Khokhar, A. Ahmad, I. Ahmad, and A. Rehman, *Pak. J. Sci. Res.* **31**, 237 (1979).

32. R. Tschesche, C. Spilles, and G. Eckhardt, *Ber.* **107**, 686 (1974).

33. O. Blanpin, M. Païs, and M. A. Quevauviller, *Ann. Pharm. Fr.* **21**, 147 (1963).

34. K. Kawai, Y. Nozawa, and Y. Ogihara, *Experientia* **33**, 1454 (1977).

35. R. A. Ravizzini, C. S. Andreo, and R. H. Vallejos, *Plant and Cell Physiol.* **18**, 701 (1977).

36. K. D. Kopple, *J. Pharm. Sci.* **61**, 1345 (1972) and references cited therein.

37. (a) R. F. Nutt, Ph.D. Thesis, University of Pennsylvania, 1981; (b) R. F. Nutt, K.-M. Chen, and M. M. Joullié, *J. Org. Chem.* **49**, 1013 (1984).

38. U. Schmidt, A. Lieberknecht, H. Griesser, and J. Häusler, *Angew. Chem. Int. Ed. Engl.* **20**, 281 (1981).

39. U. Schmidt, H. Griesser, A. Lieberknecht, and J. Talbiersky, *Angew. Chem. Int. Ed. Engl.* **20** 280 (1981).

40. F. Frappier, F. Rocchiccioli, F.-X. Jarreau, and M. Païs, *Tetrahedron* **34**, 2911 (1978).

41. F. Rocchiccioli, F.-X. Jarreau, and M. Païs, *Tetrahedron* **34** 2917 (1978).

42. J. Marchand, F. Rocchiccioli, M. Païs, and F.-X. Jarreau, *Bull. Soc. Chim. Fr.,* 4699 (1972).

43. (a) J. C. Lagarias, R. A. Houghten, and H. Rapoport, *J. Am. Chem. Soc.* **100**, 8202 (1978); (b) J. C. Lagarias, Ph.D. Thesis, University of California, Berkeley, 1979.

44. I. Suzuki, *Bull. Chem. Soc. Japan* **35**, 540 (1962).

45. L. A. La Planche and M. T. Rogers, *J. Am. Chem. Soc.* **86**, 337 (1964).

46. L. Pauling, *The Nature of the Chemical Bond,* 2nd Ed. Cornell University Press, Ithaca, NY, 1948, p. 207.

47. J. Dale and K. Titlestad, *J. C. S. Chem. Commun.* 656 (1969).

48. D. Goff, J. C. Lagarias, W. C. Shih, M. P. Klein, and H. Rapoport, *J. Org. Chem.* **45**, 4813 (1980).

49. J. Häusler and U. Schmidt, *Liebigs Ann. Chem.,* 1881 (1979).

50. R. F. Nutt, D. F. Veber, and R. Saperstein, *J. Am. Chem. Soc.* **102**, 6539 (1980).

51. U. Schmidt. A. Lieberknecht, H. Griesser, and J. Talbiersky, *J. Org. Chem.* **47**, 3261 (1982).

52. U. Schmidt, A. Lieberknecht, H. Griesser, and J. Häusler, *Liebigs Ann. Chem.* 2153 (1982).

53. U. Schmidt, H. Bökens, A. Lieberknecht, and H. Griesser, *Tetrahedron Lett.* **22**, 4949 (1981).

54. U. Schmidt, H. Bökens, A. Lieberknecht, and H. Griesser, *Liebigs Ann. Chem.,* 1459 (1983).

55. U. Schmidt, A. Lieberknecht, H. Bökens, and H. Griesser, *J. Org. Chem.* **48**, 2680 (1983).

56. D. W. Brooks, L. D. Lu, and S. Masamune, *Angew. Chem. Int. Ed. Engl.* **18**, 72 (1979).

57. U. Schmidt and U. Schaubacher, *Angew. Chem. Int. Ed. Engl.* **22**, 152 (1982).

58. U. Schmidt, A. Lieberknecht, U. Schaubacher, T. Beuttler, and J. Wild, *Angew. Chem. Int. Ed. Engl.* **21**, 770 (1982).

59. E. W. Warnhoff, in *The Alkaloids,* Vol. 1, Specialist Periodical Reports, The Chemical Society, London, 1971, p. 444.

60. G. P. Stack, Ph.D. dissertation, University of Michigan, 1976.

61. J. R. Bettrell and P. Maitland, *J. Chem. Soc.,* 5211 (1961).

62. U. Schmidt and E. Ohler, *Angew. Chem. Int. Ed. Engl.* **16**, 327 (1977).

63. U. Schmidt and E. Prantz, *Angew. Chem. Int. Ed. Engl.* **16**, 328 (1977).

64. G. E. Krejcarek, B. W. Dominy, and R. G. Lawton, *J. Chem. Soc. Chem. Commun.,* 1450 (1968).

65. Robotti, K. M., Ph.D. dissertation, University of Michigan, 1980.

66. R. F. Nutt and M. M. Joullié, *J. Am. Chem. Soc.* **104**, 5852 (1982).

67. M. M. Bowers-Nemia and M. M. Joullié, unpublished results.

68. M. M. Bowers-Nemia, J. Lee, and M. M. Joullié, *Synth. Commun.* **13**, 1117 (1983).

69. O. Mitsunobu, *Synthesis,* 1 (1981).

70. M. M. Bowers-Nemia and M. M. Joullié, *Heterocycles* **20**, 817 (1983).

71. A. Kirfel, G. Will, R. Tschesche, and H. Wilhelm, *Z. Naturforsch.* **31b**, 279 (1976).

72. A. Kirfel and G. Will, *Z. Krist.* **142**, 368 (1975).

73. M. Takai, K. Kawai, Y. Ogihara, Y. Iitaka, and S. Shibata, *J. Chem. Soc. Chem. Commun.,* 653 (1974).

74. M. Takai, Y. Ogihara, Y. Iitaka, and S. Shibata, *Chem. Pharm. Bull. (Japan),* **24**, 2181 (1976).

75. R. Tschesche, E. U. Kaussmann, and H.-W. Fehlhaber, *Ber.* **105**, 3094 (1972).

76. I. M. Hais and K. Macek, *Handbuch fur Papierchromatographie*, Vol. I., G. Fischer, Ed., Jena, (East Germany) 1963, p. 597.

77. R. Tschesche, D. Hillebrand, and I. R. C. Bick, *Phytochemistry* **19**, 1000 (1980).

78. R. Tschesche, E. Frohberg, and H.-W. Fehlhaber, *Ber.* **103**, 2501 (1970).

79. R. Tschesche, L. Behrendt, and H.-W. Fehlhaber, *Ber.* **102**, 50 (1969).

80. U. Henke and R. Tschesche, *J. Chromat.* **120**, 477 (1976).

81. J. Marchand, M. Païs, and F. X. Jarreau, *Bull. Soc. Chim. Fr.* **10**, 3742 (1971).

82. J. Marchand, M. Païs, X. Monseur, and F.-X. Jarreau, *Tetrahedron* **25**, 937 (1969).

83. K. Levsen, H. K. Wipf, and F. W. McLafferty, *Org. Mass. Spectrom.* **8**, 117 (1974).

84. R. Tschesche, I. Khokhar, C. Spilles, G. Eckhardt, and B. K. Cassels, *Tetrahedron Lett.* **34**, 2941 (1974).

85. E. Haslinger, *Tetrahedron* **34**, 685 (1978).

86. E. Haslinger and W. Robien, *Monatsh. Chem.* **110**, 1011 (1979).

87. E. Haslinger, *Monatsh. Chemie* **109**, 523 (1978).

88. M. Takai, Y. Ogihara, Y. Iitaka, and S. Shibata, *Chem. Pharm. Bull. (Japan)* **23**, 2556 (1975).

89. J. C. Lagarias, W. H. Yokoyama, J. Bordner, W. C. Shih, M. P. Klein, and H. Rapoport, *J. Am. Chem. Soc.* **105**, 1031 (1983).

90. M. Païs, F.-X. Jarreau, M. Gonzalez Sierra, O. A. Mascaretti, E. A. Ruveda, C. Chang, E. W. Hagaman, and E. Wenkert, *Phytochemistry* **18**, 1869 (1979).

91. D. M. Hindenlang, M. Shamma, G. A. Miana, A. H. Shah, and B. K. Cassels, *Liebigs Ann. Chem.* 447 (1980).

92. B. H. Lipshutz, R. W. Hungate, and K. E. McCarthy, *Tetrahedron Lett.* **24**, 5155 (1983).

93. B. H. Lipshutz and M. C. Morey, *J. Org. Chem.* **48**, 3745 (1983).

94. B. H. Lipshutz, R. W. Hungate, and K. E. McCarthy, *J. Am. Chem. Soc.* **105**, 7703 (1983).

Chapter Four

Cannabis Alkaloids

Mahmoud A. ElSohly

Research Institute of Pharmaceutical Sciences
School of Pharmacy
University of Mississippi
University, Mississippi 38677

CONTENTS

1. INTRODUCTION

Cannabis sativa L. (Cannabaceae) is the medicinal plant from which the crude drugs marijuana, hashish, and hash oil are obtained. Although there are references in the literature to the presence of more than one species of *Cannabis* [1, 2] for example, *C. sativa, C. indica,* and *C. ruderalis* it is strongly argued that there is only one species, namely, *C. sativa* [3, 4], with two subspecies (subspecies *sativa* and subspecies *indica*) and four varieties (sativa, spontanica, indica, and kafiristanica) [4].

The plant has been known in medicine for thousands of years and the drug was official in the USP (United States Pharmacopeia) from 1820 to 1942. Conditions for which *Cannabis* was recommended included rheumatism, beriberi, neuralgia, asthma, malaria, hysteria, constipation, insomnia, tetanus, and epilepsy, to name a few. In 1942, the drug was denied admission into the USP, indicating the lack of acceptable medical use in the United States. Currently, *Cannabis* is classified as a Schedule 1 drug, which means that it is used clinically only in research, after filing the proper IND (Investigative New Drug) application with the FDA (Food and Drug Administration).

The study of the chemistry of *Cannabis* goes back to the nineteenth century with many attempts to isolate and characterize the psychologically active constituent(s). The first successful report in that direction was carried out in 1896 by Wood et al. [5] who isolated a "red oil" from the ether extract of the resin of Indian hemp (Charas). This material was named cannabinol, and it was 1940 before the correct structure was determined as a result of the work carried out by Cahn [6] and Adams et al. [7]. Cannabinol, although psychologically active, is only one-fifth as active as Δ^9-tetrahydrocannabinol (Δ^9-THC), which is the major cannabinoid in drug type *Cannabis*. It is believed that cannabinol is an artifact produced by oxidation of Δ^9-THC and hence exists in appreciable amounts only in old *Cannabis* samples. Δ^9-THC, on the other hand, was first isolated as its acetate by Wollner et al. [8] and the correct chemical structure was assigned 22 years later by Prof. Mechoulam's group [9].

Today, 426 chemical components are known to exist in the plant material [10, 11] belonging to 18 different classes of compounds (Table 1).

The major active constituent of *Cannabis* was thought to be an alkaloid. Early investigations of *Cannabis* were directed toward the isolation of an alkaloid as the active principle. This could be attributed to the fact that other drugs at the time, as for example opium and cinchona, were known to contain alkaloids as the active constituents and in part to the ease of isolation and crystallization of alkaloids in general. The first report of an alkaloid in *Cannabis* (1876) was by Preobraschensky [12] who reported the presence of nicotine as the chief active constituent in Indian hemp. However, in 1881 Siebold and Bradbury [13] reported the absence of nicotine in *C. indica* and isolated trace amounts of a volatile alkaloid which they named cannabinine. Two years later, Hay [14] reported the isolation of an alkaloid with strychnine-type activity which he named tetanocannabine. These reports have not been confirmed by other

Table 1. Chemical Constituents of *Cannabis*

Chemical Class	Number of Compounds Known to Exist
1. Cannabinoids	62
a. Cannabigerol (CBG) type	6
b. Cannabichromene (CBC) type	4
c. Cannabidiol (CBD) type	7
d. Δ^9Tetrahydrocannabinol (Δ^9-THC) type	9
e. Δ^8-Tetrahydrocannabinol (Δ^8-THC) type	2
f. Cannabicyclol (CBL) type	3
g. Cannabielsoin (CBE) type	3
h. Cannabinol (CBN) type	6
i. Cannabinodiol (CBND) type	2
j. Cannabitriol (CBT) type	6
k. Miscellaneous types	10
l. Other cannabinoids	4
2. Nitrogenous compounds (Alkaloids)	20
a. Quaternary bases	5
b. Amides	1
c. Amines	12
d. Spermidine alkaloids	2
3. Amino acids	18
4. Proteins, glycoproteins, and enzymes	9
5. Sugars and related compounds	34
a. Monosaccharides	13
b. Disaccharides	2
c. Polysaccharides	5
d. Cyclitols	12
e. Aminosugars	2
6. Hydrocarbons	50
7. Simple alcohols	7
8. Simple aldehydes	12
9. Simple ketones	13
10. Simple acids	20
11. Fatty acids	12
12. Simple esters and lactones	13
13. Steroids	11
14. Terpenes	103
a. Monoterpenes	58
b. Sesquiterpenes	38
c. Diterpenes	1
d. Triterpenes	2
e. Miscellaneous compounds of terpenoid origin	4
15. Noncannabinoid phenols	20
16. Flavonoid glycosides	19
17. Vitamins	1
18. Pigments	2

From C. E. Turner, M. A. ElSohly and E. G. Boeren, Constituents of *Cannabis sativa* L. XVII. A Review of the Natural Constituents. *Journal of Natural Products*, **43**, 169-234 (1980). Reproduced by permission of the American Society of Pharmacognosy.

investigators, and no definitive structure work was done on these alkaloids. The presence of nicotine was again refuted in 1886 by Kennedy [15].

The first report of the presence of an alkaloid in *Cannabis,* which has been confirmed in recent years, was by Jahns in 1887 [16].

The initial belief that the active constituent was an alkaloid subsided after the isolation of cannabinol in 1896 by Wood et al. [5]. The concentrated ether extract of the resin of Indian hemp (Charas) was distilled at room pressure followed by reduced pressure to give the "red oil," which was named cannabinol. The physiological symptoms produced by the red oil were "peculiar to *C. indica.* . . ." This work marked the beginning of the chemistry of cannabinoids as we know it today. The interest in *Cannabis* alkaloids, however, did not stop and a wide variety of compounds were isolated and characterized. This chapter will review those alkaloidal constituents of *Cannabis,* their isolation techniques, and their spectral characteristics. In addition, other nitrogen-containing nonalkaloidal constituents will be briefly reviewed.

2.   ALKALOIDS ISOLATED FROM *CANNABIS*

Unlike most alkaloid-containing plants, *Cannabis* does not contain specific types of alkaloids. The alkaloids in *Cannabis* could be classified under the following major groups.

1.   *Quaternary ammonium bases:* Choline **(1)**, neurine **(2)**, trigonelline **(3)**, and isoleucine betaine **(4)**.
2.   *β-Arylethylamine alkaloids:* Hordenine **(6)**.
3.   *Spermidine-type alkaloids:* Cannabisativine **(7)** and anhydrocannabis-ativine **(8)**.
4.   *Amides: N-(p*-hydroxy-*β*-phenylethyl)-*p*-hydroxy-(*trans*)-cinnamide **(9)**.
5.   *Volatile amines:* There are six primary amines and four secondary amines in this group (Table 3).

2.1.   Quaternary Ammonium Bases

2.1.1.   *Choline* (1).   Choline was first isolated from Indian *Cannabis* in 1887 by Jahns [16]. The aqueous extract was evaporated, the residue was dissolved in ethanol, and choline precipitated from ethanol as its platinum salt and recrystallized from water. In his report, Jahns indicated that choline might be identical to Siebold and Bradbury's "Cannabinine" [13] and Hay's "Tetanocannabine" [14]. Choline was later isolated by many investigators including Merz and Bergner [17], Salemink et al. [18], and Mole and Turner [19]. While Salemink et al. [18] isolated choline from the seeds, Mole and Turner [19] isolated it from the roots. Details of the most recent isolation of choline will be given in the next section.

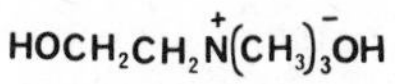

$$\text{HOCH}_2\text{CH}_2\overset{+}{\text{N}}(\text{CH}_3)_3\overline{\text{OH}}$$

1

$$\text{CH}_2{=}\text{CH}{-}\overset{+}{\underset{\text{CH}_3}{\overset{\text{CH}_3}{\text{N}}}}{-}\text{CH}_3\overline{\text{OH}}$$

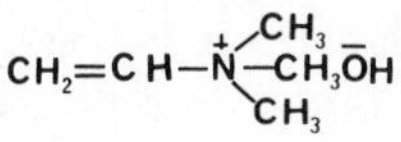

2

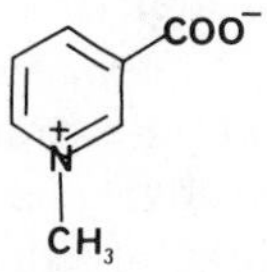

3

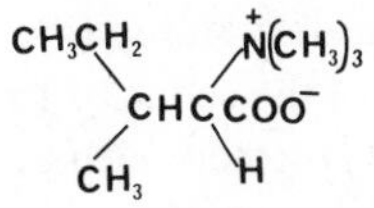

4

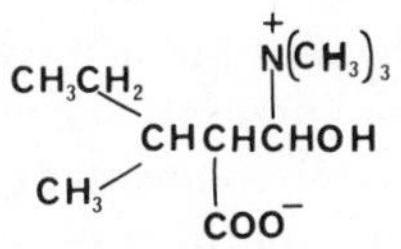

5

2.1.2. *Neurine* (2). Neurine was isolated, along with choline (**1**), from the roots of a Mexican *Cannabis* grown in Mississippi [19]. The ground roots were extracted consecutively with hexane, chloroform, ethanol, water, and 5% hydrochloric acid. The aqueous extract, when concentrated, gave a brown syrup. Fractionation was accomplished by eluting through a strong anion exchange column (OH⁻ form) with water followed by 10% acetic acid. The water eluate was concentrated to yield a brown gum having an amine base odor. Further fractionation of the gum on a strong cation exchange column (H⁺ form) eluting with water, 10% ammonium hydroxide, and 5% ammonium carbonate gave a tan crystalline solid from the ammonium carbonate fraction. Triturating the residue with ethanol gave, after removal of solvent, a "crystalline oil." Chromatography of this material on alumina gave two compounds. The first was eluted with ethanol and the second was eluted with ethanol–ammonium hydroxide (9:1). Further purification by column chromatography on alumina followed by preparative TLC on alumina provided neurine (**2**) as a nearly colorless, crystalline solid that was very hygroscopic. The second compound eluted was choline, which was purified by preparative TLC using alumina.

2.1.3. *Trigonelline* (3). Trigonelline was reported to have been isolated from *Cannabis* as its platinum salt by Schulze and Frankfurt in 1894 [20], although no isolation procedure was given. This work was later confirmed by Merz and Bergner [17] who isolated trigonelline and choline from three *Cannabis* samples grown in Germany. The samples were extracted with petroleum–ether, ether, ethanol, and water. The water extract was acidified with sulfuric acid, extracted with ether, and treated with several precipitating reagents. The two alkaloids were again isolated by Bercht and Salemink [21] from the seeds of French *Cannabis*. The ground seeds were extracted with ethanol (96%) and precipitation of the alkaloids was carried out using Kraut's reagent. The bases were then chromatographed on cellulose powder and the structures of the two alkaloids were determined by R_f values, melting points (mp), and infrared (IR) spectra.

2.1.4. L-(+)-Isoleucine-betaine (4).

This quaternary base was isolated by Bercht et al. [22] from the seeds of French-type *Cannabis* (Fibrimon). The fresh ground seeds were extracted with ethanol and the extract acidified with acetic acid. The concentrated extract was then exhaustively partitioned with benzene, and after filtration of the aqueous layer an excess of Kraut's reagent was added. The reddish precipitate was suspended in water and then decomposed by H_2S. After filtration, the filtrate was concentrated, shaken with freshly prepared AgCl, and again filtered. The residue obtained from evaporation of the filtrate was chromatographed on cellulose column using *n*-butanol saturated with water. The fraction containing a Dragendorff's positive spot on cellulose plate (R_f 0.45 using the same system) was further purified on a basic alumina column using methanol as the solvent. The compound with R_f 0.30 (basic alumina plate with methanol as solvent) was finally purified over $CaHPO_4$ column, which was first washed with *n*-pentane followed by chloroform. The yield of isoleucine-betaine was about 0.34 mg/kg of the seeds. The chemical structure was determined by spectral analysis and finally proven by synthesis.

Racemic isoleucine-betaine was prepared by treating 2-bromo-3-methyl-pentanoic acid with trimethylamine in a sealed tube at 35°C for 5 days followed by workup of the reaction mixture. L-(+)-Isoleucine-betaine on the other hand was prepared starting from L-(+)-isoleucine which was converted to the corresponding 2-bromo-3-methylpentanoic acid. The latter was converted to L(+)-isoleucine-betaine following the same procedure as for the racemic mixture. Alternatively, L-(+)-isoleucine-betaine was prepared from L-(+)-isoleucine by methylation of the *N*,*N*-dimethyl derivative using methyl iodide in weakly basic ethanolic solution. The *N*,*N*-dimethyl derivative was prepared by shaking acetic acid solution of L-(+)-isoleucine with formaldehyde and platinum oxide in an atmosphere of hydrogen.

The betaine base assigned structure **5**, which was isolated by Lousberg and Salemink [23] from the seeds of Fibrimon *Cannabis*, had the same physical data as L-(+)-isoleucine-betaine. Thus, Lousberg and Salemink's compound [23] is actually L-(+)-isoleucine-betaine **(4)**.

2.2. β-Arylethylamine Alkaloids

Hordenine **(6)** is the only alkaloid in this group isolated from *Cannabis*. It was isolated by El-Feraly and Turner in 1975 [24] from the leaves of a drug type *Cannabis*. The powdered leaves were extracted with 95% ethanol and the concentrated extract was partitioned between chloroform and 2% citric acid. The aqueous fraction was made alkaline with ammonia and then extracted with chloroform. A second acid–base partitioning step was carried out before the

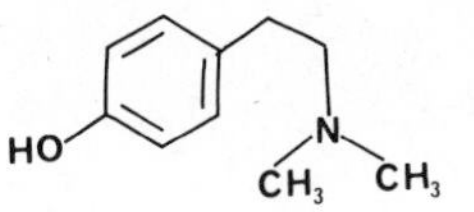

6

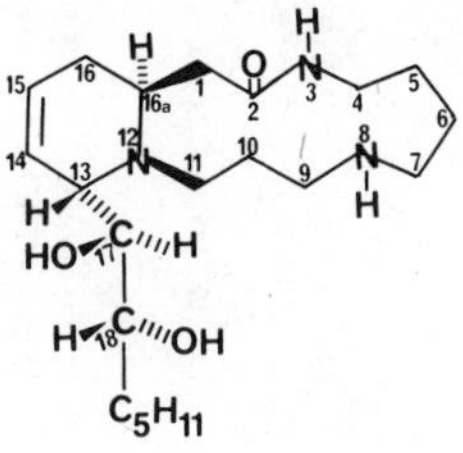

7

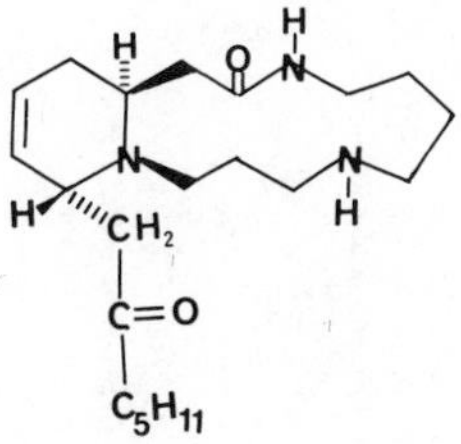

8

alkaloidal fraction was chromatographed on a silica gel column using 1% ammonia in methanol. A light yellow fraction gave rise to feathery needles of hordenine upon crystallization from acetone–hexane mixture. The chemical structure was proven on the bases of spectral evidence (Table 2) and direct comparison with authentic hordenine.

Following this report, ElSohly and Turner [25] examined *Cannabis* variants grown from seeds of different geographic origin, including Afghanistan, Australia, Czechoslovakia, Hungary, India, Jamaica, Lebanon, Mexico, South Africa, Spain, Thailand, Turkey, and Russia, and found that hordenine is common among all variants.

2.3. Spermidine-Type Alkaloids

Two alkaloids in this class were isolated from *Cannabis,* namely, cannabisativine (7) and anhydrocannabisativine (8). The isolation of these alkaloids from *Cannabis* marks the first reported existence of this class in a higher plant. Alkaloids of this system (palustine and palustridine) have been previously

Table 2. Physical and Spectral Data on *Cannabis* Alkaloids

Compound	Formula	Molecular Weight	Melting Point	$[\alpha]_D$	Infrared	Ultraviolet	Mass Spectrometry	^{1}H NMR	References
Choline (1)	$C_5H_{15}NO_2$	121.18			ν max 3400, 2960, 2925, 1480, 1410, 1350, 1135, 1090, 1000, 940, and 860 cm^{-1}			δ D$_2$O (60 MHz) 3.34 (s, 9H), 3.64, (m, 2H), and 4.24 (m, 2H)	18, 19
Neurine (2)	$C_5H_{13}NO$	103.17			ν max 3430, 2965, 2930, 1665, 1470, 1260, 965, and 895 cm^{-1}			δ D$_2$O (60 MHz) 6.74 (m, 1H), 5.87 (m, 2H), and 3.42 (s, 9H)	19
Trigonelline (3)	$C_7H_7NO_2$	137.14	218°C						17, 20, 21
L-(+)-isoleucine betaine (4)	$C_9H_{19}NO_2$	173.26	Hygroscopic not crystallizable	$[\alpha]_D^{20} + 18°.0$ (c, 1 CHCl$_3$)	ν max 3340, 2960, 1625, 1490, 1450, 1390, 1030, 970, and 860 cm^{-1}			δ D$_2$O (60 MHz) 1.02 (t, $J = 7.3$ Hz, 3H), 1.06 (d, $J = 7.0$ Hz), 1.58 (m, $J = 7.0$ Hz, 2H), 2.11 (m, 1H), 3.24 (s, 9H), and 3.67 (d, $J = 1.1$ Hz, 1H)	22
Hordenine (6)	$C_{10}H_{15}NO$	165.24	118–119°C		ν max (CHCl$_3$) 3600, 3040, 2960, 2870, 2838, 2795, 1620, 1603, 1520, 1475, 1380, 1255, 1175, 1145, 1103, 1055, 1040, 1009, 870, 845, and 830 cm^{-1}	λ max (MeOH) 274 nm (ϵ 1261) and 211 (1100)	M$^+$ 165 (14%), 58 (100)	δ CDCl$_3$ (60 MHz) 6.94 (d, $J = 9$ Hz, 2H), 6.59 (d, $J = 9$ Hz, 2H), 2.69 (t, distorted), and 2.36 (s, 6H)	24
Cannabisativine (7)	$C_{21}H_{39}N_3O_3$	381.56	167–168°C	$[\alpha]_D^{25} = +55.1°C$ (c 0.53, CHCl$_3$)	ν max 1628 cm^{-1}		M$^+$ m/z 381, 310, 280, 250, 171, 208, 129, and 72	δ CDCl$_3$ (100 MHz) 5.96 (s, 2H), —CH=CH—, 9.6 (s, broad, CO—NH)	26, 27
Anhydro-cannabisativine (8)	$C_{21}H_{37}N_3O_2$		Not crystallizable	$[\alpha]_D^{22} + 18.7°$ (c 0.1 MeOH)	ν max (KBr) 3290, 3020, 2925, 2860, 1715, 1661, 1642, 1615, 1540, 1460, 1365, 1210, 1124, 1100, and 1050 cm^{-1}		M$^+$ m/z 363 (12%), 348 (1), 292, (4), 264 (22), 250 (68), 208 (55), 198 (60), 192 (25), 171 (20), 84 (60), 80 (40), 70 (100), and 43 (60)	δ CDCl$_3$ (60 MHz) 5.8 (m, 2H, —CH=CH—), 9.6 (s, broad, 1H, CO—NH)	30
N-(p-hydroxy-β-phenylethyl) p-hydroxy-(trans)-cinnamide (9)	$C_{17}H_{17}NO_3$	283.12	252–254°C (dec.)		ν max 3200, 1650, 1590, 1510, 1230, 980, and 830 cm^{-1}	λ max (EtOH) 227 nm (log ϵ 4.1), 296 (4.1), and 312 (4.0)			31

reported in the genus *Equisetum* (Equisetaceae-Horsetail). Thus these two alkaloids could have a significant value in the chemotaxonomy of *Cannabis*.

2.3.1. Cannabisativine (7).

This alkaloid was first reported in an ethanolic extract of the roots of Mexican *Cannabis* [26, 27], and the structure was determined by X-ray crystallography [26] in 1975. Cannabisativine was later isolated from the leaves and small stems of a Thailand variant [28]. Subsequently, it was found to occur in all variants examined by ElSohly and Turner [25] with quantitative differences.

Isolation of the alkaloid from the roots [27] was carried out by partitioning the concentrated methanol extract between water and chloroform. The chloroform fraction was then partitioned between hexane and methanol–water (9:1) and the aqueous-methanol residue was chromatographed over silicic acid column. Cannabisativine was eluted with 8% methanol–chloroform and crystallized from acetone.

A different procedure was used for the isolation of this alkaloid from the leaves. The powdered leaves were extracted with 95% ethanol, and the concentrated extract was partitioned between chloroform and 2% citric acid. Cannabisativine was then isolated from the acidic fraction following the same procedure of partitioning and chromatography as previously described under hordenine.

Hordenine and cannabisativine were not separated on silica gel plates using 4% ammonia in methanol as a solvent system but were cleanly separated using the organic phase (lower) of a chloroform–acetone–ammonia (1:1:1) system [25]. The physical data of cannabisativine are given in Table 2 and the X-ray data showed that there was intramolecular hydrogen bonding between N(3) and N(8) resulting in constriction of the 13-membered ring and between OH at C(18) and N(12). In addition, the relative configuration of C(17) and C(1) are *trans* to each other and C(11) is *cis* to C(1) and the OH groups on C(17) and C(18) are erythro. Conformation differences thus existed between cannabisativine and palustrine.

2.3.2. Anhydrocannabisativine (8).

Shortly after the reported isolation and structure of cannabisativine ElSohly et al. reported the isolation and characterization of another spermidine alkaloid [29, 30]. The isolation procedure was similar to that previously described under cannabisativine and hordenine.

The alkaloid did not crystallize and neither did the picrate salt. However, the structure analysis was an easy task because of the close structural similarity to cannabisativine (7). Final proof was by conversion of cannabisativine (7) to anhydrocannabisativine (8) by heating with oxalic acid at 180°C for 1 h. The assignment of the position of the keto group was based on the mass spectral analysis. Figure 1 shows the fragmentation pattern for both spermidine alkaloids.

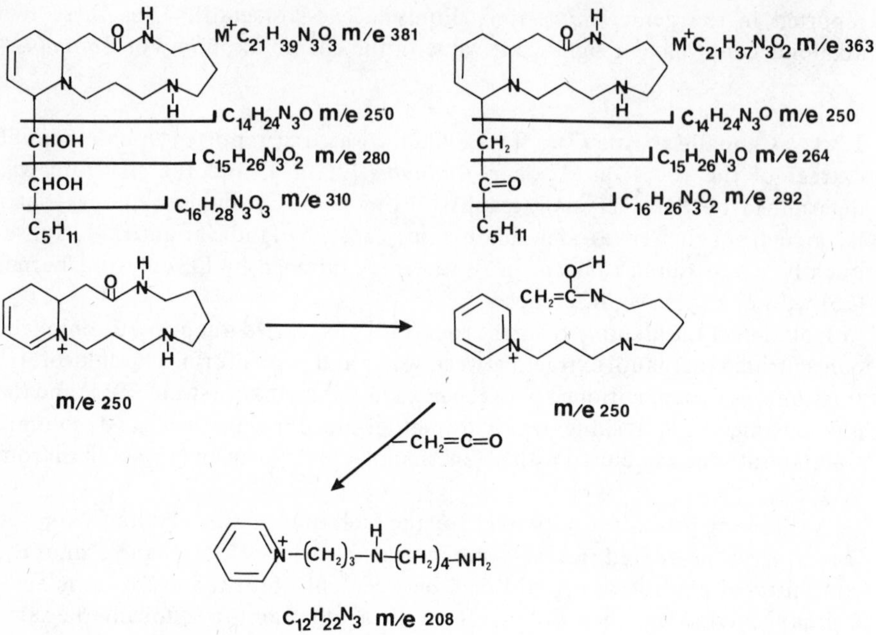

Figure 1. Fragmentation pattern of the spermidine alkaloids of *Cannabis*.

2.4. Amides

There is only one *Cannabis* alkaloid in which its only nitrogen is an amido function, namely, N-(p-hydroxy-β-phenylethyl)-p-hydroxy-(*trans*)-cinnamide (**9**). This alkaloid was reported in the roots of Mexican *Cannabis* [31] and so far it

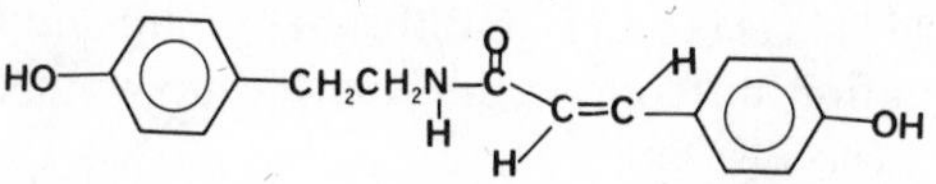

9

has not been reported in any of the above ground plant material. Although the alkaloid does not possess a basic nitrogen, it was isolated from the basic alkaloidal fraction obtained from acid–base partitioning of the concentrated ethanolic extract of the roots. Purification was carried out by chromatography over a silicic acid column followed by crystallization from chloroform-methanol. Chemical characterization was based on spectral evidence (Table 2) and through comparison with synthetic material. The alkaloid was synthesized by condensation of p-hydroxy-*trans*-cinnamic acid and tyramine in the presence of dicyclohexylcarbodiimide.

Table 3. Volatile Amines of *Cannabis*

Compound	Reference
Piperidine	32, 21
Methylamine	33
Ethylamine	33
n-Propylamine	33
n-Butylamine	33
iso-Butylamine	33
sec-Butylamine	33
Dimethylamine	33
Diethylamine	33
Pyrrolidine	33

From C. E. Turner, M. A. ElSohly and E. G. Boeren, Constituents of *Cannabis sativa* L. XVII. A Review of the Natural Constituents. *Journal of Natural Products*, **43**, 169–234 (1980). Reproduced by permission of the American Society of Pharmacognosy.

2.5. Volatile Amines

Both primary and secondary amines have been reported in *Cannabis*. These are shown in Table 3. Piperidine was isolated from Japanese *Cannabis* by Obata et al. [32] from the ether extract of the leaves. On the other hand, all other amines listed in Table 3 were reported to be present in Dutch *Cannabis* [33]. The amine fraction was isolated by steam distillation of the plant material and characterization was by capillary gas chromatography. Because these are common amines, there will be no further discussion of their physical, chemical, and biological properties.

3. UNSUBSTANTIATED REPORTS OF OTHER ALKALOIDS IN *CANNABIS*

Klein et al. [34] reported the isolation of four alkaloids from the alkaloidal fraction of the dried leaves of Mexican *Cannabis*. These were named cannabamines A, B, C, and D in a decreasing order of polarity using silica gel plates and 2.5% diethylamine in ethanol as the solvent. Cannabamine D was absent in the alkaloidal fraction of the fresh leaves.

These alkaloids were separated by preparative TLC and were analyzed by high-resolution mass spectrometry. The molecular formulas were as follows: Cannabamine A, $C_{21}H_{37}N_3O_2$ or $C_{26}H_{37}N$; Cannabamine B, $C_{18}H_{21}NO_3$; Cannabamine C, $C_{14}H_{21}N_3O_3$; and Cannabamine D, $C_{17}H_{33}N_3O_2$. Cannabamine C was thought to be a β-arylethylamine alkaloid based on the mass fragmentation. No further discussion on the structure of these four alkaloids was given.

Lousberg and Salemink [33] indicated the presence of indole alkaloids in *Cannabis*. This was based on TLC examination of alkaloidal fractions of *Cannabis* extracts and spraying with reagents known to be used for indole alkaloids. Up to this date there have been no reports of actual isolation of any indole alkaloid from *Cannabis*. In the same report [33], Lousberg and Salemink tentatively identified *n*-pentylamine, isoamylamine, β-phenethylamine, cadaverine (1,5-diaminopentane), ethanolamine or histamine, and benzylamine or tyramine in the free amines fraction of hemp.

Muscarine has been reported to be a component of *Cannabis* alkaloids; however, this report was refuted by Salemink et al. [18].

The above compounds are in addition to nicotine, cannabinine, and tetanocannabine which have been discussed in the introduction of this chapter.

4. OTHER NONALKALOIDAL, NITROGEN-CONTAINING COMPOUNDS IN *CANNABIS*

This group of compounds includes amino acids, amino sugars, proteins, glycoproteins, and enzymes.

4.1. Amino Acids

Table 4 shows the different amino acids known to occur in *Cannabis*. Leucine, methionine, threonine, valine, histidine, and lysine were isolated from the tops of wild hemp by Obata et al. [32]. The tops were extracted with petroleum ether, the extract concentrated, and the concentrate extracted with 95% ethanol. Workup of the ethanol extract afforded these amino acids. Lousberg and Salemink [33] studied the amino acid composition of the hydrolyzed amino acid and smaller peptides isolated from hemp. Using an amino acid analyzer the following acids were determined: glycine, alanine, serine, threonine, asparagine and/or asparatic acid, glutamine and/or glutamic acid, lysine, arginine, histidine, proline, valine, leucine, isoleucine, phenylalanine, and tyrosine. However, none of the sulfur-containing amino acids was detected. The amino acid proline was also isolated from the ethanol extract of the leaves, stems, and flowering tops of Indian *Cannabis* [35]. The amino acids cystine, methionine, and tryptophane were reported among other amino acids in the sequence of hemp cytochrome *C* [36].

4.2. Amino Sugars

Galactosamine and glucosamine were isolated by Wold and Hillstad [37] from the light petrol extract of the leaves and stems of South African *Cannabis* grown in Oslo, Norway. Ion exchange chromatography on DEAE–cellulose and gel

Table 4. Amino Acids of *Cannabis*

Compound	Reference
Alanine	33
Aspartic acid	33
Cystine	36
Glutamic acid	33
Glycine	33
Serine	33
Arginine	33
Histidine	32
Isoleucine	33
Leucine	32
Lysine	32
Methionine	32, 36
Phenylalanine	33
Proline	33, 35
Threonine	32
Tryptophan	36
Tyrosine	33
Valine	32

From C. E. Turner, M. A. ElSohly and E. G. Boeren, Constituents of *Cannabis sativa* L. XVII. A Review of the Natural Constituents. *Journal of Natural Products*, **43**, 169–234 (1980). Reproduced by permission of the American Society of Pharmacognosy.

filtration on Sepharose 4B were used for the isolation of the two amino sugars that are thought to be part of a carbohydrate protein polymer.

4.3. Proteins and Enzymes

Table 5 shows the components of this group that have been isolated from the fruits of *Cannabis*. In addition, two glycoproteins were isolated from the leaves of South African [45] and Thai [46] material. The glycoprotein from the South African *Cannabis* contained the amino acid hydroxyproline while the Thai material did not.

5. BIOLOGICAL ACTIVITIES OF *CANNABIS* ALKALOIDS

The literautre is deficient on data pertinent to the biological activity of *Cannabis* alkaloids. This is in part because of their low concentration in the plant material and thus the difficulty in isolating sufficient quantities for testing. However, several investigators have attempted to evaluate the pharmacology of the crude

Table 5. Proteins and Enzymes of *Cannabis*

Compound	Reference
Edestin	38
Zeatin	39
Zeatin nucleoside	39
Edestinase	40
Glucosidase (emulsin)	41
Polyphenol oxidase	42
Peptidase (ereptase)	43
Peroxidase	42
Adenosine-5-phosphatase	44

From C. E. Turner, M. A. ElSohly and E. G. Boeren, Constituents of *Cannabis sativa* L. XVII. A Review of the Natural Constituents. *Journal of Natural Products,* **43,** 169–234 (1980). Reproduced by permission of the American Society of Pharmacognosy.

alkaloidal fraction. Gill et al. [47] examined the aqueous fraction of the ethanol extract of Pakistani *Cannabis* (tincture of *Cannabis* B.P.). The aqueous extract was analyzed by high-voltage paper electrophoresis; a strip of the paper was visualized by a color reagent and the remainder divided into six parts and eluted. The eluate corresponding to trigonelline was found to have an acetylcholine-antagonizing activity with a slow course of action comparable to atropine. Pure trigonelline does not have such activity. Two other components with faster mobility on the electrophoretic paper showed acetylcholinelike activity, similar to the methyl and ethyl esters of trigonelline. The authors concluded that the extract had at least one atropinic and two muscarinic substances and that the latter components could contribute to the irritant activity of the smoke. Lousberg and Salemink [33] subjected the complete basic fraction to preliminary pharmacological assays and found that it decreased body temperature in mice by the intraperitoneal route. Different doses caused gradual decrease in motility, increased convulsions, peculiar downward bowing of the neck, ataxia, and screaming. A weak analgesic activity (hot plate) was also observed. When the same investigators [33] examined the free-amines fractions, it was found to have no apparent biological action. The LD_{50} of the crude alkaloidal fraction was estimated to be 6 mg/ 10 g of mice weight. Klein et al. [34] examined the effects of three batches of extracts (each at pH 4.0 and about 0.5% total crude alkaloids) given subcutaneously, intraperitonially, or intravenously to mice. The only observed effect was slightly reduced motor activity. However, no gross signs of toxicity were observed and the extracts had no effect on pentobarbital sleeping time.

No pharmacological evaluation has been carried out on the pure alkaloids isolated from *Cannabis* except *N*-(*p*-hydroxy-β-phenylethyl)-*p*-hydroxy-(*trans*)-cinnamide **(9)** which was found to have mild analgesic activity [31].

REFERENCES

1. R. E. Schultes, W. M. Klein, T. Plowman, and T. E. Lockwood, *Bot. Mus. Leafl. (Harvard Univ.)* **23**, 337, 1974.
2. W. A. Emboders, *Econ. Bot.* **28**, 304 (1974).
3. E. Small, *Bull. Narcot.* **27**, 1 (1975).
4. E. Small, and A. Cronquist, *Taxonomy* **25**, 405 (1976).
5. T. B. Wood, W. T. N. Spivey, and T. H. Easterfield, *J. Chem. Soc.* **69**, 539 (1896).
6. R. S. Cahn, *J. Chem. Soc.* 1342 (1932).
7. R. Adams, B. R. Baker, and R. B. Wearn, *J. Am. Chem. Soc.* **62**, 2204 (1940).
8. H. J. Wollner, J. R. Matchett, J. Levine, and S. Loewe, *J. Am. Chem. Soc.* **64**, 26 (1942).
9. Y. Gaoni and R. Mechoulam, *J. Am. Chem. Soc.* **86**, 1646 (1964).
10. C. E. Turner, M. A. ElSohly, and E. G. Boeren, *J. Nat. Prod. (Lloydia)* **43**, 169 (1980).
11. M. A. ElSohly, H. N. ElSohly, and C. E. Turner, *Cannabis:* New constituents and their Pharmacological action, in *Topics in Pharmaceutical Sciences,* Elsevier Science Publishers B. V., Amsterdam 1983, D. D. Breimer and P. Speiser, editors, p.p. 429–440.
12. W. Preobraschensky, *Jahresber. Pharmacog. Toxicol.* **11**, 98 (1876).
13. L. Siebold and T. Bradbury, *Pharm. J.* **41**, 326 (1881).
14. M. Hay, *Pharm. J.* **42**, 998 (1883).
15. G. W. Kennedy, *Pharm. J.* **46**, 453 (1886).
16. E. Jahns, *Arch. Pharm. (Weihheim)* **XXV**, 479 (1887).
17. K. W. Merz and K. G. Bergner, *Arch. Pharm. (Weihheim)* **278**, 49 (1940).
18. C. A. Salemink, E. Veen, and W. A. de Kloet, *Planta Med.* **13**, 211 (1965).
19. M. L. Mole and C. E. Turner, *Acta Pharm. Jugoslav.* **23**, 203 (1973).
20. E. Schulze and S. Frankfurt, *Ber. dtsch. Chem. Ges.* **27**, 769, 1894.
21. C. A. L. Bercht and C. A. Salemink, *U.N. Secretariat Document ST/SOA/Ser.S/21,* 1969.
22. C. A. L. Bercht, R. J. J. Ch. Lousberg, F. J. E. M. Kuppers, and C. A. Salemink, *Phytochemistry* **12**, 2457 (1973).
23. M. Paris, F. Boucher, and L. Cosson, *Econ. Bot.* **29**, 245 (1975).
24. F. S. El-Feraly and C. E. Turner, *Phytochemistry* **14**, 2304 (1975).
25. M. A. ElSohly and C. E. Turner, *U.N. Secretariat Document ST/SOA/SER.S/54,* 1977.
26. H. L. Lotter, D. J. Abraham, C. E. Turner, J. E. Knapp, P. L. Schiff, Jr., and D. J. Slatkin, *Tetrahedron Lett.* **33**, 2815 (1975).
27. C. E. Turner, M. F. H. Hsu, J. E. Knapp, P. L. Schiff, Jr., and D. J. Slatkin, *J. Pharm. Sci.* **65**, 1084 (1976).
28. F. S. El-Feraly and C. E. Turner, *U.N. Secretariat Document ST/SOA/SER.S/52,* 1975.
29. M. A. ElSohly and C. E. Turner, *U.N. Secretariat Document ST/SOA/SER.S/53,* 1976.
30. M. A. ElSohly, C. E. Turner, C. H. Phoebe, Jr., J. E. Knapp, P. L. Schiff, Jr., and D. J. Slatkin, *J. Pharm. Sci.* **67**, 124 (1978).

31. D. J. Slatkin, N. J. Doorenbos, L. S. Harris, A. N. Masoud, M. W. Quimby, and P. L. Schiff, *J. Pharm. Sci.* **60**, 1891 (1971).

32. Y. Obata, Y. Ishikawa, and R. Kitazawa, *Bull. Agr. Chem. Soc. Japan* **24**, 670 (1960).

33. R. J. J. Ch. Lousberg and C. A. Salemink, *Pharm. Weekbl.* **108**, 1 (1973).

34. F. K. Klein, H. Rapoport, and H. W. Elliott, *Nature* **232**, 258 (1971).

35. M. L. Mole, Jr., and C. E. Turner, *J. Pharm. Sci.* **63**, 154 (1974).

36. D. G. Wallace, R. H. Brown, and D. Boulter, *Phytochemistry* **12**, 2617 (1973).

37. J. K. Wold and A. Hillestad, *Phytochemistry* **15**, 325 (1976).

38. D. M. Stockwell, J. M. Dechary, and A. M. Altschul, *Biochim. Biophys. Acta.* **82**, 221 (1964).

39. A. J. St. Angelo, R. L. Ory and H. S. Hansen, *Phytochemistry* **8**, 1873 (1969)

40. A. J. St. Angelo, L. Y. Yatsu, and A. M. Altschul, *Arch. Biochim. Biophys.* **124**, 199 (1968).

41. G. Leoncini, *Boll. Ist. Super. Agrar. Pisa* **7**, 603 (1931).

42. V. I. Ostapenko, *Botan. Zhur.* **45**, 114.

43. S. H. Vines, *V. Ann. Botan.* **22**, 103 (1908).

44. N. Bargoni and A. Luzzati, *Atti. accad. nazl. Lincei, Rend., Classe sci fis mat. e nat.,* **21**, 450 (1956).

45. A. Hillestad and J. K. Wold, *Phytochemistry* **16**, 1947 (1977).

46. A. Hillestad, J. K. Wold, and T. Engen, *Phytochemistry* **16**, 1953 (1977).

47. E. W. Gill, W. D. M. Paton, and R. G. Pertwee, *Nature* **228**, 134 (1970).

Chapter Five

Synthesis of *Lycopodium* Alkaloids

Todd A. Blumenkopf and Clayton H. Heathcock

Department of Chemistry
University of California
Berkeley, California 94720

CONTENTS

1. BIOSYNTHESIS

About 400 varieties of club moss (genus *Lycopodium*) have been described, but the alkaloid content has been determined for only 10% of these species [1]. From this small sample, more than 100 different basic compounds have been isolated. The structures of most of these compounds have been fully elucidated [2].

The study of *Lycopodium* alkaloids began more than 100 years ago when Bödeker isolated from *L. complanatum* a substance that seems to correspond with the alkaloid now known as lycopodine [3]. In 1938 Achmatowicz and Uzieblo examined *L. clavatum* and isolated lycopodine, to which they assigned the correct molecular formula, and two new alkaloids [4]. This report was followed by a series of papers in which Manske and Marion reported the isolation and partial characterization of a number of alkaloids from *Lycopodium* [5].

The *Lycopodium* alkaloids have little medicinal value. Several pharmacological studies performed prior to 1938 were with impure plant extracts [6]. Achmatowicz and Uzieblo reported that samples of these substances stimulate the respiratory center of mammals [4]. However, this report was shown later to be incorrect [7]. Lycopodine does produce uterine contractions and an increase in the peristable movements of small intestines of rabbits, rats, and guinea pigs and causes complete paralysis in frogs when injected into the ventral lymph sac [7]. The compound is moderately toxic and causes severe convulsions and eventual death. Several species of club moss have been used in Chinese herbal medicine since at least the seventeenth century for treatment of skin disorders and as a tonic [8].

The *Lycopodium* alkaloids thus far characterized can be classified according to the 12 principal skeletons depicted in Figure 1. Lycopodine **(1)** is the most

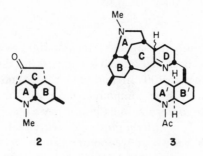

1

abundant and most widely distributed member of the family [9,10]. The lycopodines, of which lycopodine is one example, are also the most common skeletal class; more than 35 examples have been characterized. One or more lycopodines have been detected in almost every species of club moss thus far studied; in some samples, the lycopodine-type compounds constitute more than 90% of the basic extracts [1].

The simplest *Lycopodium* alkaloid is luciduline **(2)** [11]. There are no other known members of the luciduline group. The most complex are the lucidines, which contain three nitrogen atoms per molecule. This class contains three alkaloids, lucidine B **(3)**, lycolucine **(4)**, and dihydrolycolucine **(5)** [12]. It is not

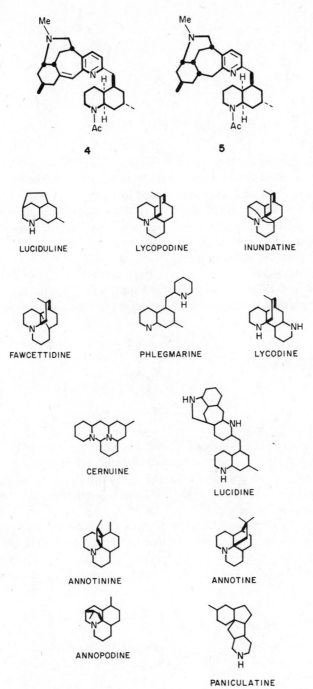

Figure 1. Principal *Lycopodium* alkaloid skeletons.

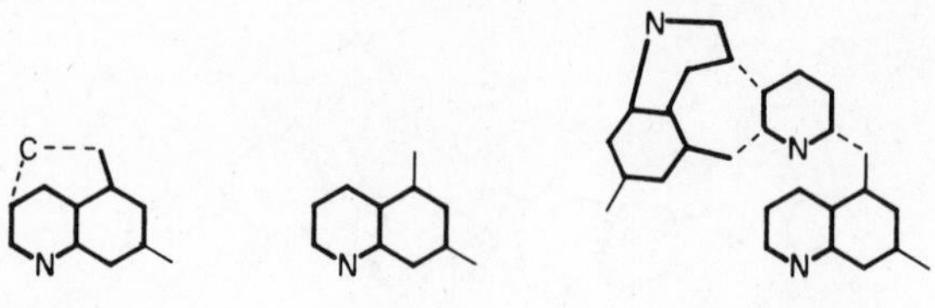

LUCIDULANE   6   LUCIDANE

Figure 2.   Structural relationship between the lucidine and luciduline skeletons.

surprising that the luciduline and lucidine classes are found in the same species, as both are readily derivable from the $C_{11}N$ fragment **6** [12]. The lucidine skeleton is composed of two such fragments joined to a piperidine ring (see Fig. 2). It is interesting that although the AB-ring systems in the luciduline and lucidine skeletons have the same absolute stereostructure, the A'B'-ring system in the lucidine family is derived from the enantiomeric $C_{11}N$ fragment.

The phlegmarine skeleton is composed of a single $C_{11}N$ unit attached to a piperidine ring. Alkaloids that have the lycodine skeleton, for example, lycodine **(7)** [13], can be derived by joining the $C_{11}N$ unit to a piperidine ring via two C—C bonds. The absolute configuration of lycodine **(7)** is the same as that found in the

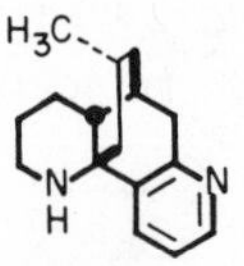

7

lucidine and luciduline AB ring systems. The absolute sense of chirality of the phlegmarine alkaloids has not yet been determined. Since these compounds have been proposed as intermediates in the biosynthesis of more complex *Lycopodium* alkaloids [14], such as the lucidines, it is important that their absolute configurations be elucidated.

The first *Lycopodium* alkaloid to be fully characterized was annotinine **(8)** [15]. This accomplishment was followed by the structural elucidation of lycopodine **(1)** [9], selagine **(9)** [16], lycodine **(7)** [13], and the obscurines [e.g., α-obscurine **(10)**] [17]. The first biosynthetic proposal was made by Conroy,

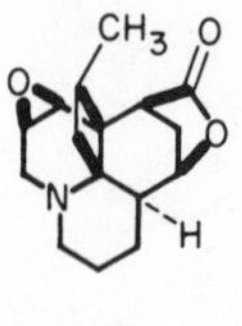

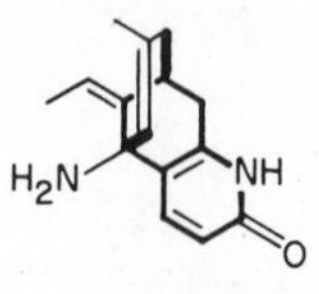

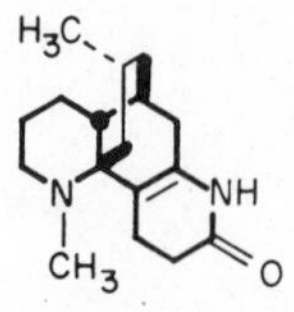

8   9   10

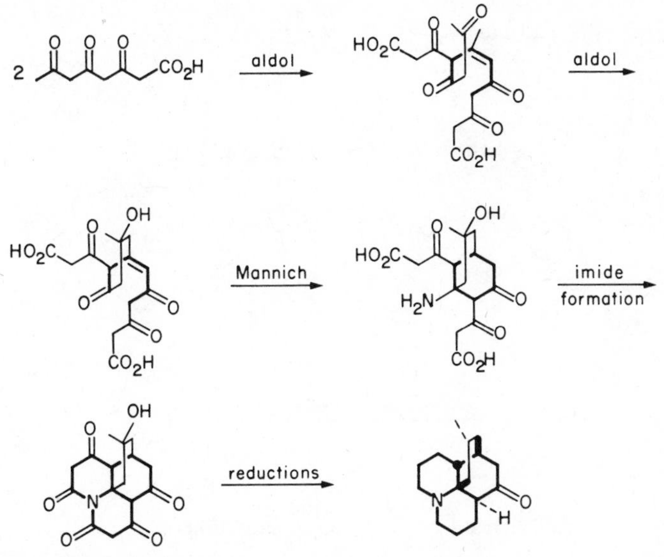

Scheme 1. Conroy hypothesis.

who suggested that the *Lycopodium* alkaloids arise from the double condensation of two eight-carbon polyacetate chains (see Scheme 1), followed by a Mannich reaction with ammonia and subsequent loss of water to give imide **11** [18]. A series of reductions would lead to lycopodine (**1**), while minor deviations would afford other known members of the family. Conroy's proposal was cautiously accepted by some early workers in the field [19], but later work proved it to be incorrect. Although Leete and Louden found that ^{14}C-labeled acetate was indeed incorporated into annotinine (**8**) by *Lycopodium annotinum*, the distribution of activity was not that which would be predicted by the polyketide hypothesis [20, 21].

Prompted by the latent symmetry in the lycodine skeleton, Spenser and MacLean postulated that the *Lycopodium* alkaloids may be formed through the dimerization to two C_8N units. Coniine (**12**), one likely candidate for such a C_8N

unit, is of polyketide origin (see Scheme 2) [22]. However, the structurally similar alkaloid pelletierine (**13**) is formed in an entirely different manner (see Scheme 3) [23]. In this case only the side chain has an acetate origin. The piperidine ring is

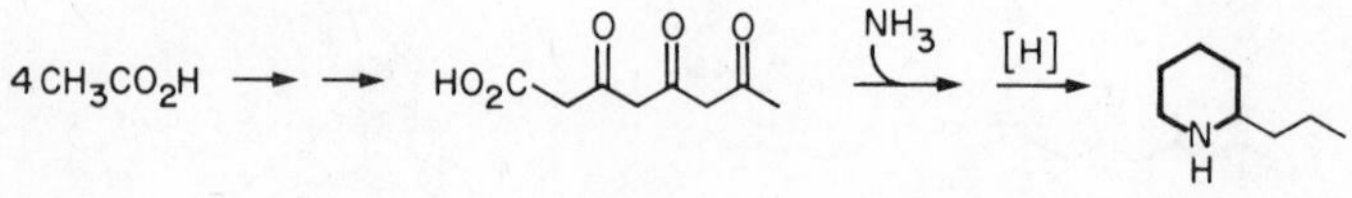

Scheme 2.   Coniine biosynthesis

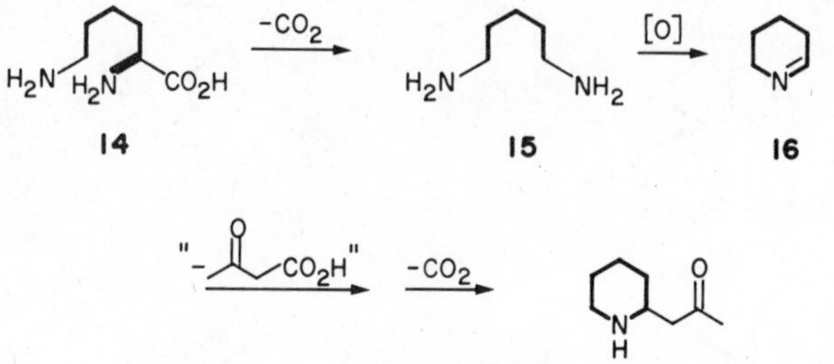

Scheme 3.   Pelletierine biosynthesis.

generated by decarboxylation of lysine **(14)** to give cadaverine **(15)**, which is subsequently converted to Δ^1-piperideine **(16)**. Condensation of this intermediate with an acetoacetate unit and subsequent decarboxylation provides pelletierine **(13)**.

To test the C_8N hypothesis, Spenser and MacLean performed a series of ^{14}C-acetate feeding experiments in intact plants and excised shoots of *Lycopodium tristachym* Pursh [24]. If the biosynthesis of lycopodine **(1)** proceeds through a coniine dimer (the polyketide route) 1-^{14}C-acetate should be incorporated at the odd-numbered carbons, whereas 2-^{14}C-acetate would give lycopodine labeled at all even-numbered carbons. The pelletierine-dimer pathway would be indicated by 1-^{14}C-acetate incorporation at C(7) and C(15), and by 2-^{14}C-acetate incorporation at C(6), C(8), C(14), and C(16) of lycopodine **(1)**.

The experimental results are not compatible with a polyketide-derived biosynthesis of lycopodine but are consistent with the pelletierine hypothesis. Further evidence for this pathway was the finding that lysine **(14)** is incorporated into lycopodine [24]. However, the ^{14}C-acetate feeding experiments indicated that lycopodine is *not* formed as a dimer of pelletierine, since only the C(9)–C(16) portion of the molecule is derived from this source. Yet, the lysine-feeding experiments indicated that both C(1)–C(5) and C(9)–C(13) arise from lysine. Furthermore, both 2-^{14}C-lysine and 6-^{14}C-lysine are incorporated into lycopodine to give the same pattern of activity [24]. This result implies that lysine is incorporated by way of a symmetrical intermediate, such as cadaverine **(15)**.

Spenser and MacLean also examined the possibility that C(9)–C(16) of lycopodine might be derived from pelletierine **(13)**, while the C(1)–C(8) fragment might arise from a related piperidine alkaloid derived from lysine, cadaverine, and Δ^1-piperideine. They considered 2-allylpiperidine **(17)** as a candidate for the

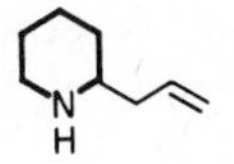

17

unknown C_8N precursor of the C(1)–C(8) portion [24]. However, feeding experiments showed that while cadaverine (15) and Δ^1-piperideine (16) are incorporated into lycopodine, 2-allylpiperidine is not [25].

The biosynthesis of cernuine (18) has also been studied [26]. As in lycopodine biosynthesis, specific incorporation of a C_5N unit derived from lysine (14) via

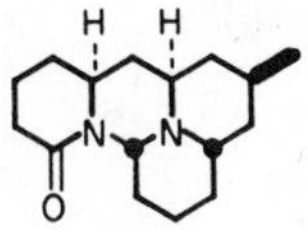

18

cadaverine (15) and Δ^1-piperideine (16) and a C_8N unit derived from pelletierine (13) was observed. Furthermore, pelletierine was shown to be an obligatory intermediate in the origin of both lycopodine and cernuine [27]. This discovery suggests that the two alkaloids are biosynthesized by related pathways, as shown in Scheme 4.

Spenser and MacLean have also tested the hypothesis that the C(1)–C(8) and C(9)–C(16) fragments of lycopodine might both arise from the piperdine-2-acetoacetate derivative 19. Decarboxylation of 19 would give pelletierine (13), which could condense with another molecule of 19 to give 20, the immediate precursor to the lycopodine-type alkaloids. Hydrolysis of the immonium salt followed by cyclization with the A-ring nitrogen would give lycopodine (1). An alternative mode of cyclization could lead to the cernuine-type alkaloids. However, piperidine-2-acetic acid is not incorporated in lycopodine biosynthesis, so some modification of this scheme is still required [28].

Braekman has put forth an appealing pathway for the biosynthesis of *Lycopodium* alkaloids [29]. The Braekman proposal is based on phlegmarine (22, only the skeletal framework of the molecule had been established), a recently

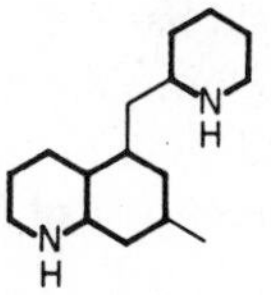

22

discovered *Lycopodium* alkaloid which could serve as an intermediate between pelletierine (13) and a variety of *Lycopodium* compounds (see Scheme 5). It was suggested that intermediate 23 could be generated from the condensation of pelletierine (13) and an acetoacetate derivative. This intermediate could cyclize

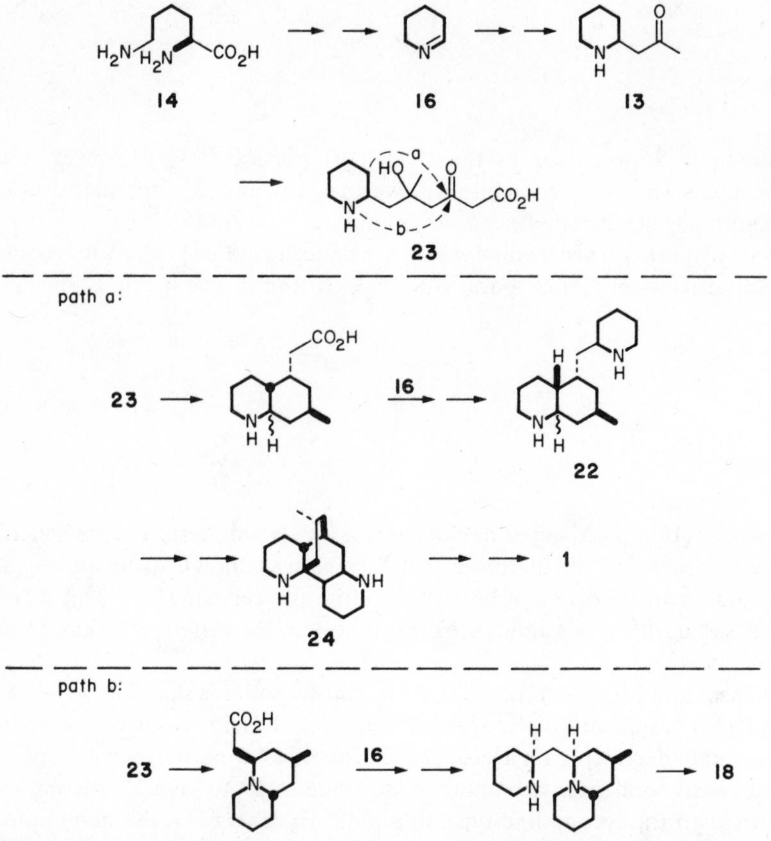

Scheme 4. Spenser–MacLean proposal for the biosynthesis of lycopodine.

by either of two pathways. Path *a* would lead to the lycodine and lycopodine-type alkaloids, via phlegmarine **22,** while path *b* would give the cernuine type alkaloids. The hypothesis is attractive, as it is in accordance with the results obtained by Spenser and MacLean [21, 28] and explains how two units of Δ^1-piperideine but only one pelletierine unit can be incorporated into these structures. The Braekman proposal has yet to be corroborated experimentally.

Several other proposals for the biosynthetic conversion of one *Lycopodium* alkaloid into another have been made. For example, lycodoline (12-hydroxyly-copodine, **25**) is the postulated precursor of fawcettimine **(26)** (Scheme 6) [19].

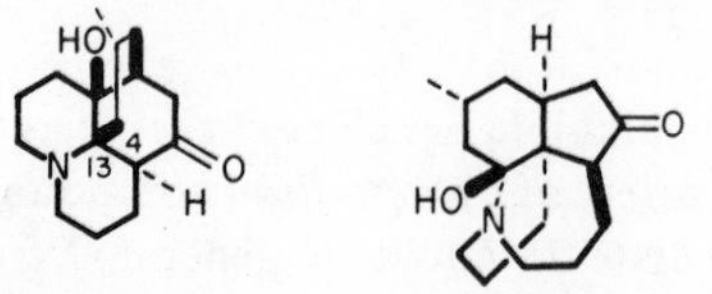

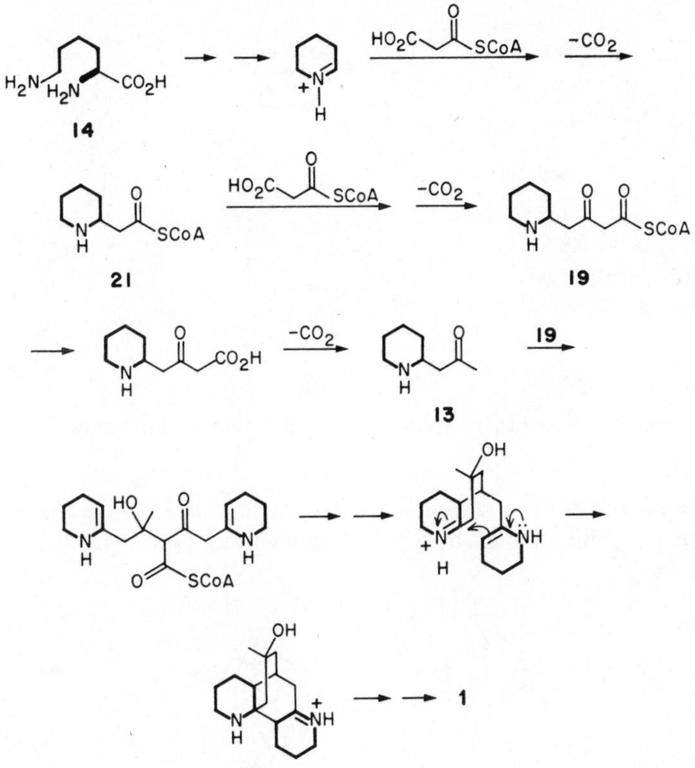

Scheme 5. Braekman proposal for the biosynthesis of lycopodine, cernuine, and lycodine.

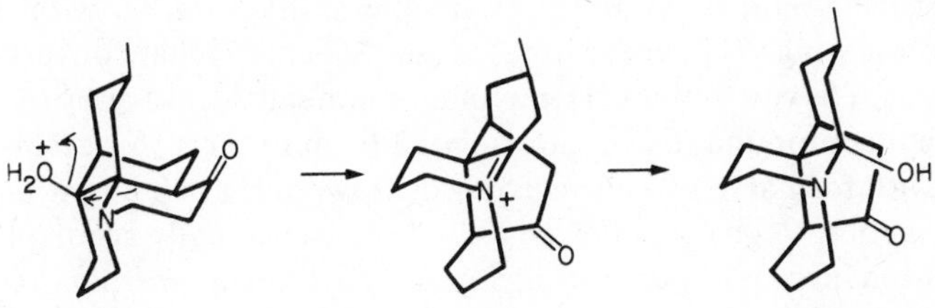

Scheme 6. Biosynthetic conversion of lycodoline into fawcettimine.

Protonation of the angular hydroxyl in lycodoline (25) followed by loss of water with concomitant migration of the C(4)–C(13) bond produces immonium salt 27. Hydration of this material gives fawcettimine (26). Although attempts to duplicate this conversion in the laboratory have been unsuccessful, Horii and co-workers have affected a similar transformation with a model system which lacks the C ring of lycodoline (compound 28) [30]. Treatment of N-protected

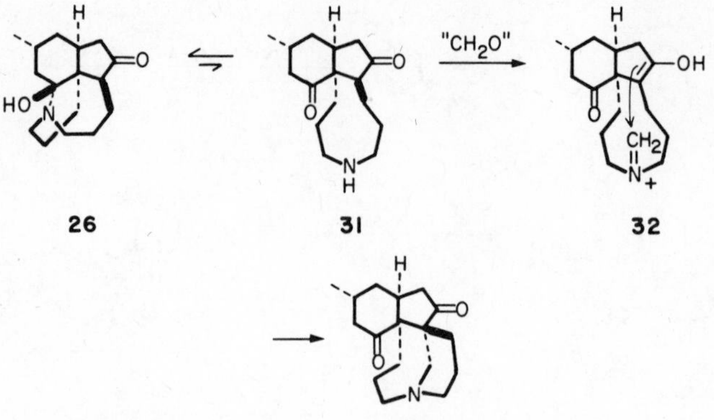

26   **31**   **32**

30

Scheme 7.   Biosynthetic conversion of fawcettimine into lycoflexine.

carbamate **28** with concentrated hydrochloric acid or 48% hydrobromic acid in acetic acid gives enamine **29** by a rearrangement analogous to that described above.

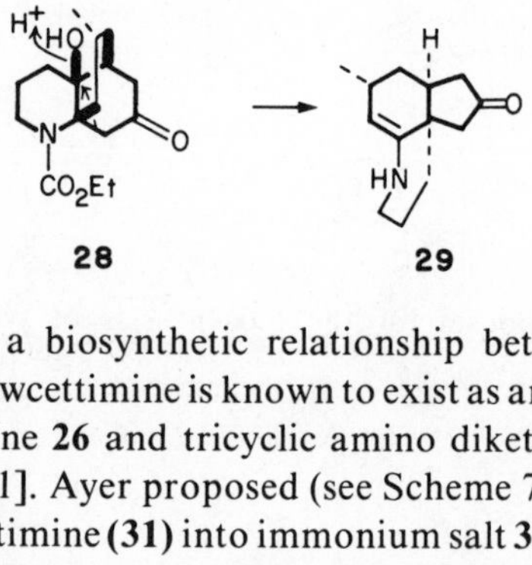

28   **29**

Ayer has suggested a biosynthetic relationship between fawcettimine and lycoflexine **(30)** [30]. Fawcettimine is known to exist as an equilibrium mixture of tetracyclic carbinolamine **26** and tricyclic amino diketone **31,** with the closed form predominating [31]. Ayer proposed (see Scheme 7) that conversion of the tricyclic form of fawcettimine **(31)** into immonium salt **32,** either by oxidation of the *N*-methyl compound (as yet not isolated from natural sources) or by condensation with formaldehyde or its equivalent, would set up an intramolecular Mannich reaction to give lycoflexine **(30)**. Ayer has actually accomplished this transformation in the laboratory using formaldehyde in hot methanol containing a trace of hydrobromic acid; under these conditions, lycoflexine is formed in approximately 50% yield.

2.   TOTAL SYNTHESIS

2.1.   Annotinine

Annotinine **(8)** was the first *Lycopodium* alkaloid to be characterized by structural and spectroscopic means [15]. The proposed structure was later confirmed by X-ray crystallographic analysis of annotinine bromohydrin **(33)** [32].

Ten years after this pioneering work, Wiesner and his colleagues published a total synthesis of annotinine [33].

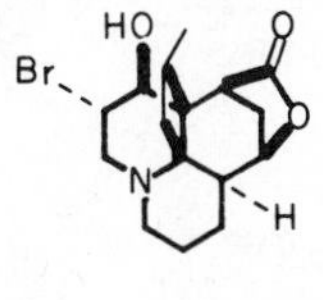

33

The synthesis begins with vinylogous amide **34** (see Scheme 8), which is readily prepared from methyl acrylate and acrylonitrile according to the method of Grob and Wilkens [34]. When compound **34** is heated with excess acrylic acid, tricyclic vinylogous imide **35** is produced in quantitative yield. Wiesner suggests that the reaction proceeds via acrylamide **36**, which is protonated on the amide oxygen to give **37**. This material may then undergo a charge-accelerated thermal cyclization to give **35**. Low yields of **35** are obtained if either acrylonitrile or ethyl acrylate is used in place of acrylic acid.

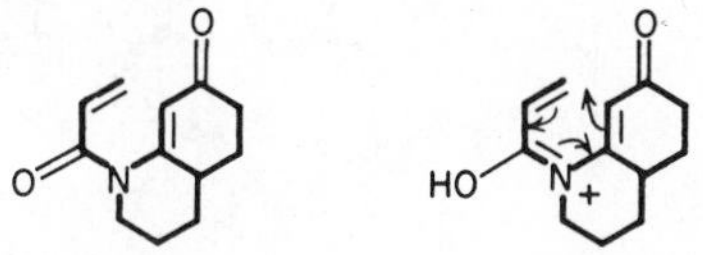

36   **37**

The four-membered D ring was assembled by 2 + 2 cycloaddition of allene to **35**. High diastereoface selectivity in this reaction is required to avoid production of B–C ring juncture diastereomers. In addition, high regioselectivity is required in order to generate the new ring with the exocyclic methylene at the desired location. Studies with the achiral model **38**, which lacks the C ring of compound **35**, showed that photoaddition with allene produces an equimolar mixture of **39** and the undesired regioisomer **40** [35]. In contrast, low-temperature photolysis

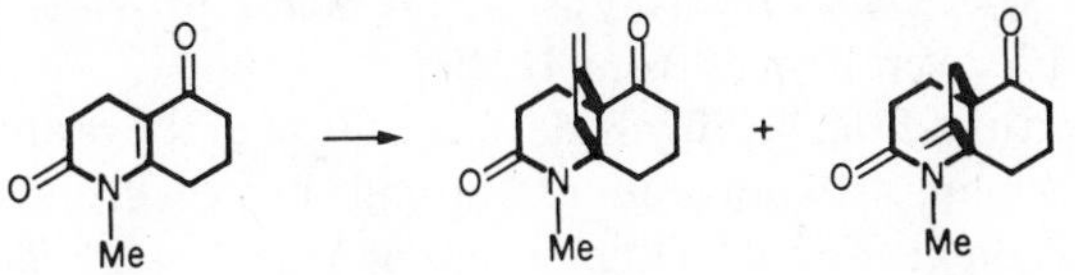

38   **39**   **40**

of **35** in the presence of a large excess of allene gives a single, crystalline adduct shown to be the desired photoadduct **41** in quantitative yield. The different behavior of **38** and **35** is apparently steric in origin, as Weisner found that analogs of **38** in which the N-methyl group is replaced by bulkier alkyl chains give predominantly photoadducts analogous to structure **39** [35].

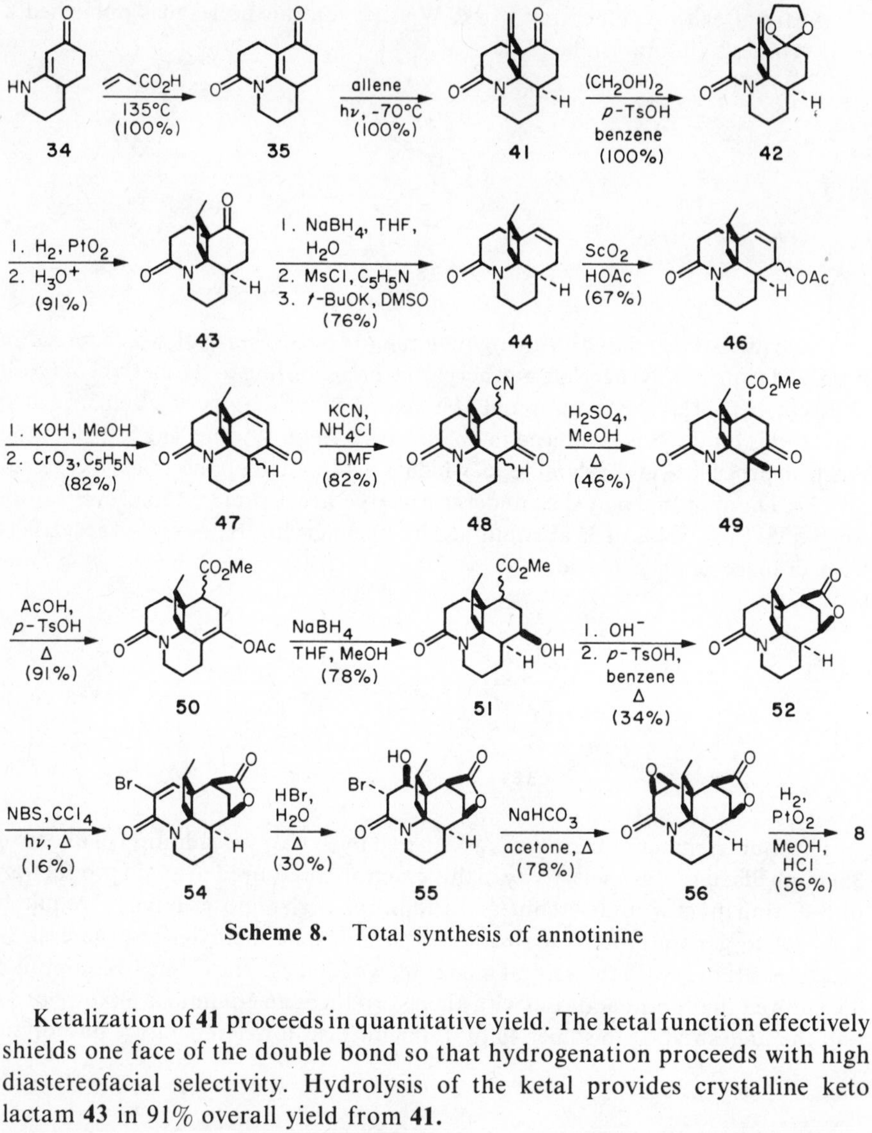

Scheme 8. Total synthesis of annotinine

Ketalization of **41** proceeds in quantitative yield. The ketal function effectively shields one face of the double bond so that hydrogenation proceeds with high diastereofacial selectivity. Hydrolysis of the ketal provides crystalline keto lactam **43** in 91% overall yield from **41**.

Reduction with sodium borohydride, mesylation of the resulting alcohol, and subsequent treatment with potassium *t*-butoxide in anhydrous DMSO gives the desired elimination product **44**. Other basic conditions resulted in the formation of substantial amounts of the rearrangement product **45** in addition to the desired olefin **44**. Functionality at C(5) was introduced by oxidation with selenium dioxide to give allylic acetate **46** in moderate yield. Hydrolysis and oxidation with CrO$_3$-pyridine provides enone **47**. It was not determined if any epimerization of the C(4) hydrogen had taken place, and its configuration was left unassigned.

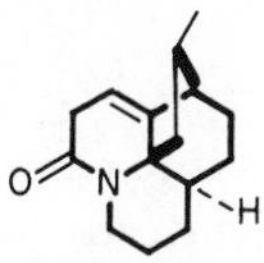

45

Hydrocyanation of 47 gives a mixture of epimeric cyano ketones 48 which was not separated but was treated with methanol and sulfuric acid to afford methyl ester 49. The spectral properties and TLC behavior of synthetic, racemic 49 were identical with optically active 49 prepared by degradation of annotinine [36], thus allowing assignment of relative configuration to the C(4) hydrogen. Compound 49 was resolved by crystallization of the brucine salts of the corresponding carboxylic acid.

The conversion of ester 49 into "annotinine lactam" (52) [37] required inversion of two asymmetric centers (α-keto and α-carboxyl hydrogens) and creation of a third with the correct relative stereochemistry. Compound 49 was converted to enol acetate 50 in high yield; the configuration of the carboxyl group was left unassigned. Reduction with sodium borohydride in a mixture of THF and methanol gives a mixture of hydroxy ester 51 and an epimeric compound assumed to be 53. The reduction process is believed to occur by

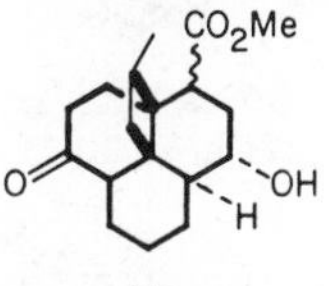

53

solvolysis of the enol actate to the corresponding enolate anion, followed by kinetically controlled protonation of the enolate (thereby providing the α-keto epimer) and subsequent reduction of the ketone carbonyl. The mixture of compounds 51 and 53 was subjected to ester hydrolysis and lactonization to give annotinine lactam (52) in 25% overall yield from enol acetate 50. The hydroxy acid corresponding to 52 does not lactonize. An alternate, less efficient conversion of 49 into annotinine lactam has also been reported [38].

Compound 52 was converted into bromo enone 54 in low yield by treatment with N-bromosuccinimide in refluxing carbon tetrachloride while irradiating with visible light. Hydration of 54 with hot aqueous hydrobromic acid produced a mixture of compounds from which bromohydrin 55 crystallized. Treatment of 55 with sodium bicarbonate in refluxing aqueous acetone affords oxoannotinine (56), which had been previously reduced with hydrogen and platinum oxide in acidic methanol to give annotinine (8) [39]. The entire synthesis requires 21 steps to convert vinylogous amide 34 into optically active annotinine and proceeds in 0.07% overall yield, not including the material lost in the resolution. The route has several interesting features, such as the photoaddition to form ring D, but is plagued by low-yielding reactions toward the end of the synthesis.

2.2.   Lycopodine

While Wiesner and his colleagues were completing their annotinine synthesis, several other research groups were investigating routes to the most common of the *Lycopodium* alkaloids—lycopodine (1). Several early studies were aimed at the preparation of isomers of the hexahydrojulolidine ring system (e.g., 57, 58, and 59). Compound 59 corresponds in stereochemistry to the ABC ring skeleton of lycopodine.

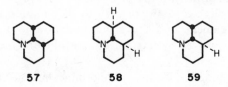

57   **58**   **59**

Mandell and co-workers [40] examined the copper chromate-catalyzed reductive cyclization of the readily available cyano ketones 60 and 61, a strategy

60: R = CN
61: R = CO$_2$Et

for the preparation of perhydroquinolines first developed by Leonard [41]. Although an effective entry into the *cis,trans*-hexahydrojulolidine system did not result from this study, it was found that hydrogenation of dicyano ketone 60 over copper chromate in acetic acid gives exclusively the *trans,trans*-isomer 58.

Wenkert also reported an efficient synthesis of isomer 58 (see Scheme 9) [42]. Selective hydrogenation of pyridinium salt 64, prepared from methyl nicotinate (62) and bromo ketal 63 [43], gives tetrahydropyridine 65 [44]. Acidic hydrolysis of the ketal and intramolecular Michael reaction leads to quinolizidine 66, the relative stereostructure of which was not determined. Claisen cyclization of the keto ester, followed by conversion of the initially formed β-diketone with methanolic HCl provides the β-methoxy-α,β-unsaturated ketone 67. Reduction of this substance with lithium aluminum hydride, and subsequent acidic workup provides enone 68. Saturation of the double bond and removal of the carbonyl group yields *trans,trans*-hexahydrojulolidine (58).

Several years later, Wenkert published further studies on the synthesis of hydrojulolidines [45]. Michael addition of methyl acetoacetate to enone 68 affords diketone 69 after hydrolysis and decarboxylation (see Scheme 10). Oxidation of this material with *m*-chloroperoxybenzoic acid gives an *N*-oxide that undergoes the Potier variant of the Polonovski reaction [46, 47] to give vinylogous amide 70 [48]. Bohlmann had earlier prepared the related ester 71 by

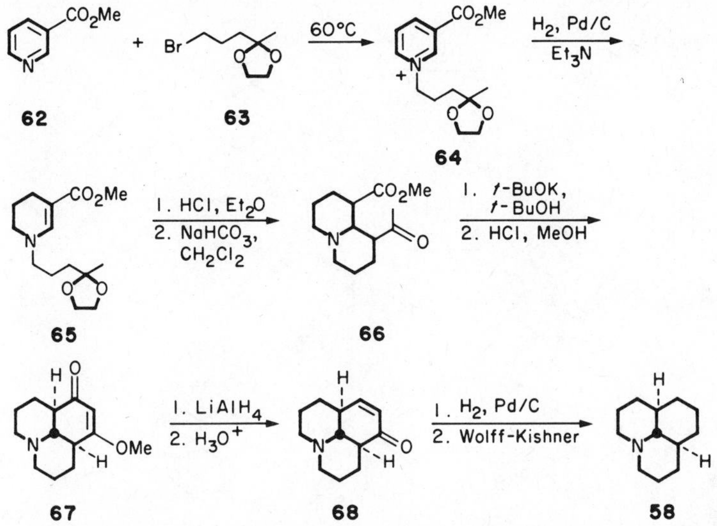

Scheme 9. Wenkert hexahydrojulolidine synthesis.

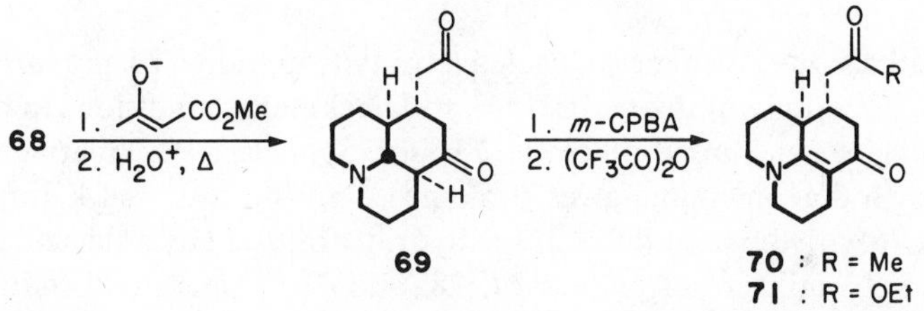

Scheme 10. Wenkert–Bohlmann approach to the lycopodine skeleton.

condensation of dehydroquinolizidine and diethyl glutaconate [49]. However, in neither system was formation of the 12-epilycopodine D ring ever achieved. In the case of acetonyl compound **70**, the failure of a D-ring formation may be due to a low population of the conformer required for cyclization. Because of the double bond, tetrahydrojulolidine **70** is constrained to a conformation in which the angular hydrogen is axial and the acetonyl group is equatorial. In order for the acetonyl group to achieve the axial position necessary for intramolecular Michael addition to the enone, it is necessary for the molecule to adopt a conformation in which all three rings are boats; consequently, tetracyclic dione **72** is not formed [50].

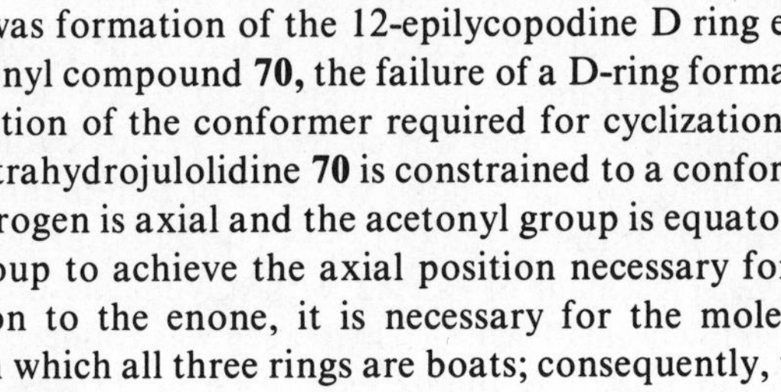

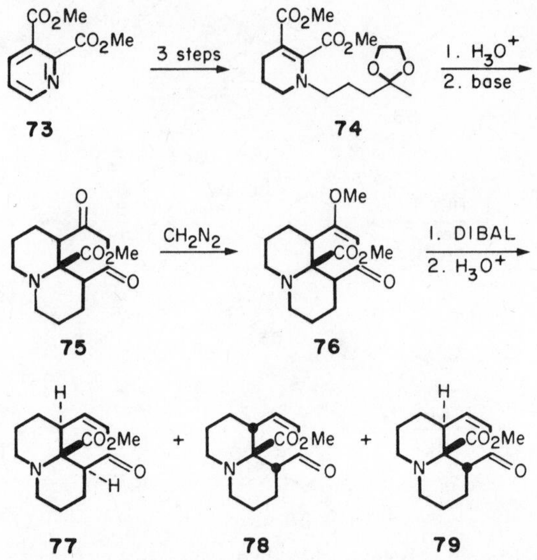

Scheme 11.   Wenkert's approach to the lydopodine skeleton.

In a related study, Wenkert subjected tetrahydropyridine **74,** prepared in three steps from dimethyl quinolate **(73)** [46], to Dieckmann conditions to obtain the symmetrical, *trans,trans*-β-diketone **75** (see Scheme 11). Treatment of this material with diazomethane gives β-methoxy-α,β-unsaturated ketone **76.** Reduction of **76** with diisobutylaluminum hydride followed by mild acid treatment gives a 1:5:13 mixture of enones **77, 78,** and **79.** The major product of this conversion **(79)** has the correct ring juncture relative stereochemistry for elaboration into *Lycopodium* alkaloids. No further work along this line has been reported.

Wiesner and co-workers accomplished the first synthesis of the complete lycopodine skeleton (see Scheme 12) [51]. The julolidine phenol **80,** prepared from *m*-anisidine and 1-bromo-3-chloropropane in 30% yield, was reduced to obtain vinylogous amide **81,** which was converted to vinylogous imidate **82** by *O*-alkylation with methyl iodide. Unlike Wenkert or Bohlmann, Wiesner chose to attach the D ring by forming the C(13)—C(14) bond prior to forming the C(7)—C(8) linkage. Treatment of salt **82** with allylmagnesium bromide gave amino ketone **83** in low yield. The C(12) hydrogen was epimerized by a series of equilibrations on alumina. The final closure of ring D was achieved by an HBr-induced Prins reaction, which resulted in complete conversion to a 54:46 mixture of isomeric olefins **84.** The bromine was removed by reduction with sodium amalgam and the double bond saturated to obtain **85,** a tetracyclic model of the lycopodine skeleton.

An attempt to extend this promising method to the synthesis of lycopodine itself was thwarted by an unexpected 1,6-hydride shift in the Prins reaction (see Scheme 13). Vinylogous amide **87,** prepared in two steps from dihydroorcinol

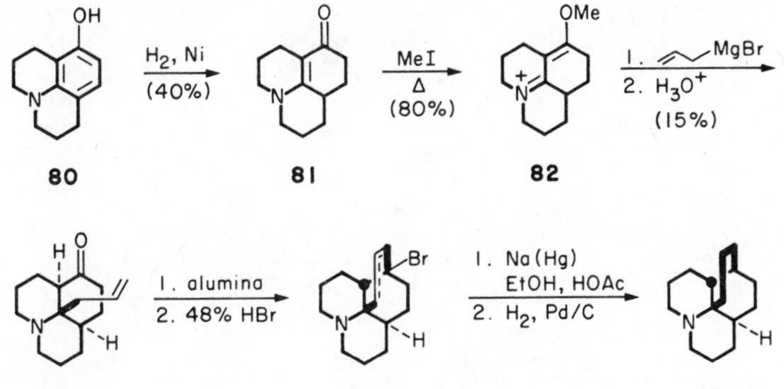

Scheme 12. Wiesner's synthesis of the lycopodine skeleton.

(86), was converted to enol ether **88** as in the model study. In the subsequent step, it was attempted to form ring C (rather than ring D, as in the model study) by the intramolecular Prins reaction. Exposure of enol ether **88** to strong acid does generate the desired tricyclic cation **89**. However, this material undergoes a hydride shift to give the more stable tertiary carbocation **90**, which deprotonates to give a mixture of double bond isomers **91**. This rearrangement leaves no functionality in ring B for introduction of the C(5) carbonyl, although its occurrence does confirm the assumed stereochemistry of the Grignard addition.

Colvin and co-workers have investigated a rather different approach in which a preformed bicyclo [3.3.1]nonane derivative serves as the progenitor of rings C and D of lycopodine (see Scheme 14) [52]. Compound **92** [53] is converted in four steps into benzyl carbamate **93**. The carbamate is cleaved by HBr in acetic acid to give an amino ketone, which is isolated as the hydrobromide salt **(94).** Treatment of **94** with pyruvic acid in the presence of phosphorus oxychloride and triethylamine provides the corresponding pyruvamide, which is cyclized with

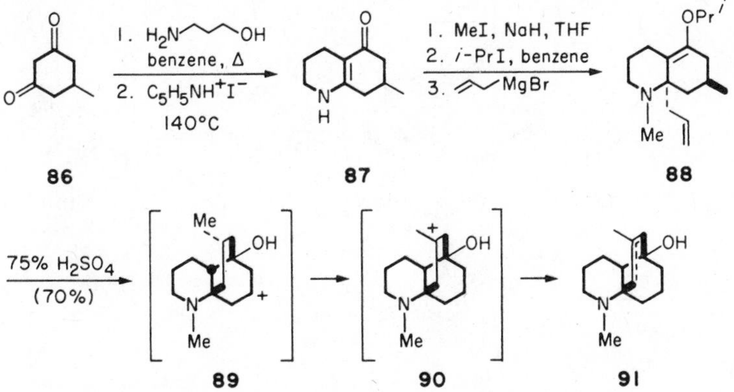

Scheme 13. Wiesner's approach to the synthesis of lycopodine.

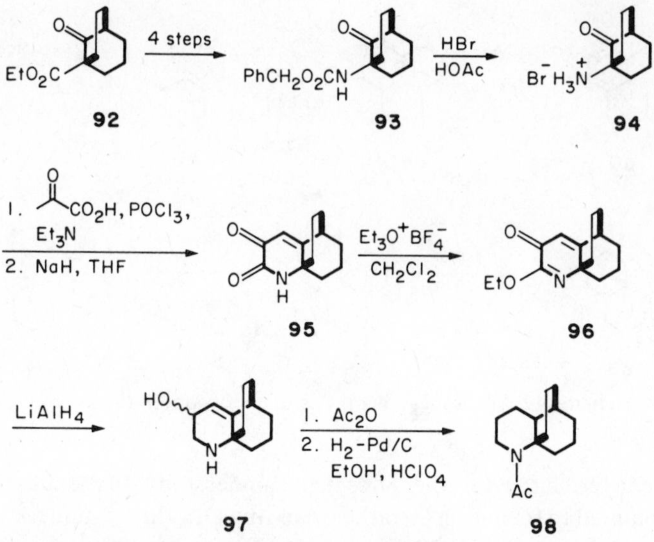

Scheme 14. Colvin's approach to the lycopodines.

sodium hydride to obtain enone lactam **95**. Alkylation with Meerwein's salt gives **96**, which is reduced to obtain allylic alcohol **97**. Exhaustive acetylation, followed by hydrogenolysis of the allylic acetate and reduction of the double bond gives acetamide **98**, a model for the ABD ring system of lycopodine. However, this study does not really address the establishment of configuration at C(12), as amide **98** contains only one asymmetric carbon. That is, the two enantiomers of **98** correspond to lycopodine and 12-epilycopodine.

In a related study, Horii and co-workers prepared amino dienone **103** from substituted bicyclononene **99** by a sequence of reactions essentially identical to that employed in Colvin's approach (see Scheme 15) [54]. Although compound **103** has appropriate functionality for elaboration into the ABD ring system of lycopodine, no further work on this synthesis has been reported.

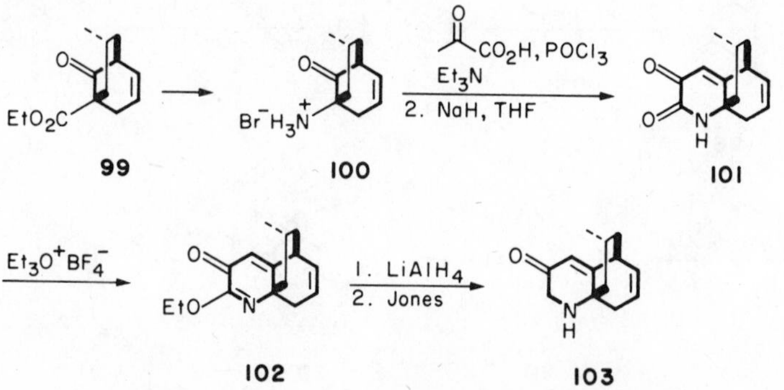

Scheme 15. Horii's approach to lycopodine.

The total synthesis of lycopodine (1) presents a significant challenge to synthesis. One of the more interesting features of its structure is the fact that a single carbon atom, C(13), is common to all four rings. A number of successful and unsuccessful routes have been investigated in order to assemble the tetracyclic framework around this central carbon. Five general strategies that have evolved for the synthesis of lycopodine or of the basic lycopodine skeleton are depicted in Figure 3 [50]. The work of Colvin and of Horii, just discussed, are examples of the second mode of ring construction, in which the heterocyclic A and B rings are built onto the preexisting BD carbocyclic ring system. The most popular synthetic strategy utilizes hydrojulolidine derivatives onto which the D ring is added. Examples of this approach include the work of Bohlmann, Wenkert, and the Prins reaction methodology of Wiesner. The only successful total synthesis of lycopodine (1) which employs this general strategy was reported by Ayer in 1968 (see Scheme 16) [55].

As in Wiesner's route, Ayer constructed the C(13)—C(14) bond of ring D by nucleophilic addition of an organometallic reagent to an immonium salt (105) derived from a functionalized julolidine. However, because Ayer and co-workers used a chiral Grignard reagent, a diastereomeric mixture resulted (106), thus necessitating an isomer separation later in the synthesis. The *cis,cis*-hydro-julolidine 106 was epimerized at C(4) by a four-step sequence involving enone 107 as an intermediate. Methyl ether 108 was cleaved with boron tribromide, and the resulting mixture of diastereomeric alcohols was separated chromatographically. Alcohol 109 was protected as the acetate, and the resulting product was oxidized to obtain lactam 110 [56]. The latter transformation was necessary in order to reduce the nitrogen basicity so as to avoid the formation of quaternary salts when the primary acetate was converted to mesylate 111. An interesting selectivity is

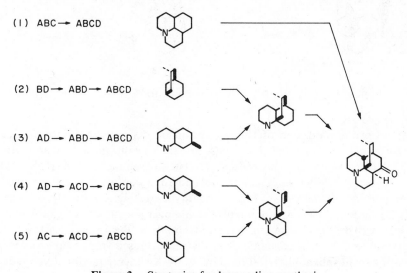

(1) ABC → ABCD

(2) BD → ABD → ABCD

(3) AD → ABD → ABCD

(4) AD → ACD → ABCD

(5) AC → ACD → ABCD

Figure 3. Strategies for lycopodine synthesis.

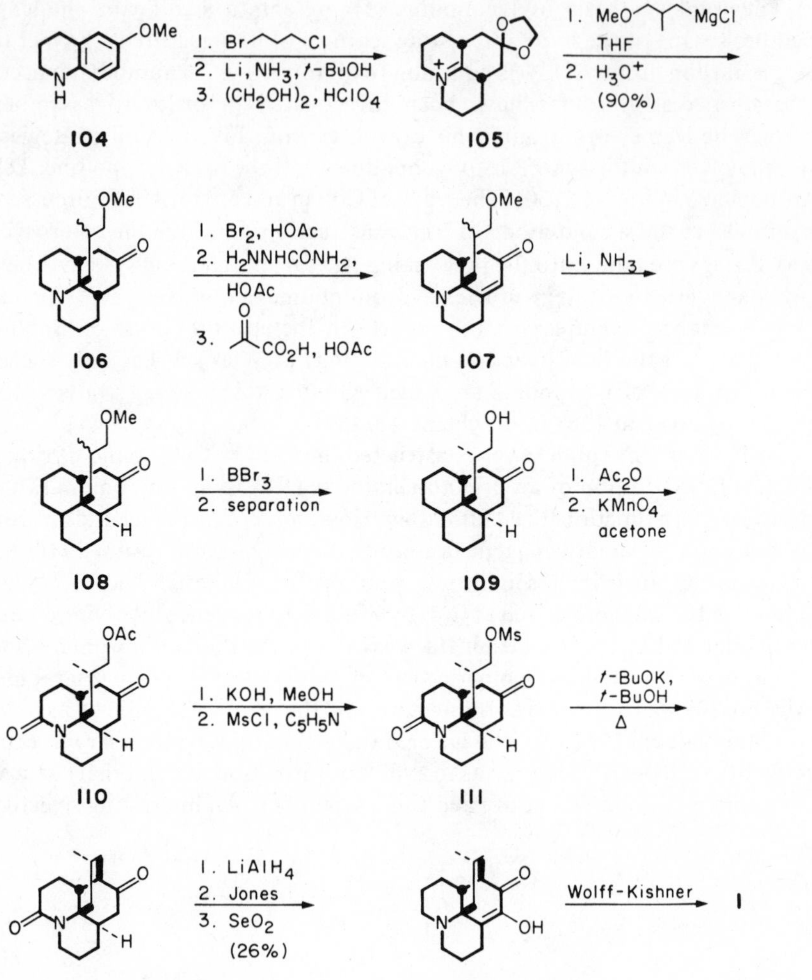

Scheme 16. Ayer synthesis of lycopodine.

observed in the oxidation of **109**; oxidation occurs only at the future C(9). Treatment of keto mesylate **111** with base brings about intramolecular alkylation at C(7) to give the tetracyclic product **112**. The C(4) epimerization (**106 → 108**) was necessary in order to place the molecule in a conformation amenable to intramolecular alkylation (see Fig. 4). Again, interesting selectivity is obtained, since alkylation at C(7) predominates heavily over alkylation at C(5). A minor unidentified product from this reaction could be the isomer arising from alkylation at C(5). The carbonyl group was transposed from C(6) to C(5) by oxidation to diosphenol **113**, followed by Wolff–Kishner reduction. Lycopodine

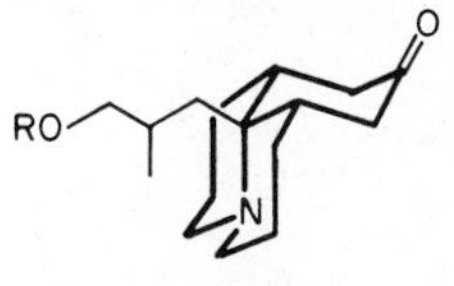

106 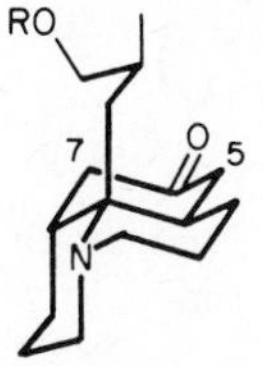   108

Figure 4. Predominant conformations of compounds **106** and **108**.

(1) and anhydrodihydrolycopodine (**114**), another natural product, were produced in yields of 26 and 40%, respectively. The Ayer synthesis gives lycopodine in 18 steps and 0.07% overall yield from compound **104**.

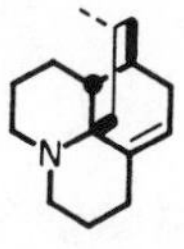

114

Stork's synthesis of lycopodine is an example of the third mode of ring assembly (see Scheme 17) [57]. Symmetrical diketone **116**, readily prepared from *m*-methoxybenzaldehyde (**115**), is converted into enone **117**. Conjugate addition of methylmagnesium iodide occurs exclusively *trans* to the *p*-methoxybenzyl side chain. The pyrrolidine enamine of **118** is heated with acrylamide to give a separable mixture of hexahydroquinolone **119** and its isomer **126**. The mixture of **119** and **126** results from a lack of selectivity in enamine formation; the desired quinolone **119** may be isolated in 20–25% yield.

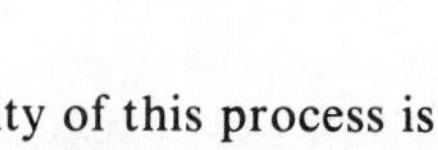

126

Although the regioselectivity of this process is poor, compound **119** can be made in large quantity because of the ready availability of the starting materials and the ease with which these reactions may be adapted for large-scale operation. Furthermore, the regioselectivity problem was later circumvented by an alternate route to intermediate **119** (see Scheme 18), in which the order of introduction of the methyl and methoxybenzyl side chains is reversed [58]. Conjugate addition of *m*-methoxybenzylmagnesium bromide to 5-methylcyclo-

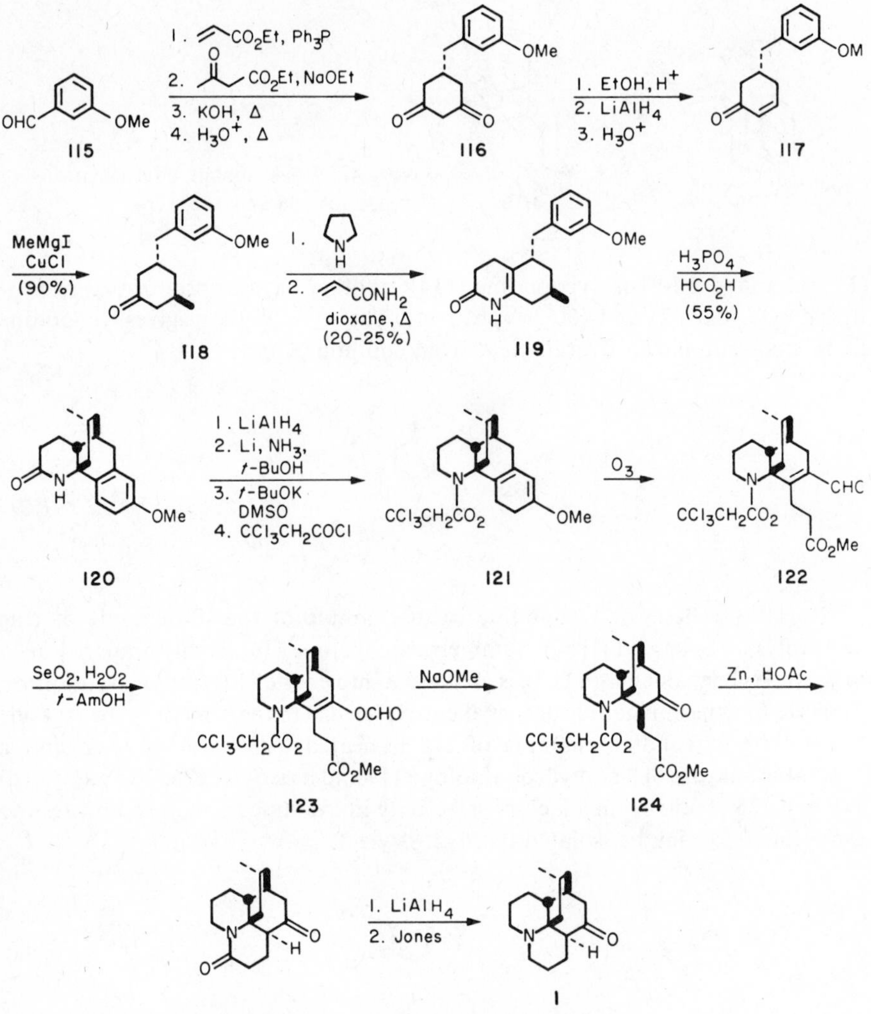

Scheme 17. Stork's lycopodine synthesis.

hexenone results in generation of the specific enolate **127**. The normally unreactive magnesium enolate is alkylated directly and without loss of structural specificity by the addition of a polar co-solvent (HMPT) and a reactive alkylating agent (allyl bromide). Functionalization of the allyl side chain and treatment with ammonia gives hexahydroquinolone **119**.

Cyclization to form ring B was accomplished by an intramolecular Friedel–Crafts reaction. Two epimeric immonium ions (**128** and **129**) may result from the reversible protonation of **119** (see Fig. 5). However, it was reasoned that the rate of cyclization of **128** would be markedly higher than that of **129**. In the former

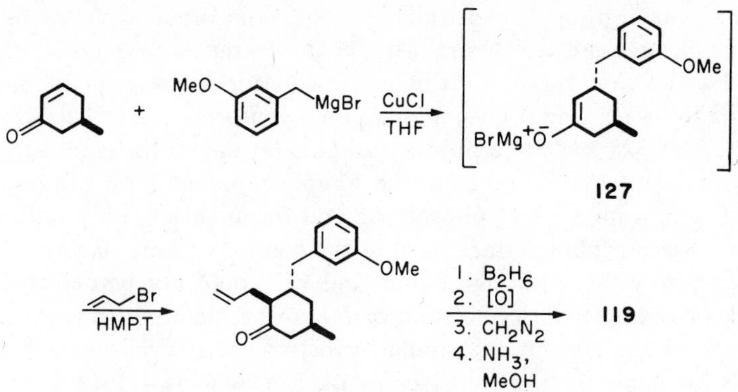

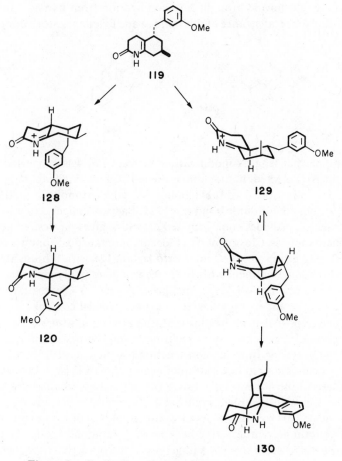

Scheme 18. Adaptation of the Stork lycopodine synthesis.

Figure 5. Cyclization conformations of quinolone **119**.

cation the benzyl group is constrained to the axial orientation necessary for cyclization through a relatively strain-free chairlike transition state. In the case of cation **129,** the side chain is held in the equatorial position; in order for cyclization to occur, ring B must adopt a boat conformation. Cyclization from enamide cation **128** leads to a product suitable for the synthesis of lycopodine, while cyclization from **129** leads to the 12-epilycopodine series. Indeed, upon treatment with a mixture of phosphoric and formic acids, only one product arising from electrophilic attack *para* to the methoxy group, lactam **120,** was formed in 55% yield. Although compound **130** could not be detected, some by-products arising from cyclization *ortho* to the methoxy substituent were isolated. Stork points out the similarity between this cyclization and early proposals for the biosynthesis of lycopodine (see Scheme 4) [58].

It was expected that Birch reduction of amine **131** would provide enone **132,** which could be readily ozonized to give keto ester **124.** Unfortunately, only the corresponding α,β-unsaturated ketone was obtained from this reduction; the double bond could not be brought into conjugation from its tetrasubstituted position. Using the recalcitrance of this bond to an advantage, Stork designed the

131   132

following somewhat lengthy modification. Lactam **120** was converted in four steps into the *N*-protected, homoannular diene **121,** resulting from migration of the other, less substituted, double bond. Selective ozonolysis of the more electron-rich enol ether double bond gives **122.** Baeyer–Villiger oxidation of the aldehyde function provides enol formate **123,** which gives the desired keto ester **124** upon methanolysis. Closure of ring C occurs spontaneously upon removal of the *N*-protecting group. Reduction of keto lactam **125** with lithium aluminum hydride followed by treatment with Jones reagent gives (±)-lycopodine **(1).** The Stork synthesis requires 18 steps and proceeds in 1.3% overall yield. The synthesis is notable for its high degree of stereochemical control.

Wiesner has reported an interesting synthesis of a stereoisomer of lycopodine, 12-epilycopodine [59]. As with Stork's approach, this synthesis employs the third general mode of lycopodine ring construction (see Scheme 19). Lactam **134** is prepared by acid-catalyzed cyclization of cyano dione **133** [60]. Low-temperature photolysis of the *N*-benzyl derivative **135** in the presence of allene provides cyclobutane **136** with complete regio- and diastereoface selectivity, albeit in rather low yield. This step is analogous to the formation of compound **41** in the successful annotinine synthesis (see Scheme 8). Compound **136** is protected as the ethylene ketal to avoid Baeyer–Villiger reaction during the subsequent olefin

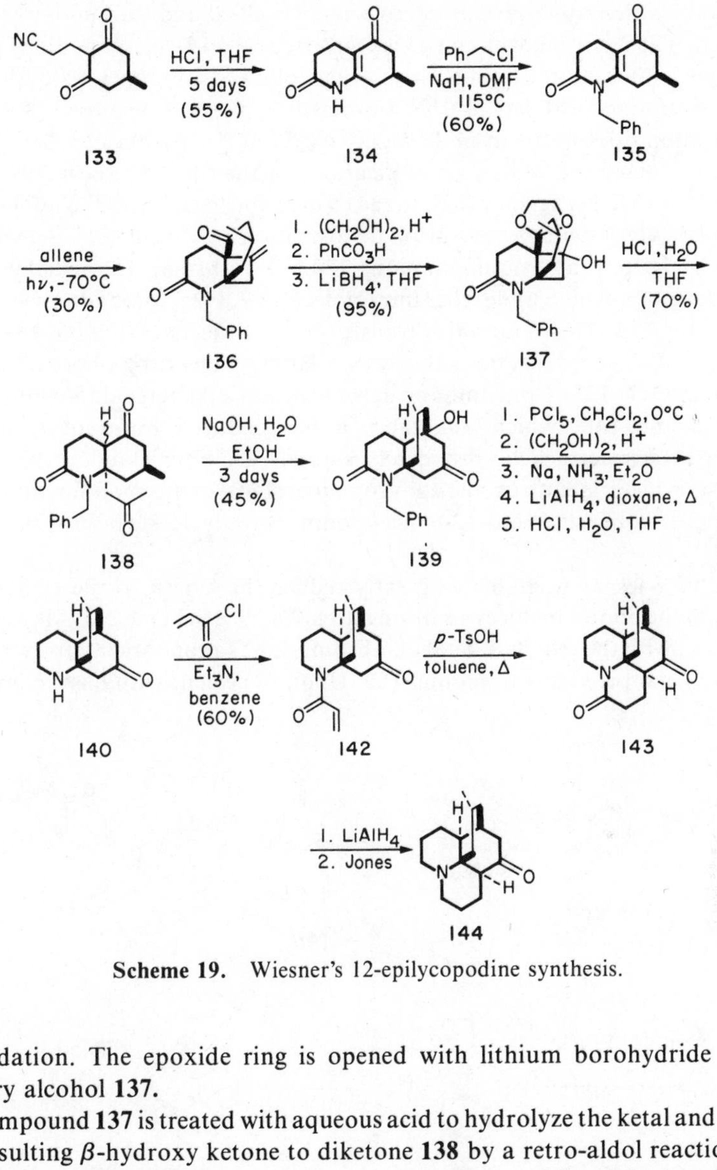

Scheme 19.   Wiesner's 12-epilycopodine synthesis.

epoxidation. The epoxide ring is opened with lithium borohydride to give tertiary alcohol **137**.

Compound **137** is treated with aqueous acid to hydrolyze the ketal and convert the resulting β-hydroxy ketone to diketone **138** by a retro-aldol reaction. This material is treated with dilute base to close ring B, providing β-hydroxy ketone **139**. The resulting product has the 12-epi configuration, presumably because the intermediate perhydroquinolone **138** adopts the *cis*-configuration under the basic conditions of the intramolecular aldol reaction.

The synthesis of 12-epilycopodine is completed with fairly standard reactions. β-Hydroxy ketone **139** is converted in five steps to amino ketone **140**. Ring C is added by acylation of the secondary amine with acryloyl chloride to obtain

amide **142,** which undergoes intramolecular Michael addition upon treatment
with acid. The resulting lactam **143** is converted into 12-epilycopodine **(144)** by
reduction with lithium aluminum hydride followed by Jones oxidation. Although compound **144** has not been isolated from natural sources, the 12-epi
configuration does occur in more highly oxidized lycopodines [61].

Wiesner later reported two modifications of the foregoing synthesis which
improve its overall efficiency [62]. In the refined approach, ring C is added to the
preformed AD precursor **134** prior to formation to ring B (see Scheme 20).
Alkylation of tetrahydroquinolindione **134** with 6-bromo-1,2-hexadiene provides allene **145,** which undergoes intramolecular 2 + 2 photoaddition to afford
cyclobutane **146.** This material is transformed into tertiary alcohol **147** by the
identical sequence of reactions that was used for converting photoadduct **136**
into compound **137.** Concomitant deketalization and retroaldolization gives
diketo lactam **148,** which undergoes a base-induced intramolecular aldol
reaction to give tetracyclic β-hydroxy ketone **149.** Four standard operations
convert this material into (±)-12-epilycopodine **(144).** In this modification, cyano
dione **133** is converted into 12-epilycopodine in only 12 steps in 4.3% overall
yield.

Finally, Wiesner was able to greatly reduce the length of the synthesis by
bypassing the photochemical addition altogether (see Scheme 21). Alkylation of
lactam **134** with the ethylene ketal of 6-bromo-2-hexanone and hydrolysis of the
resulting ketal provides compound **150.** Upon treatment with base, compound

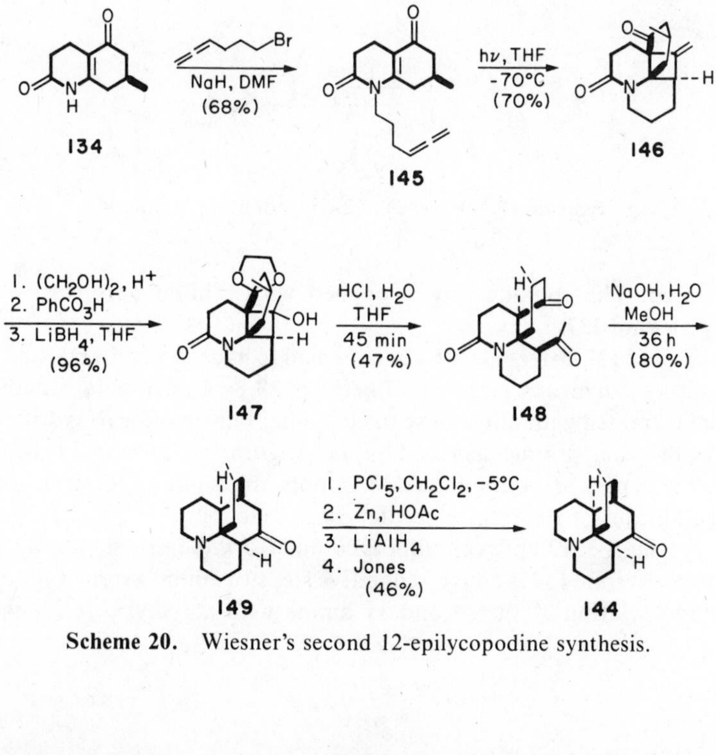

Scheme 20.   Wiesner's second 12-epilycopodine synthesis.

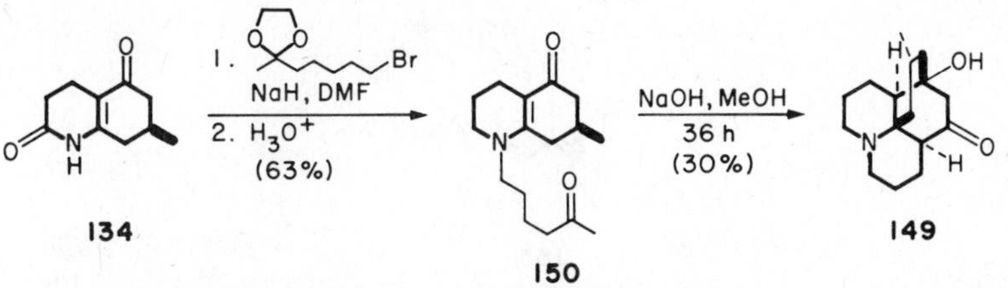

Scheme 21.   Wiesner's third 12-epilycopodine synthesis.

150 undergoes a remarkable transformation involving sequential intramolecular Michael and aldol additions to provide tetracyclic β-hydroxy ketone **149**. Thus, the seven-step sequence required for the conversion of **134** into aldol adduct **149** is condensed to only two reactions. The overall yield from cyano dione **133** to 12-epilycopodine is only slightly improved (4.8%), but the entire synthesis requires only eight steps from readily available starting materials.

Lycopodine (**1**) is an α-amino ketone and can therefore be envisioned as a Mannich reaction product [63]. Three research groups have examined the application of this reaction to the lycopodine problem.

The first attempt to utilize a Mannich reaction for the synthesis of lycopodine was an unsuccessful effort from the University of Wisconsin [64]. It is the only example of the fifth mode of ring assembly (see Scheme 22). Imino ether **152** is prepared by C-alkylation of **152** with the ethylene ketal of 4-iodo-2-butanone. Reaction of **152** with Nazarov's reagent [65] in the presence of a catalytic amount of p-toluenesulfonic acid gives **153**. Unfortunately, all attempts to induce a Mannich-type cyclization of this material failed. It is suggested [64] that the vinylogous amide **153** is simply too stable, resulting in a high activation barrier for ring closure. From another perspective, the immonium ion **155** is likely to be rather acidic, so that the equilibrium concentration of this obligatory intermediate is probably low. Attempts to cyclize the ketone obtained by hydrolysis of

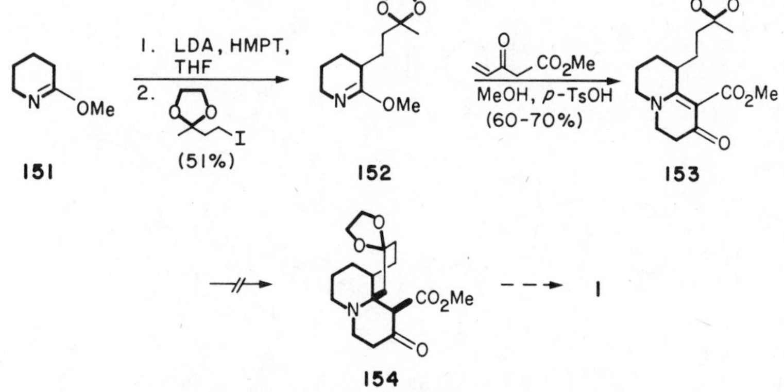

Scheme 22.   Kunz's approach to lycopodine.

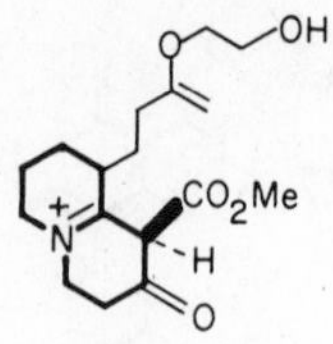

155

153 by base-catalyzed intramolecular Michael reaction were also unsuccessful. The latter failure is somewhat surprising, in light of the successful Michael closure of **150** (see Scheme 21). However, in that case, an unfavorable equilibrium might be overcome by the ensuing aldol cyclization.

Heathcock, Kleinman, and Binkley successfully employed the intramolecular Mannich reaction for the synthesis of lycopodine [66, 67]. Retrosynthetic analysis reveals that the direct Mannich precursor to lycopodine would be the bicyclic amino diketone **156** (see Fig. 6). However, the 12-membered ring of compound **156** would pose a considerable synthetic challenge. The plan is simplified by postponing ring-C closure until after the Mannich step. If one realizes that ring C could be readily closed from a tricyclic intermediate such as **157** or **158,** the synthesis plan reduces to the intramolecular Mannich condensations of amino diketones **159** and **160,** respectively. Both of these possibilities have been examined and successfully applied to the synthesis of lycopodine.

Cyano dione **133** [61] is converted via chloro enone **161** into cyano enone **162** [68] by reactions readily adaptable to large-scale procedures (see Scheme 23). Three methods were developed for the attachment of the acetonyl side chain. The most efficient of these methods is the Sakurai reaction [69, 70], whereby a methallyl group is stereoselectively introduced *trans* to the C(5) methyl group; subsequent ozonolysis gives **163.** After protection of the two carbonyl groups, the cyanoethyl side chain is refunctionalized into an appendage that can later be

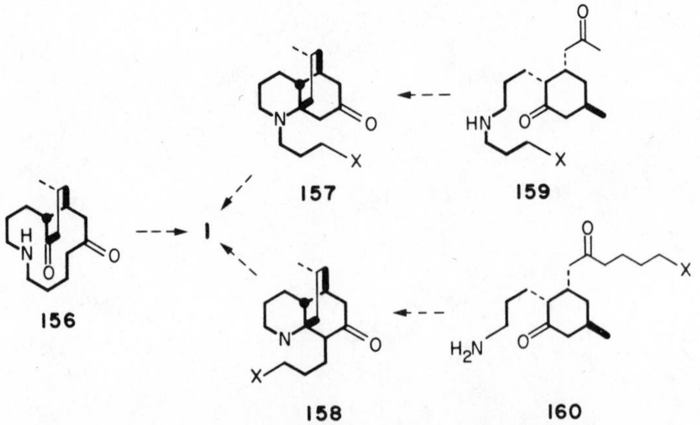

Figure 6.   Mannich-based retrosynthetic analysis of lycopodine.

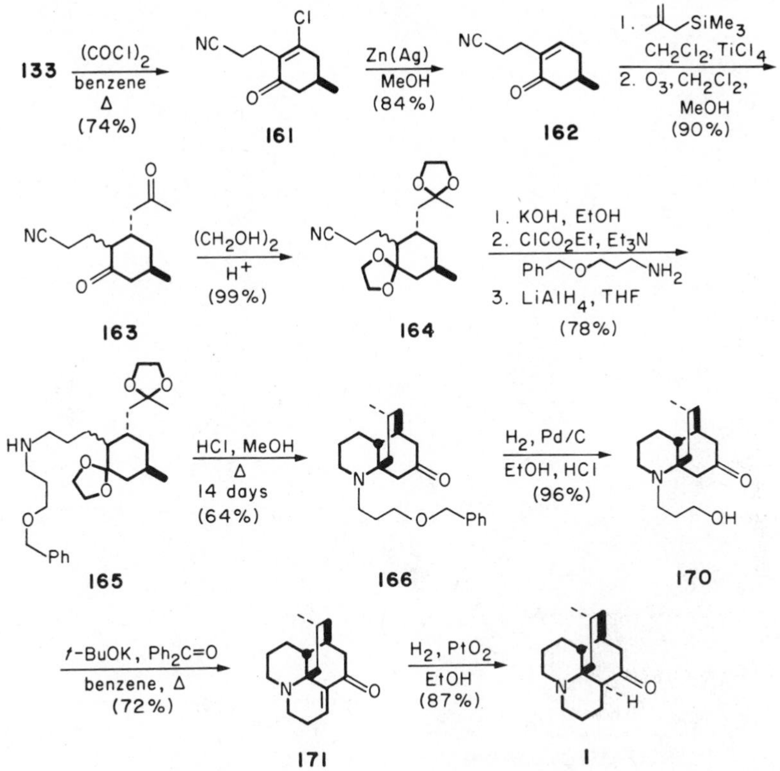

Scheme 23. First Berkeley lycopodine synthesis.

used to form ring C. Alkaline hydrolysis of nitrile **164** provides an acid which is condensed with 3-benzyloxypropylamine. The resulting amide is reduced with lithium aluminum hydride to give amino diketal **165,** the crucial Mannich reaction precursor.

As **165** is a mixture of C(2) epimers, two diastereomeric immonium ions (**166a** and **166b**) are produced upon ketal hydrolysis (see Fig. 7). As in the Stork synthesis, cation **166a** is suitably oriented for cyclization to the desired product **167,** while **166b** is not. Treatment of amino diketal **165** with acidic methanol provides a single tricyclic amino ketone **166** in 64% yield. None of the corresponding bridgehead epimer was detected in the product of this reaction.

Initial attempts to convert Mannich product **166** into lycopodine (1) by intramolecular alkylation of bromide **169** were unsuccessful, apparently due to

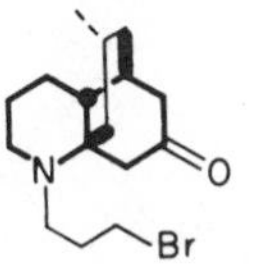

169

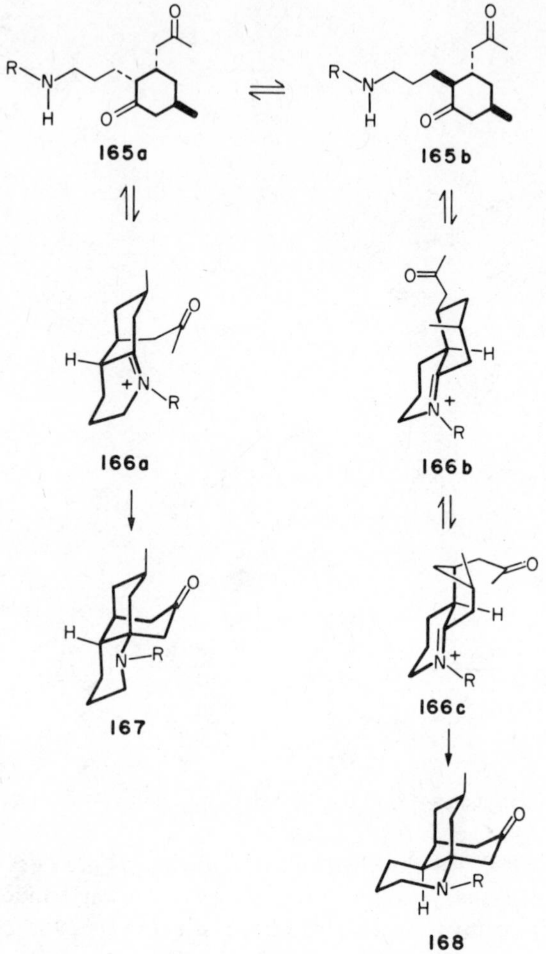

Figure 7. Cyclization conformations in the intramolecular Mannich condensation of **165**.

poor orbital overlap between the enolate π-system and the carbon–bromine bond [66, 67]. This problem was circumvented by using an intramolecular aldol condensation to close the final C(3)—C(4) bond. Hydrogenolysis of the benzyl ether affords alcohol **170**. Oppenauer oxidation [71] with subsequent aldolization and dehydration gives enone **171**, which is hydrogenated to give lycopodine **(1)**. The synthesis proceeds in high overall yield (16%) and requires only 12 steps for the conversion of cyano dione **133** into (±)-lycopodine **(1)**. Compared to previous lycopodine syntheses, the route is distinguished by an efficient use of existing functionality in the retrosynthetic design.

In a later paper, Heathcock and Kleinman reported a more convergent refinement of the synthesis [72]. In this modification, the elements of ring C are

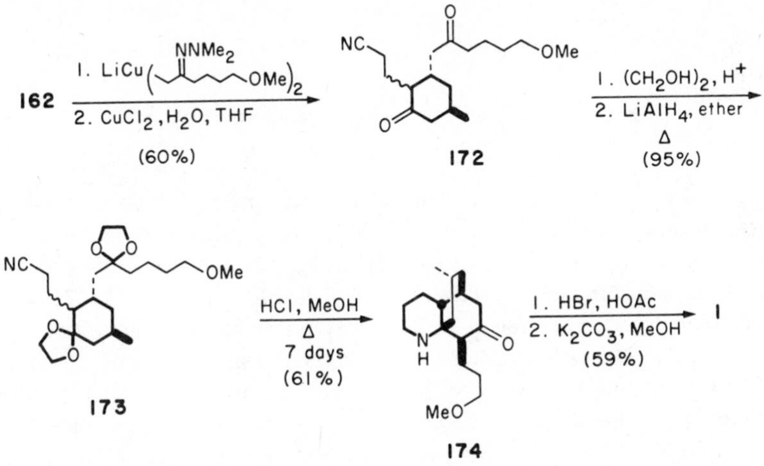

Scheme 24. Second Berkeley lycopodine synthesis.

attached to the acetonyl side chain rather than to the amino side chain, thus leaving formation of the C(1)—N bond as the last step (see Scheme 24). The cuprate reagent derived from the lithium enolate of the N,N-dimethylhydrazone of 6-methoxyhexan-2-one is added to cyano enone **162** to give cyano dione **172**, after hydrolysis of the hydrazone. All the carbons and the nitrogen of lycopodine are assembled in this reaction. Protection of the carbonyls, followed by reduction with lithium aluminum hydride gives amino diketal **173**. The intramolecular Mannich reaction provides tricyclic amino ketone **174**. Treatment of **174** with HBr in acetic acid cleaves the methyl ether to a primary bromide which, after basic workup, spontaneously cyclizes to lycopodine. This modified synthesis gives the racemic natural product in only seven operations from cyano dione **133**, albeit in lower overall yield (13%) than in the original snythesis. As in the first synthesis, no isomer separations are required.

The most recent synthesis of lycopodine, reported by Schumann and co-workers in Berlin [73] takes an approach similar to that used in the original Berkeley synthesis [66, 67]. Exhaustive lithium aluminum hydride reduction of cyano dione **133** provides the unsaturated amino alcohol **175** (see Scheme 25).

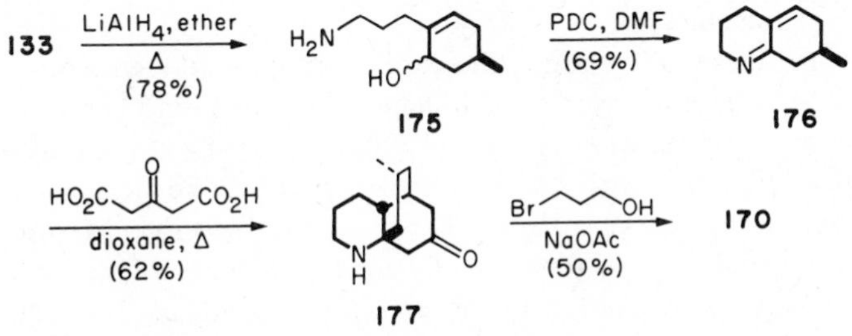

Scheme 25. Schumann synthesis of lycopodine.

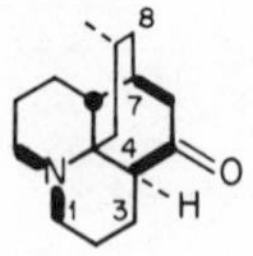

Figure 8.   Strategic bond analysis of lycopodine.

Oxidation with pyridinium dichromate gives an enone which spontaneously cyclizes to unsaturated imine **176**. When compound **176** is treated with acetonedicarboxylic acid in refluxing dioxane, 1,4- addition followed by ring closure takes place to give tricyclic amino ketone **177**. Compound **177** is the only stereoisomer produced in this reaction, presumably because the initial Michael addition to the α,β-unsaturated imine places the acetonyl side chain in the axial configuration. Alkylation with 3-bromopropanol affords alcohol **170** and completes a formal total synthesis of lycopodine **(1)**, as this compound had previously been converted into the natural product by the Heathcock group [66, 67]. This synthesis provides the shortest route (six steps) from cyano dione **133** to lycopodine **(1)**, although the overall yield (10%) is slightly lower than those achieved in the Berkeley syntheses.

As a final comment on these synthetic endeavors, it is interesting to note the application of Corey's strategic bond analysis to lycopodine [74]. The disconnections that result from an application of the Corey heuristics are shown as heavy lines in Figure 8. Of the five successful approaches to lycopodine, only two have employed one of the final ring closures suggested by this analysis; Stork's synthesis and the second Berkeley route form the C(1)—N bond last. In contrast, the first Berkeley synthesis and Schumann's synthesis form the C(3)—C(4) bond last, and Ayer closes the C(7)—C(8) bond to complete the lycopodine framework. Three of the routes suggested by Corey have yet to be successfully applied. Thus, the lycopodine problem may still bear fruitful investigation.

2.3.   Lycodine

Lycodine **(7)** is the simplest and most common example of the dinitrogen lycodine skeletal class. The only total synthesis yet reported for this compound was accomplished by Heathcock and Kleinman [67, 72]. As in their previous lycopodine syntheses, the critical step involves an intramolecular Mannich reaction (see Scheme 26).

For the synthesis of lycodine, the cuprate reagent derived from the lithium enolate of *N, N*-dimethylhydrazone of hept-6-en-2-one is added to cyano enone **162** to give adduct **178**, after hydrolysis of the hydrazone moiety. By a sequence of operations identical with that employed in the second Berkeley lycopodine synthesis (see Scheme 24), diketone **178** is transformed into Mannich product **179**. As usual, only one stereoisomer is produced in the Mannich cyclization. Ozonolysis of the vinyl group in the presence of acid, used to protonate the nitrogen and prevent amine oxidation, gives a solution presumably containing

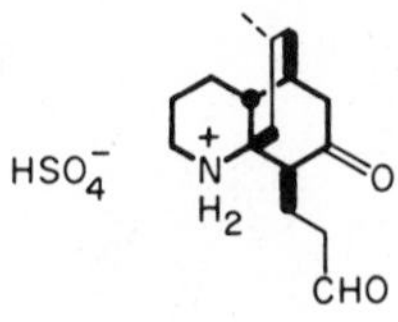

180

aldehyde **180**. Excess hydroxylamine hydrochloride is then added to form the oxime. Cyclization and dehydration of the resulting N-hydroxydihydropyridine affords (±)-lycodine **(7)** in good yield. The synthesis is highly efficient and provides lycodine in seven operations from the precursor **(133)** to cyano enone **162**, in 15% overall yield.

2.4.   Lycodoline and Anhydrolycodoline

Heathcock and Kleinman have also applied the Mannich technology in a synthesis of lycodoline **(25)** [67, 75]. The critical feature of this molecule is the presence of the sensitive tertiary alcohol function, which precludes the use of strongly acidic conditions to promote the key Mannich cyclization, and necessitated the development of new conditions to affect the Oppenauer oxidation used (as in their original lycopodine synthesis) in the formation of ring C.

The synthesis begins with cyano diketal **164**, which is reduced to the corresponding amine with lithium aluminum hydride. Hydrolytic cleavage of the two ketal groups results in cyclization to the exceedingly air-sensitive octahydroquinoline **181** (see Scheme 27). This substance undergoes facile autoxidation upon exposure to oxygen to give a tertiary hydroperoxide, which is immediately subjected to catalytic hydrogenolysis to afford hydroxy imine **182**, in which the angular hydroxyl has been introduced *trans* to the acetonyl side chain. A small

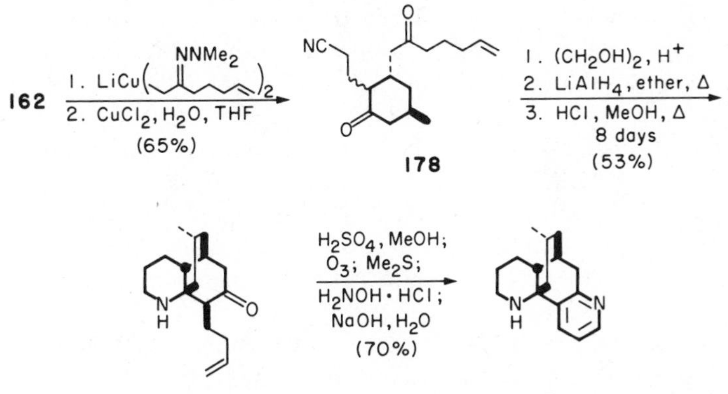

Scheme 26.   Synthesis of lycodine.

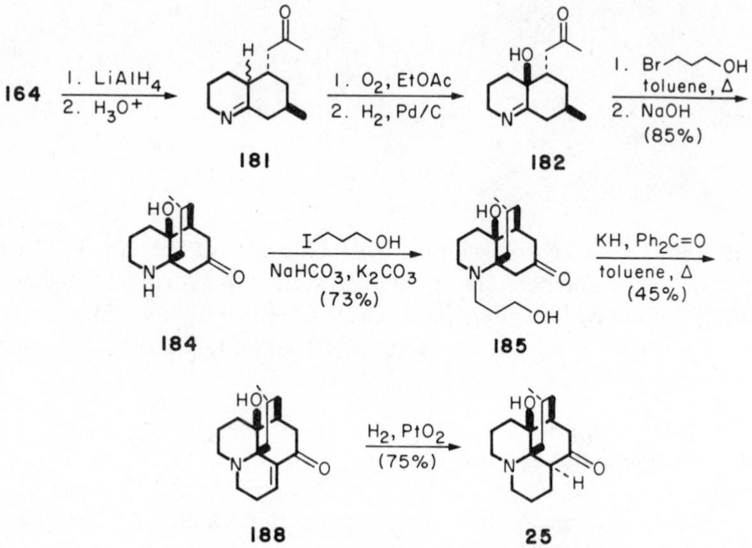

Scheme 27. Synthesis of lycodoline.

amount of hemiketal **183**, arising from oxygen attack *cis* to the keto side chain is also formed.

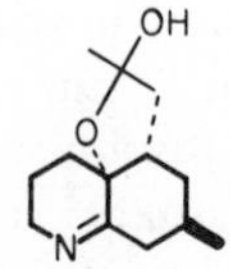

183

The Mannich cyclization was affected under unusual conditions. Treatment of **182** with methanolic HCl results in the formation of complex mixtures, whereas heating the preformed hydrobromide salt of **182** gives no reaction. However, when a solution of **182** and 3-bromo-1-propanol in toluene is heated at 110°C, tricyclic Mannich product **184** is produced in high yield. It has been suggested that polymerization of the bromo alcohol serves as a source for the gradual delivery of HBr. A low HBr concentration affords a greater opportunity for acid-catalyzed enolization to occur in a molecule that does not possess a protonated nitrogen. Rapid cyclization should then occur upon subsequent imine protonation. Treatment of amine **184** with 3-iodo-1-propanol in refluxing actone gives diol **185**.

Attempts to oxidize **185** under Oppenauer conditions (*t*-BuOK, benzo-phenone, toluene) used successfully on the related substrate **170** in the first Berkeley lycopodine synthesis (see Scheme 23) resulted only in amine dealkyla-tion. It is presumed that loss of acrolein occurs in a reverse Michael reaction from intermediate aldehyde **186.** Intramolecular proton transfer from the hydroxyl group, which is known to be hydrogen bonded to the amine, makes the leaving group in this reverse Michael reaction alkoxide **187,** rather than an amide anion, so that this side reaction is more facile than with substrate **170.** The side reaction

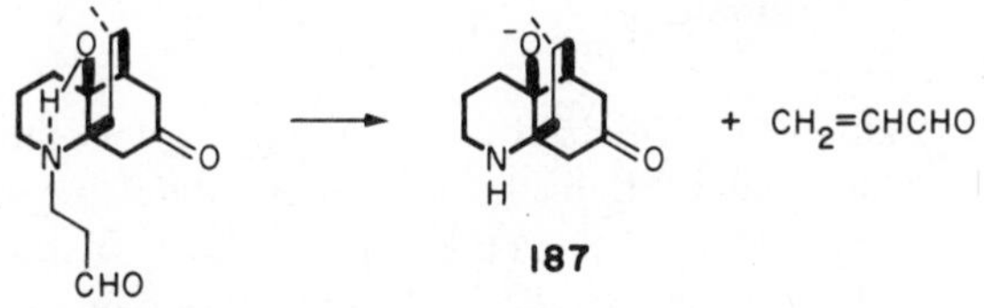

is circumvented in a straightforward manner by using a stronger base, potassium hydride, in the Oppenauer oxidation so that both the primary and tertiary hydroxyls are deprotonated; retro-Michael reactions are thus suppressed as expulsion of an amide-alkoxide dianion would be required. Under these conditions, dehydrolycodoline **(188)** is produced in moderate yield. Catalytic hydrogenation of **188** gives (±)-lycodoline **(25).** The synthesis requires 10 steps from cyano dione **133** and proceeds in 8% overall yield.

An unsuccessful approach to the synthesis of lycodoline from the Horii group [30, 76] bears some resemblance to their earlier approach to lycopodine [54]. Bicyclononene benzyl carbamate **189** is treated with excess dimethylsulfonium methylide to give epoxide **19,** resulting from kinetically controlled attack of the reagent on the less hindered face of the carbonyl (see Scheme 28). Horii and co-workers also suggest that the stereochemical outcome of this addition may be influenced by interaction of the olefin π-electrons in **189** with the positively charged end of the ylide dipole, as shown in the hypothetical intermediate **191.**

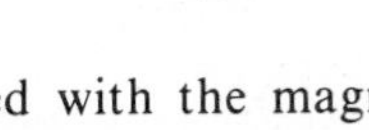

The epoxide ring is opened with the magnesium salt of ethyl hydrogen malonate to give lactone **192,** in which the carbamate has also undergone transesterification. Hydrolysis and decarboxylation of the carbamate and ester gives amino lactone **193,** which is converted to hydroxy lactam **194** in high yield by treatment with a catalytic amount of triton-B. By standard methods, lactam **194** is readily transformed into N-protected amino ketone **28,** which contains all the requisite functionality for rings A and B of lycodoline **(25).** However, acrylamide **195,** prepared from the amino ketone resulting from hydrolysis of **28,**

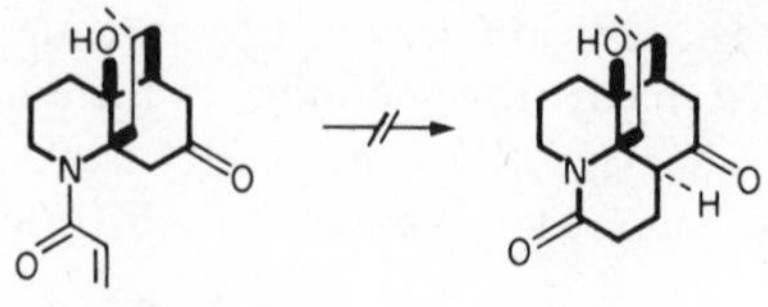

195   **25**

fails to undergo intramolecular Michael addition to give lycodoline (**25**) [77]. The failure of this intramolecular Michael addition is interesting in light of the successful closure employed by Wiesner and co-workers in the 12-epilycopodine synthesis (see Scheme 19). Model studies and examination of molecular models reveal that a higher degree of orbital overlap between the acrylamide and ketone enolate pi-systems is achieved in the 12-epi series (such as in Wiesner's compounds) than in Horii's substrates.

The problem was eventually circumvented by successfully converting carbamate **28** into anhydrolycodoline (**196**) (see Scheme 29). Dehydration of tertiary alcohol **195** gives olefin **197**, which also does not undergo intramolecular Michael addition under the acidic conditions used by Wiesner in the 12-epilycopodine synthesis. However, the desired cyclization occurs under basic conditions; treatment of acrylamide **197** with a catalytic amount of sodium ethoxide and dicyclohexyl 18-crown-6 in dimethylformamide yields exclusively the tetracyclic lactam **198**. The success of this cyclization depends heavily on the use of a crown ether–sodium ethoxide complex in dimethylformamide, rather

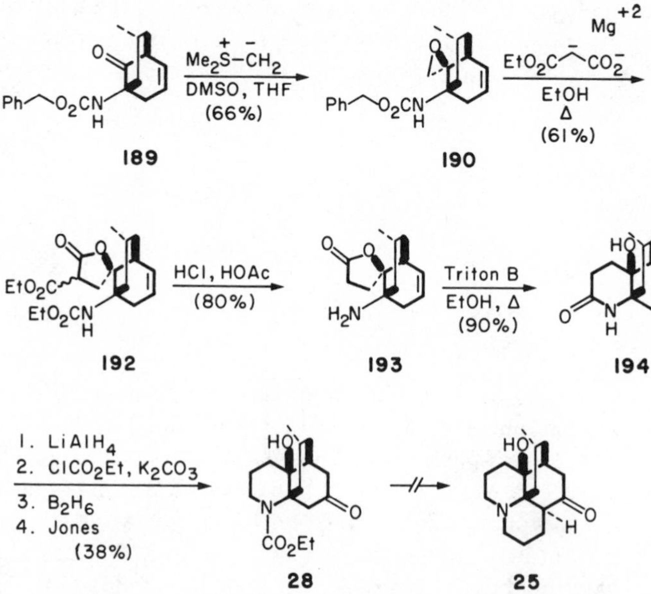

Scheme 28.   Horii's attempted lycodoline synthesis.

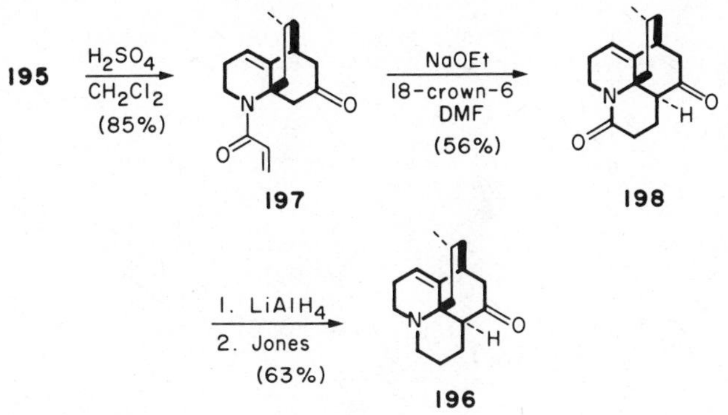

Scheme 29. Horii's total synthesis of anhydrolycodoline.

than ethanol; otherwise substantial amounts of the ethanol adduct **199** is produced.

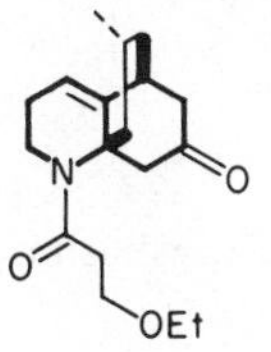

199

Reduction of lactam **198** with lithium aluminum hydride, followed by oxidation with Jones reagent, gives racemic anhydrolycodoline **(196).** The Horii synthesis represents the only recorded total synthesis of alkaloid **196**. It could probably be extended to a total synthesis of lycopodine, since the enantiomerically homogeneous version of **196** has previously been hydrogenated to a mixture of lycopodine and 12-epilycopodine [78].

2.5. Luciduline

Luciduline **(2)** is the only known representative of the luciduline skeletal class. The first total synthesis of luciduline was accomplished by Evans and Scott in 1972 (see Scheme 30) [79]. Bicyclooctenone **200,** readily obtained from the Diels–Alder reaction of 2,5-dihydroanisole with the ketene equivalent 2-chloro-acrylonitrile [80], was treated with isopropenylmagnesium bromide to give tertiary allylic alcohol **201**. When this compound is heated at 250°C Cope rearrangement occurs, giving **202** as a mixture of enol ether regioisomers and methyl epimers. The enol ether is selectively transformed into a dioxolane and the resulting mixture of epimers crystallized to obtain the desired isomer **203**. Equilibration of the undesired diastereomer provides additional amounts of **203**.

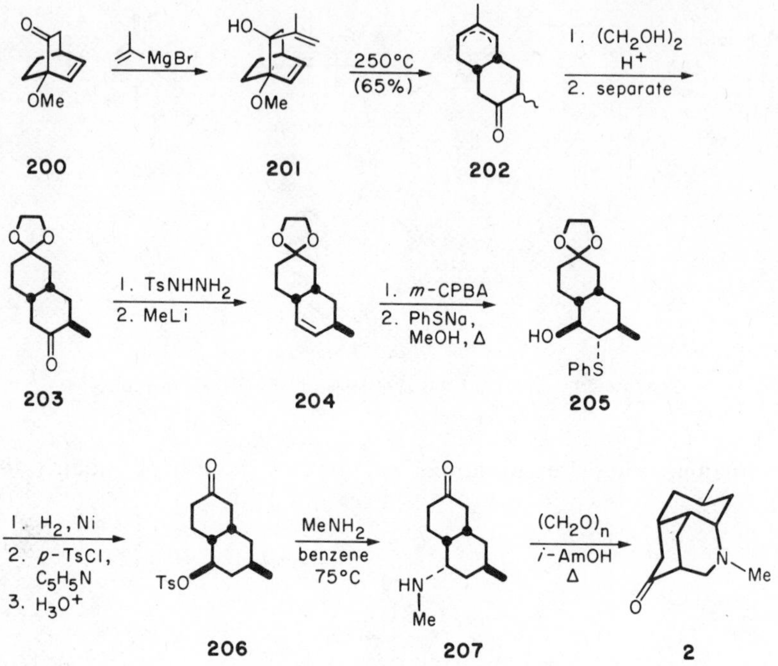

Scheme 30. Evans luciduline synthesis.

Compound **203** is subjected to the Dauben–Shapiro modification of the Bamford–Stevens reaction [81] to obtain exclusively unsaturated ketal **204**. Olefin **204** is stereoselectively epoxidized with *m*-chloroperoxybenzoic acid. In principle, epoxidation could occur from either *cis*-octalin conformer, **204a** or **204b**.

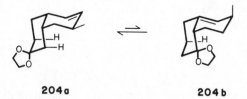

204a   **204b**

In the more stable conformer **204a**, attack *trans* to the methyl group is blocked by the two axial hydrogens in ring A. In the less stable conformer **204b**, the indicated methylene group impedes epoxidation from the bottom. The oxirane ring is opened regioselectively with sodium thiophenoxide, yielding compound **205**. Desulfurization, tosylation, and deketalization provides keto-tosylate **206**, which is treated with excess methylamine to obtain amino ketone **207**. Evans postulates that this displacement occurs by a pathway involving intramolecular alkylation of *gem*-diamine **208**, which subsequently hydrolyzes to provide **207**.

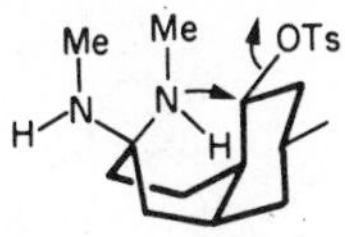

208

As evidence for this theory, Evans notes that treatment of the ethylene ketal of **206** with methylamine gives mostly elimination products. The synthesis is completed by heating amino ketone **207** with paraformaldehyde to afford (±)-luciduline (**2**). The route is fairly straightforward, providing luciduline in 11 steps overall from ketone **200**.

A novel and highly efficient luciduline synthesis has been reported by Oppolzer and Petrzilka [82]. The key step employs an intramolecular 1,3-dipolar cycloaddition to assemble the heterocyclic ring and introduce the oxygen function. Lewis-acid promoted Diels–Alder reaction of 5-methylcyclohexenone and butadiene gives a 60:40 mixture of *cis*- and *trans*-octalones **209** and **210** (see Scheme 31). Although a mixture of ring juncture isomers is produced, addition takes place entirely *trans* to the C(5)-methyl group.

Since *cis*-octalone **209** undergoes oxime formation more rapidly than *trans*-isomer **210**, the mixture is treated with 0.6 equivalents of hydroxylamine hydrochloride to obtain a mixture of oxime **211** and unreacted ketone **210**, which are easily separated by recrystallization. Reduction of **211** with sodium cyanoborohydride occurs stereoselectively to give hydroxylamine **212**. When this substance is heated with paraformaldehyde in the presence of a dehydrating

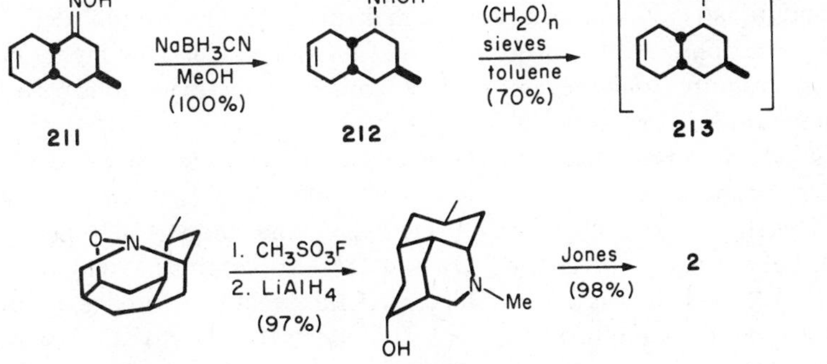

Scheme 31.   Oppolzer luciduline synthesis.

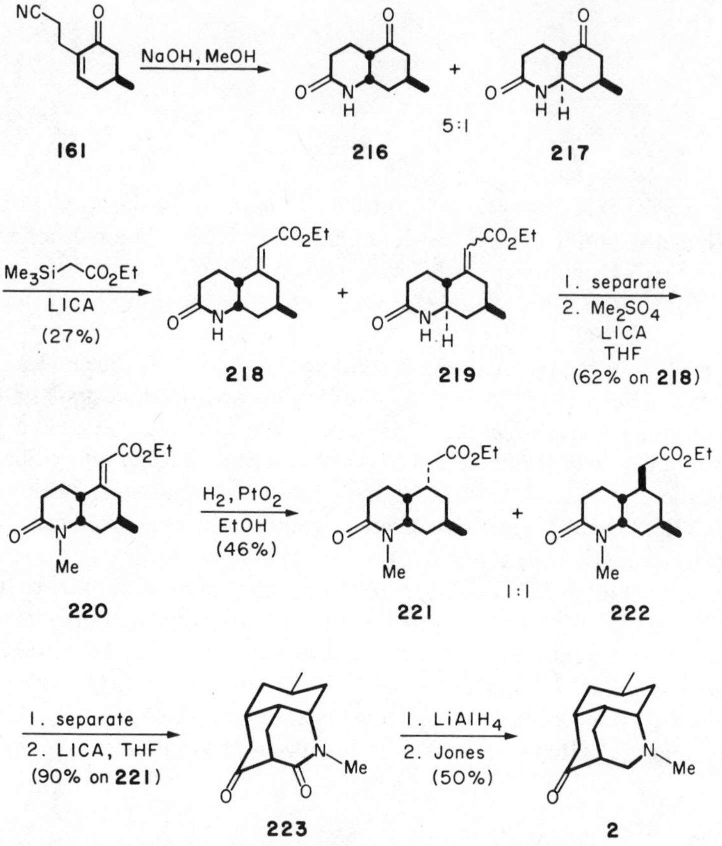

Scheme 32. MacLean luciduline synthesis.

agent, bridged isoxazolidine **214** is formed in good yield. The reaction occurs by way of the transient nitrone **213,** which undergoes intramolecular cycloaddition to the double bond. Reductive cleavage of the N—O bond is achieved by methylation with methyl fluorosulfonate followed by treatment of the resulting ammonium salt with lithium aluminum hydride to give amino alcohol **215.** Subsequent Jones oxidation provides (±)-luciduline **(2).** This highly efficient synthesis requires only seven steps to convert 5-methylcyclohexenone into luciduline in 26% overall yield.

MacLean has developed a synthesis of luciduline **(2)** which was designed to intercept proposed intermediates in the biosynthesis of *Lycopodium* alkaloids [83]. Hydrolysis of 2-cyanoethyl-5-methylcyclohexenone **(162)** gives a 5:1 mixture of *cis-* and *trans*-lactams **216** and **217,** in a combined yield of 36% (see Scheme 32) [84]. It was not established if the reaction is under equilibrium control, and the 5:1 mixture of **216** and **217** reflects their relative thermodynamic stabilities, or if **216** predominates for a kinetic reason.

Peterson olefination with ethyl trimethylsilylacetate on the mixture of lactams gives exclusively the *E* isomer from the *cis*-lactam **216**, but both *E* and *Z* isomers from the *trans*-compound **217**. Lactam **218** is separated and converted to its *N*-methyl derivative **220**, which is hydrogenated. Examination of **220** by ^{1}H NMR spectroscopy shows that the compound exists primarily in conformation **220a**. Steric hindrance by the axial methyl group is expected to result in hydrogenation of this conformer from the bottom face of the double bond, leading to isomer **222**. However, in the other conformation, **220b**, hydrogenation is also expected to occur in such a manner as to deliver hydrogen from the equatorial direction, leading to isomer **221**. In the event, catalytic hydrogenation

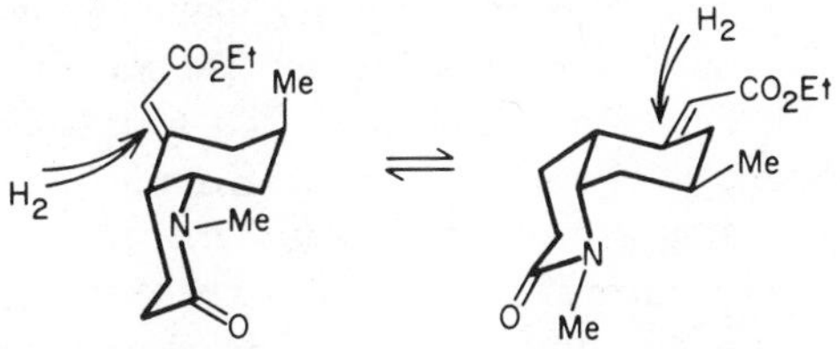

220a   **220b**

gives a 50:50 mixture of **221** and **222**. Treatment of **221** with lithium isopropylcyclohexylamide induces cyclization to give luciduline lactam **223**, the spectroscopic properties of which were identical with an optically active sample prepared by oxidation of luciduline with potassium permanganate [11]. Reduction of **223** with lithium aluminum hydride followed by treatment with Jones reagent gives (±)-luciduline (**2**) in seven steps from cyanoenone **162**. Although the synthesis contains several low-yield steps and requires isomer separation at two points, the goal of this project was fulfilled; MacLean was able to prepare proposed biosynthetic intermediates **221** (see Scheme 5) and luciduline lactam (**223**).

2.6. Phlegmarine

Braekman first reported the isolation of phlegmarine in 1978, and postulated that it may serve as a key intermediate in the biosynthesis of *Lycopodium* alkaloids (see Scheme 5) [29]. He assigned general structure **22** as the skeleton of the compound, but was unable to assign any stereochemical relationships. Three years later, MacLean published a total synthesis which established the relative configurations of four of the five asymmetric centers in the molecule [85].

It was initially assumed that phlegmarine has stereostructure **224**, in which C(5), C(7), C(9), and C(10) have the same configuration as the corresponding centers in luciduline (**2**) and the lucidine alkaloids; in addition, the stereochemistry at C(5), C(7), and C(10) is the same as that found in lycopodine (**1**) and the majority of *Lycopodium* alkaloids. In a systematic fashion, MacLean set out

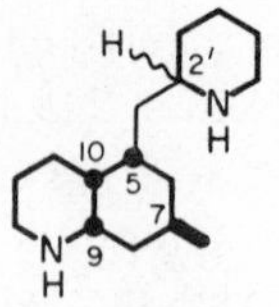

224

to prepare all possible stereoisomers of the tetrahydroquinoline portion of the molecule. He soon found that the original assumption had been incorrect and that **224** is not the structure of phlegmarine.

The phlegmarine skeleton was prepared in a manner similar to that used in MacLean's earlier luciduline synthesis. Isomer **230** was the first prepared, as shown in Scheme 33. Peterson olefination of *cis*-lactam **216** [83] with 2-trimethylsilylpyridine gives a 4 : 1 mixture of *Z* and *E* double-bond isomers **225** and **226,** in contrast to the analogous reaction with ethyl trimethylsilylacetate (see Scheme 32). The major isomer **225** is hydrogenated to give only lactam **227** in which both the methyl and pyridylmethyl groups occupy equatorial positions. Hydogenation of the *E* isomer **226** affords a mixture of **227** and **229,** as in

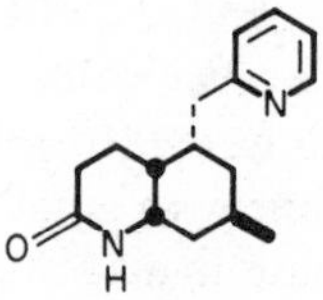

229

hydrogenation of **220** in MacLean's luciduline synthesis. After quaternization of the pyridine nitrogen, the aromatic ring is hydrogenated to obtain a 50:50 mixture of C(2′) epimers **228**. Reduction with lithium aluminum hydride, followed by acetylation of the resulting amine **(230),** gives the acetamide derivative **231**. However, neither C(2′) epimer is identical with *N*-α-acetyl-*N*-β-methylphlegmarine derived from natural sources.

MacLean next investigated the synthesis of isomers containing the *trans*-decahydroquinoline ring system (see Scheme 34). Dissolving metal reduction of compound **134** affords *trans*-keto lactam **232,** which undergoes highly stereoselective Peterson olefination with 2-trimethylsilylpyridine to produce a 50:1 mixture of *E* and *Z* adducts **233** and **234**. Isomer **233** is hydrogenated in poor yield to give a 50:50 mixture of pyridylmethyl side chain epimers, contaminated by material resulting from partial reduction of the pyridine ring. To effect separation, it is necessary to convert the hydrogenation product mixture to the corresponding pyridine *N*-oxides **235** and **236**. Treatment of pure *N*-oxide **235** with phosphorus trichloride regenerates the pyridine ring system, which is subsequently quaternerized and hydrogenated to give amino lactam **237** as an inseparable mixture of C(2′) diastereomers. Reduction of this material with

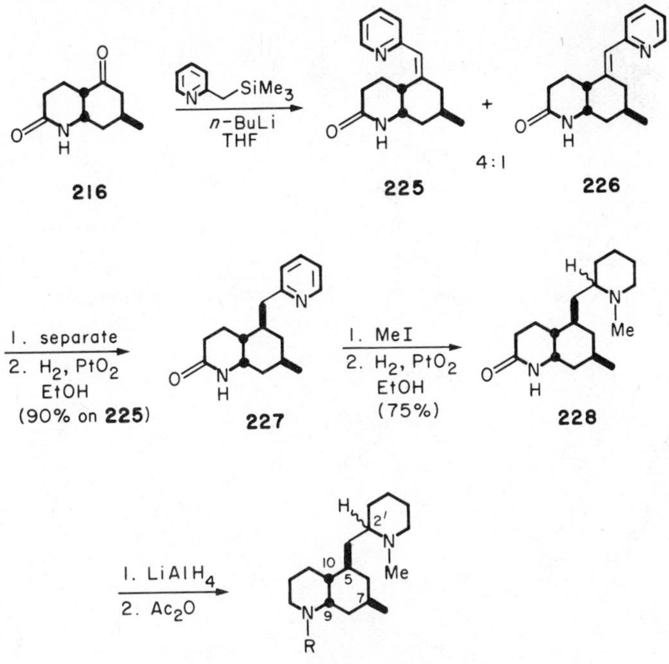

Scheme 33. MacLean's first synthesis of phlegmarine isomers.

lithium aluminum hydride, and acetylation of the resulting diamine, provides epimeric acetamides **238** and **239**, which are separable by chromatography. One of these epimers [the C(2′) stereochemistry could not be assigned] was shown to be identical in its TLC behavior in several solvent systems with a sample of *N*-α-acetyl-*N*-β-methylphlegmarine derived from natural material. Unfortunately, insufficient amounts of natural material were available to allow spectroscopic comparison. Assuming that the natural product does have the relative configuration of **238** or **239**, the MacLean synthesis establishes that it has the same relative stereochemistry at C(5), C(7), C(9), and C(10) as is found in lycopodine and most other *Lycopodium* alkaloids. The synthesis of phlegmarine derivatives provides a method for the preparation of labeled compounds for biosynthetic studies. However, no such reports have yet appeared.

2.7. Cernuine

Cernuine **(18)** is an interesting tetracyclic amino lactam that contains two tertiary nitrogens but, unlike most *Lycopodium* alkaloids, no tetra-substituted carbon atoms. At first glance, cernuine appears to bear little structural relationship to lycopodine **(1)**. However, as discussed in an earlier section, Spenser has

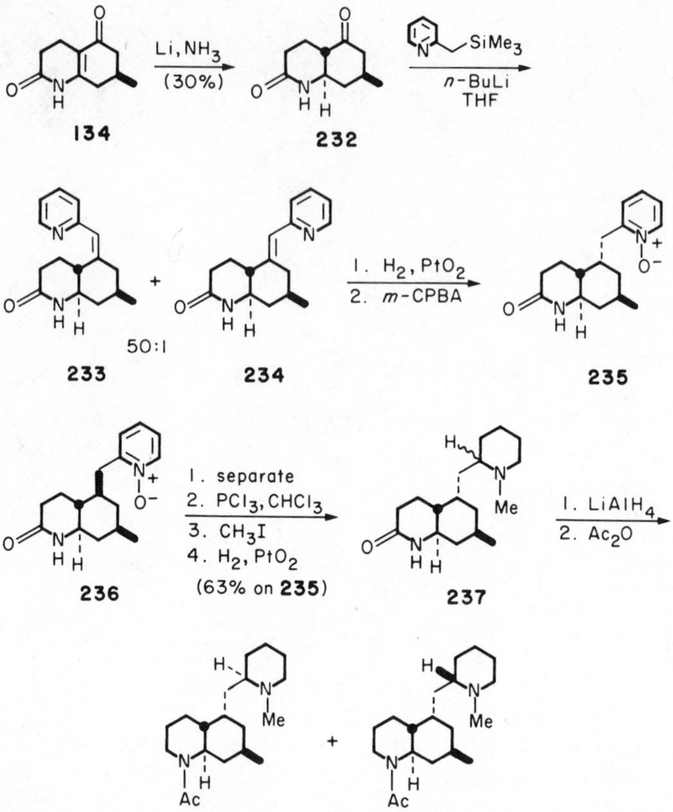

Scheme 34. MacLean's second synthesis of phlegmarine isomers.

interrelated biosynthetic pathways for cernuine and lycopodine [26, 27]. No work on the synthesis of cernuine itself has yet appeared, but two routes to dihydrodeoxyepiallocernuine **(240)**, a degradation product of natural lyco-cernuine **(241)**, have been published.

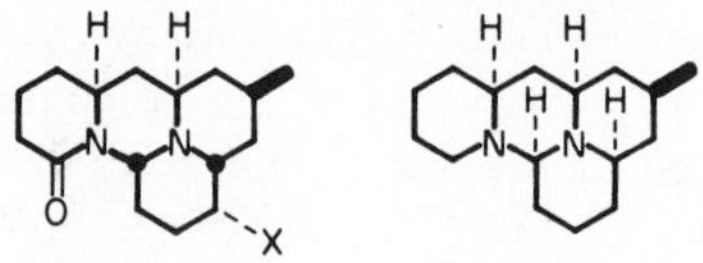

In conjunction with his description of the stereostructure of cernuine **(18)** and lycocernuine **(241)** [86, 87], Ayer reported the first synthesis of **240** (see Scheme 35) as support for his structural assignments [88]. Alkylation of lithiated 2,4,6-collidine **(242)** with allyl bromide, followed by hydroboration and protection of

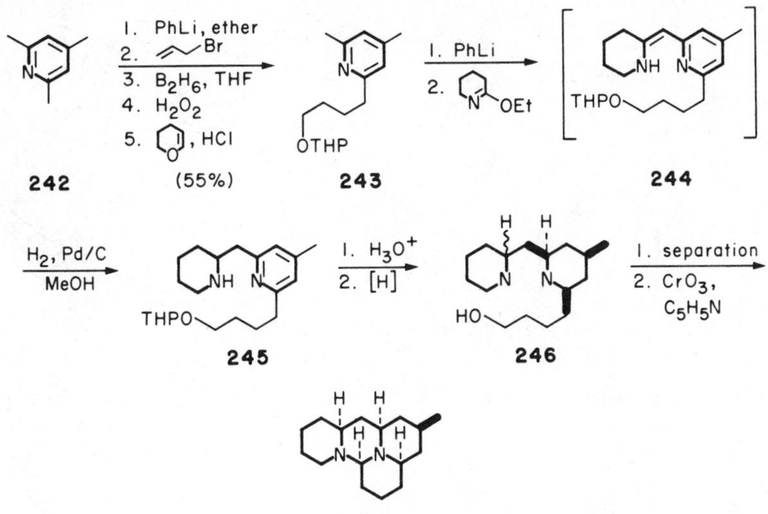

Scheme 35. Ayer's synthesis of dihydrodeoxyepiallocernuine.

the resulting alcohol, give tetrahydropyranyl ether **243**. Metallation of the C(6)-methyl group, followed by condensation with 2-ethoxy-Δ^1-piperidene, provides an unstable product **244** which is immediately hydrogenated to obtain diamine **245**. Catalytic hydrogenation at high pressure (H_2,Rh/C, EtOH, AcOH) or dissolving metal reduction (Na, isoamyl alcohol) gives the fully reduced system **246**, as an inseparable mixture of C(5) epimers, in which all the appendages on the trisubstituted piperidine are in equatorial orientations. Formation of the thermodynamically most stable product is generally expected from dissolving metal reductions; formation of the same material by catalytic hydrogenation may result from isomerizations caused by dehydrogenation–hydrogenation alpha to the nitrogen, eventually leading to the most stable product [89]. The diastereomeric diamines are separated as their N-formyl derivatives, which are then hydrolyzed to obtain the pure diamines.

The structures of the diamino alcohols **246** were assigned on the basis of the products formed in the subsequent oxidative cyclization (see Fig. 9). It was argued that cyclization of intermediate **247** could occur from either face of the immonium ion yielding two products. However, cyclization of epimeric cation **248** would be hindered from the top face and should lead to only the most stable all-*trans*-product **240**. Indeed, when the separated isomers of **246** were oxidized with CrO$_3$-pyridine, one gave two isomers and one gave only a single product, which was therefore assigned structure **240**. The structural assignment is supported by the observation of intense Bohlmann bands in the infrared spectrum of **240** [90]. The synthetic material was spectroscopically identical with a sample of dihydrodeoxyepiallocernuine derived from natural lycocernuine.

Compound **240** has also been synthesized from pelletierine (**13**), a biosynthetic precursor of cernuine (**18**) [91]. N-Acetylpelletierine (**249**) is treated with

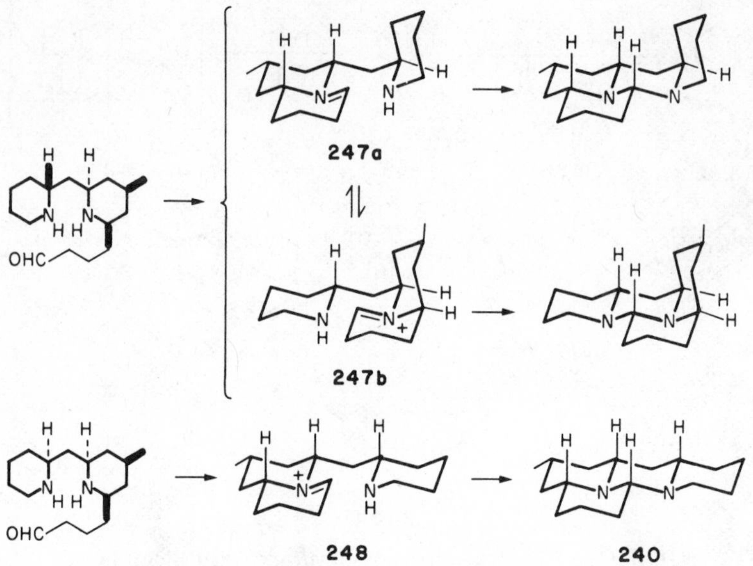

Figure 9. Cyclization reactions to the cernuine skeleton.

Meerwein's reagent and the resulting imidate salt **250** cyclized to give un-saturated amide **251** in 62% overall yield (see Scheme 36). Treatment of acetamide **249** with base gives vinylogous amide **254**.

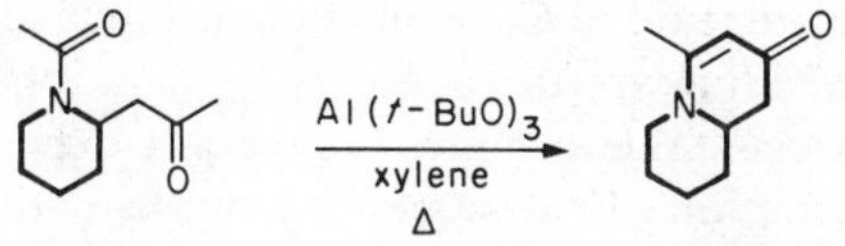

Catalytic hydrogenation of **251**, followed by condensation with α-picolyl-lithium and reduction of the resulting enamine double bond furnishes diamine **252**, presumably as a mixture of diastereomers. The pyridine ring is hy-drogenated to give **253**, which is cyclized to **240** by a Hoffmann–Loeffler–Freytag reaction. Chlorination of **253**, *in situ* with *N*-chlorosuccinimide, followed by irradiation gives (±)-dihydrodeoxyepiallocernuine **(240)** in 30% yield. Addi-tional stereoisomers are presumably produced in this synthesis, but they were not isolated and characterized.

2.8. Serratinine and 8-Deoxyserratinine

The fawcettidine skeletal system is a modification of the hydrojulolidine ring system, in which a Wagner–Meerwein shift of the C(4)—C(13) bond has formed a link between C(4) and C(12). The five-membered ring, thus formed, is the

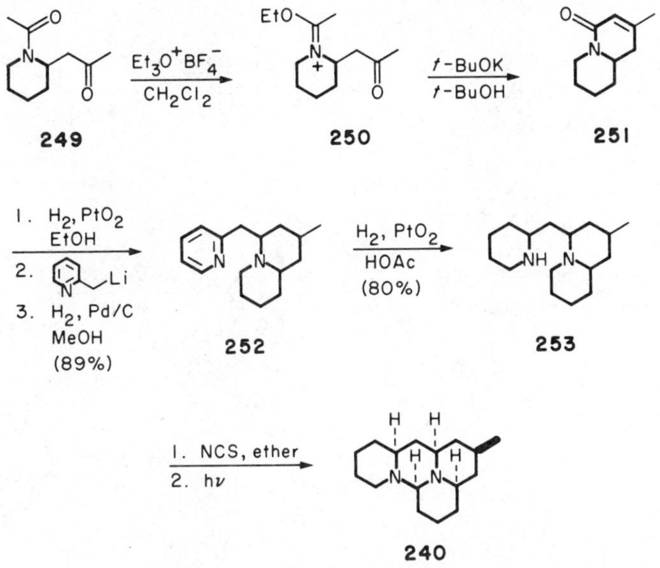

Scheme 36. Ban synthesis of dihydrodeoxyepiallocernuine.

distinguishing feature of this class of compounds. In serratinine (266) this framework has undergone further change by migration of the C(13)—N bond to C(4). This compound, with its six contiguous stereocenters, poses a considerable synthetic challenge.

Inubushi and co-workers have devised a synthesis of (±)-serratinine [92] which starts with benzoquinone 255 [93] (see Scheme 37). Diels–Alder reaction of 255 with butadiene provides bicyclic system 256 in 39% yield, along with several uncharacterized by-products. If the same reaction is carried out with an analog of 255 having a saturated ester side chain, addition occurs at both quinone double bonds. Thus, the side chain double bond serves to activate the proper position on the quinone; it is reduced later in the synthesis. Saturation of the enedione double bond, reduction of the resulting diketone, and acetylation provides octalin 257. The hydride reduction proceeds in only 40% yield to the desired diol, suggesting that other stereoisomers are also formed in this reaction. This sequence of reactions places the methyl group and the neighboring acetoxy in the incorrect configuration. Osmium tetroxide hydroxylation of the isolated double bond, hydrogenation of the conjugated alkene, and cleavage of the newly formed vicinal diol with periodate gives dialdehyde 258.

Cyclization of dialdehyde 258 occurs when it is chromatographed on alumina. However, closure under these conditions occurs in the undesired manner to give exclusively unsaturated aldehyde 267, presumably by attack of the less hindered enolate on the remaining aldehyde carbonyl. Cyclization via an enamine derivative (piperidinium acetate, benzene, 60° C) gives the same result. However, when the cyclization is carried out under Mannich-type conditions (excess

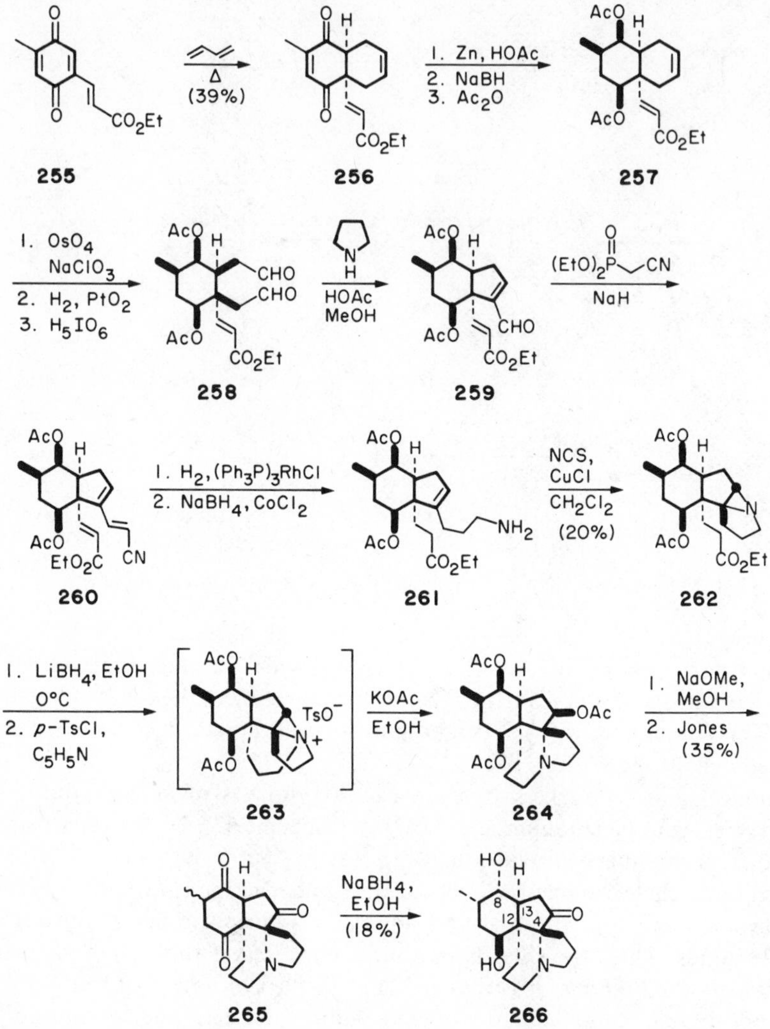

Scheme 37. Inubushi's serratinine synthesis.

pyrrolidine, excess acetic acid, methanol, 0° C), a mixture of desired aldol product **259** and regioisomer **267** is produced in an 8 : 1 ratio. It seems likely that the desired condensation proceeds via intermediate immonium salt **268** [94].

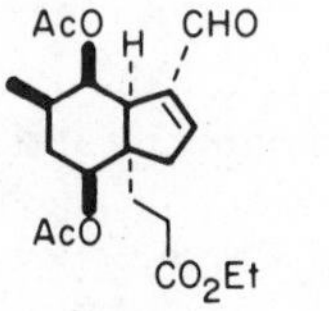

267

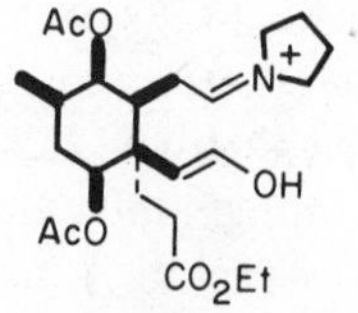

268

Treatment of aldol adduct **259** with diethyl cyanomethylphosphonate furnishes conjugated nitrile **260** as a 50:50 mixture of E and Z diastereomers. Selective hydrogenation of the less substituted double bond with Wilkinson's catalyst, followed by selective reduction of the cyano group (NaBH$_4$, CoCl$_2$, methanol) gives amine **261**. This material is now employed in a highly innovative construction of the heterocyclic ring system. Chlorination of the amine and subsequent α-elimination gives a nitrene which adds to the double bond to give aziridine **262** in 20% yield; a small amount of the isomeric aziridine resulting from addition to the top face of the cyclopentene is also produced. Selective reduction of the ethyl ester (in the presence of two acetoxy groups) with lithium borohydride gives a primary alcohol, which is tosylated to induce intramolecular alkylation affording aziridinium salt **263**. This highly strained compound is converted to triacetate **264** upon treatment with potassium acetate in ethanol. Attempts at selective hydrolysis of this compound were unsuccessful. However, if all three esters are saponified, the resulting triol may be oxidized to triketone **265** (a mixture of methyl epimers). Selective reduction of the cyclohexanedione carbonyls gives a mixture of (±)-serratinine **(266)** and (±)-8-episerratinine in yields of 18 and 20%, respectively. The faster rate of reduction of cyclohexanones compared to cyclopentanones is well precedented [95]. Ayer has suggested [2j] that the cyclopentanone carbonyl may be protected by formation of an internal hemiketal, such as **269**. Such selective reductions have been observed in related

269

systems [96]. The synthesis of racemic serratinine requires 17 steps from benzoquinone **255** and proceeds in an overall yield of less than 0.1%. The synthesis provides several fascinating examples of functional group selectivity.

Inubushi and co-workers applied the experience gained in the foregoing synthesis to the synthesis of 8-deoxyserratinine **(281)** [97]. 2-Allyl-5-methyl-cyclohexenone **(270)**, conveniently prepared in four steps from dihydroorcinol **(86)**, undergoes Diels–Alder reaction with butadiene from the side opposite the C(5) substituent [98] to give *cis*-octalone **271**, albeit in low yield (see Scheme 38). The stereoselectivity in this reaction allowed the authors to avoid the methyl epimerization that was required in the serratinine synthesis. Ketalization, hydroboration of the less substituted double bond, and protection of the resulting primary alcohol provides compound **272**. Lemieux–Johnson oxidation of the cyclohexene double bond provides dialdehyde **273**, which is cyclized under Mannich-like conditions, as in the earlier work, to give desired unsaturated aldehyde **274** and its unwanted regioisomer in a 25:1 ratio. The aldehyde side

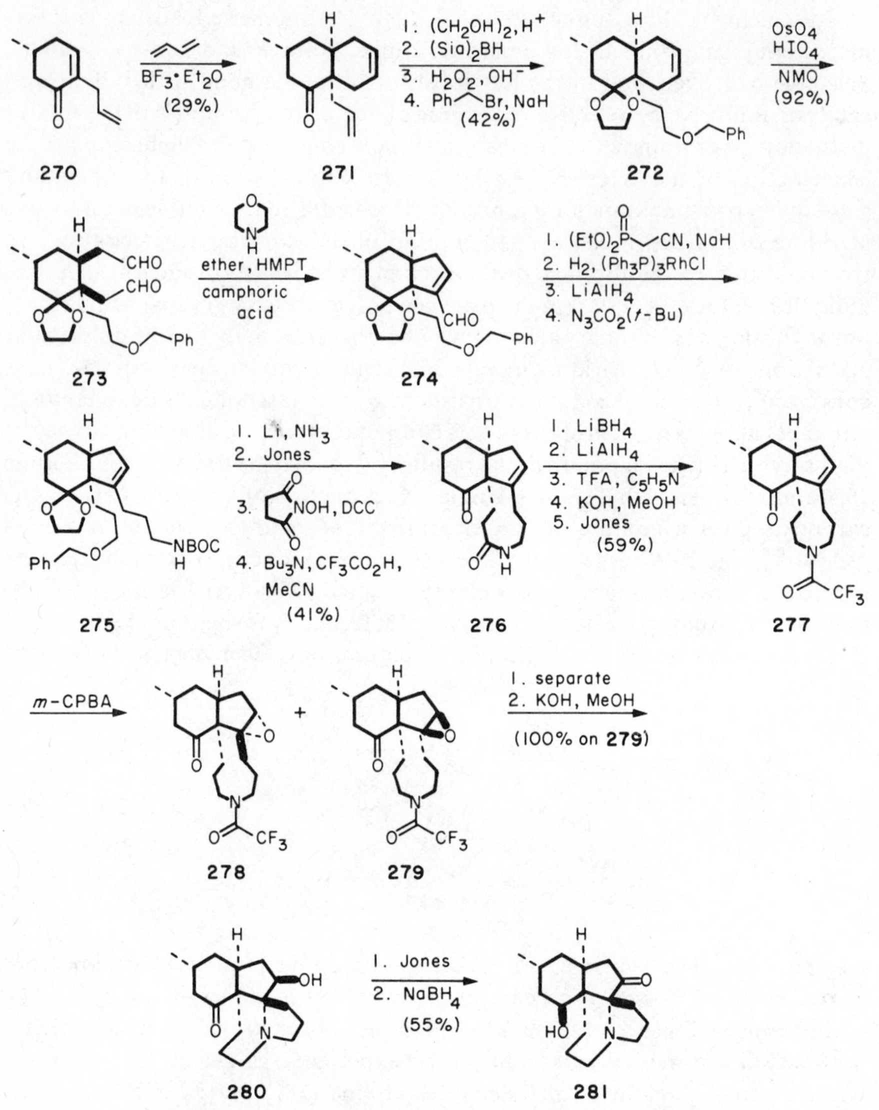

Scheme 38.   Inubushi's synthesis of 8-deoxyserratinine.

chain is homologated as before, and the resulting primary amine is protected as the *t*-butyoxycarbonyl derivative.

Construction of the heterocyclic ring system is accomplished by forming a nine-membered amide ring spanning C(4) and C(12) and subsequently joining the nitrogen to C(4). The two side chains are coupled by conversion of the C(11) appendage into a propionic acid moiety, which is coupled with the C(4) aminopropyl appendage by high-dilution lactamization to give **276**. The yield of this four-step sequence is good (41%), particularly in light of the fact that one of the steps involves formation of a nine-membered ring.

The cyclohexanone ketal is hydrolyzed during ring closure and it cannot be reinstalled, probably because of steric hindrance from the adjacent quaternary center. It is therefore necessary to reduce the cyclohexanone carbonyl. The lactam is then reduced with lithium aluminum hydride, the resulting amino alcohol protected on nitrogen, and the hydroxyl reoxidized with Jones reagent [99]. Epoxidation of the cyclopentene double bond affords two diastereomeric epoxides, 278 and 279, in yields of 58 and 40%, respectively. Upon base-catalyzed hydrolysis of the minor epoxide 279, the resulting amine opens the epoxide at C(4) to provide the tetracyclic alcohol 280 in quantitative yield. Jones oxidation and stereoselective reduction of the cyclohexanone carbonyl from the convex face of the molecule provides (±)-8-deoxyserratinine (281). The synthesis is rather long, requiring 25 steps to prepare the target compound from dihydro-orcinol (86) in less than 0.1% overall yield.

2.9. Fawcettimine

A slight variation of Inubushi's route to 8-deoxyserratinine (281) provides a synthesis of the related alkaloid fawcettimine (26) [97]. Compound 278, the major isomer produced by epoxidation of 278 (see Scheme 38) is treated with boron trifluoride etherate to furnish allylic alcohol 282 (see Scheme 39) in good yield. While one might anticipate that treatment of isomeric epoxide 279 under these conditions would yield an allylic alcohol epimeric with 282, in fact a complex mixture of unidentified products is produced from this isomer. Compound 282 is oxidized to the α,β-unsaturated ketone, which is subjected to catalytic hydrogenation to saturate the double bond. Reduction occurs from the convex face of the molecule, establishing the correct relative stereochemistry at C(4). Hydrolytic removal of the N-protecting group affords (±)-fawcettimine (26). The synthesis requires 26 steps and produces the natural product in about 0.1% overall yield.

Burnell, who first isolated fawcettimine (26) [100], has published an approach to its synthesis (see Scheme 40) [101]. Cyano dione 133 is alkylated with propargyl bromide to give 283 as a mixture of diastereomers. Partial reduction of

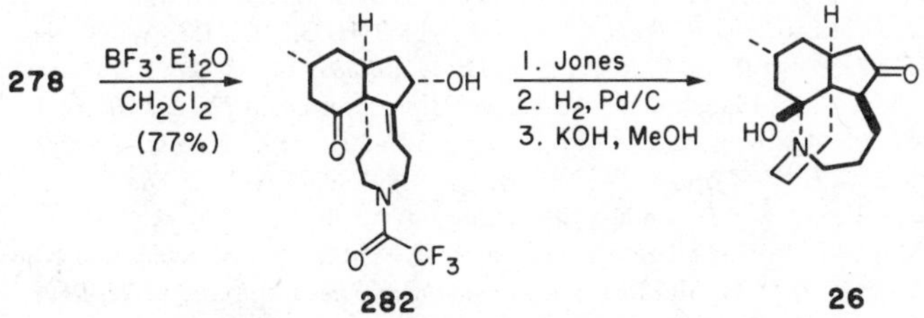

Scheme 39. Inubushi's fawcettimine synthesis.

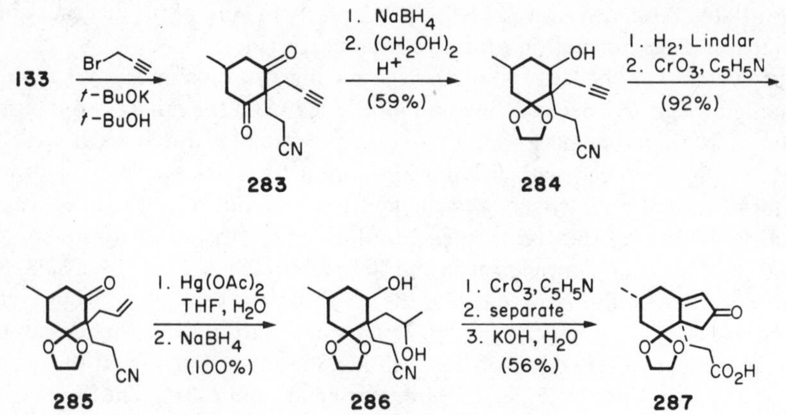

Scheme 40. Burnell's approach to fawcettimine.

283, followed by ketalization of the remaining carbonyl provides **284.** Hydration of the acetylene was planned to provide an acetonyl side chain for a subsequent intramolecular aldol ring closure to form the bicyclic system. However, hydrolysis of the protecting group occurs during hydration of the triple bond, resulting in cleavage of the nonenolizable 1,3-diketone. Compound **284** may be partially hydrogenated under the influence of Lindlar's catalyst, but oxymercuration of the resulting alkene is also unsuccessful. Oxidation with Collins reagent provides unsaturated ketone **285,** which is subjected to oxymercuration to furnish diol **286,** after reduction with sodium borohydride. Oxidation to the corresponding diketone gives a mixture from which the desired diastereomer may be obtained by chromatography. Treatment of this material with refluxing alkali gives hydrindenone carboxylic acid **287.** In this approach, eight steps are required to convert cyano dione **133** into target compound **287.** Compound **287** incoporates only two of fawcettimine's stereocenters, and it is not clear how the critical C(3)—C(4) bond would be introduced.

REFERENCES

1. J. C. Braekman, L. Nyembo, and J. J. Symoens, *Phytochemistry* **19,** 803 (1980).

2. For reviews of *Lycopodium* alkaloid structure and chemistry, see (a) L. F. Small, in *Organic Chemistry, An Advanced Treatise.* 2nd Ed., Vol. 2, H. Gilman, Ed., Wiley, New York, 1943, pp. 1166–1258; (b) R. H. F. Manske, in *The Alkaloids,* Vol. 5, R. H. F. Manske, Ed., Academic, New York, 1955, pp. 295–300; (c) R. H. F. Manske, in *The Alkaloids,* Vol. 7, R. H. F. Manske, Ed., Academic, New York, 1960, pp. 505–507; (d) K. Wiesner, *Fortschr. Chem. Org. Naturst.* **20,** 271 (1962); (e) K. W. Bentley, The Alkaloids, Part II, in *The Chemistry of Natural Products,* Vol. 7, K. W. Bentley, Ed., Wiley-Interscience, New York, 1965, pp. 202–214; (f) D. B. MacLean, in *The Alkaloids,* Vol. 10, R. H. F. Manske, Ed., Academic, New York, 1968, pp. 305–382; (g) D. B. MacLean, in *Chemistry of The Alkaloids,* S. W. Pelletier, Ed., Van Nostrand Reinhold Co., New York, 1970, pp. 469–502; (h) D. B. MacLean, in *The Alkaloids,* Vol. 14, R. H. F. Manske, Ed., Academic, New York, 1973, pp. 348–406; (i) W. A. Ayer, in

MTP International Review of Science, Series 1, Vol. 9, D. H. Hey and K. Wiesner, Eds., University Park Press, Baltimore, MD, 1973, pp. 1–25; (j) W. A. Ayer, *Alkaloids (London)* **6**, 252 (1976); (k) R. V. Stevens, in *The Total Synthesis of Natural Products,* Vol. 3, J. ApSimon, Ed., Wiley, New York, 1977, pp. 489–515; (l) W. A. Ayer, *Alkaloids (London)* **8**, 216 (1978); (m) D. R. Dalton, *The Alkaloids: The Fundamental Chemistry—A Biogenetic Approach,* Dekker, New York, 1979, pp. 97–109, 131–144; (n) W. A. Ayer, *Alkaloids (London)* **10**, 205 (1980); (o) W. A. Ayer, *Alkaloids (London)* **11**, 199 (1981); (p) G. A. Cordell, *Introduction to Alkaloids: A Biogenetic Approach,* Wiley, New York, 1981, pp. 170–195; (q) Y. Inubushi and T. Harayama, *Heterocycles* **15**, 611 (1981); (r) P. Welzel, *Nachr. Chem. Tech. Lab.* **31**, 462 (1983).

3. K. Bödeker, *Justus Liebigs Ann. Chem.* **208**, 363 (1881).

4. O. Achmatowicz and W. Uzieblo, *Roczniki Chem.* **18**, 88 (1938).

5. R. H. F. Manske and L. Marion, *Can. J. Res.* **20B**, 87 (1942); L. Marion and R. H. F. Manske, **20B**, 153 (1942); R. H. F. Manske and L. Marion, **21B**, 92 (1943); L. Marion and R. H. F. Manske, **22B**, 1 (1944); R. H. F. Manske and L. Marion, **22B**, 53 (1944); L. Marion and R. H. F. Manske, **22B**, 137 (1944); R. H. F. Manske and L. Marion, **24B**, 57 (1946); L. Marion and R. H. F. Manske, **24B**, 63 (1946); R. H. F. Manske and L. Marion, *J. Am. Chem. Soc.* **69**, 2126 (1947); L. Marion and R. H. F. Manske, *Can. J. Res.* **26B**, 1 (1948); R. H. F. Manske, *Can. J. Chem.* **31**, 894 (1953).

6. P. N. Arata and R. Canozoneri, *Gazz. Chim. Ital.* **22**, 146 (1892); J. A. Dominguez, *Rev. Centro. Estud. Farm. Bioqium.* **20**, 534 (1931); A. Orechoff, *Arch. Pharm.* **272**, 673 (1934); J. Muszynski, *Arch. Pharm.* **273**, 452 (1935).

7. (a) H. M. Lee and K. K. Chen, *J. Am. Pharm. Assoc.* **34**, 197 (1945); (b) G. Marier and R. Bernard, *Can. J. Res.* **E26**, 174 (1948).

8. C. Yang, *Zhongyao Tongbao* **6**, 12 (1981); *CA* **96**; 177926q (1982).

9. (a) D. B. MacLean, R. H. F. Manske, and L. Marion, *Can. J. Res.* **B28**, 460 (1950); (b) L. R. C. Barclay and D. B. MacLean, *Can. J. Chem.* **34**, 1519 (1956); (c) D. B. MacLean and W. A. Harrison, *Can. J. Chem.* **37**, 1757 (1959); (d) W. A. Harrison and D. B. MacLean, *Chem. Ind. (London),* 261 (1960); (e) W. A. Harrison, M. Curcumelli-Rodostamo, D. F. Carson, L. R. C. Barclay, and D. B. MacLean, *Can. J. Chem.* **39**, 2086 (1961); (f) D. Rogers, A. Quick, and M.-U. Haque, *J. Chem. Soc., Chem. Commun.,* 522 (1974).

10. The numbering used for lycopodine is that corresponding to biosynthetic suggestions (*vide infra*).

11. W. A. Ayer, N. Masaki, and D. S. Nkunika, *Can. J. Chem.* **46**, 3631 (1968).

12. W. A. Ayer, L. M. Browne, Y. Nakahara, M. Tori, and L. T. J. Delbaere, *Can. J. Chem.* **57**, 1105 (1979).

13. W. A. Ayer and G. G. Iverach, *Can. J. Chem.* **38**, 1823 (1960).

14. L. Nyembo, A. Goffin, C. Hootelé, and J. C. Braekman, *Can. J. Chem.* **56**, 851 (1978).

15. K. Weisner, W. A. Ayer, L. R. Fowler, and Z. Valenta, *Chem. Ind. (London),* 564 (1957).

16. Z. Valenta, H. Yoshimura, E. F. Rogers, M. Ternbah, and K. Wiesner, *Tetrahedron Lett.,* 26 (1960).

17. W. A. Ayer and G. G. Iverach, *Tetrahedron Lett.,* 19 (1960).

18. H. Conroy, *Tetrahedron Lett.,* 34 (1960).

19. Y. Inubushi, H. Ishii, B. Yasui, and T. Harayama, *Tetrahedron Lett.,* 1551 (1966); *Chem. Pharm. Bull.* **16**, 101 (1968).

20. E. Leete and M. C. L. Louden, 145th National Meeting of the American Chemical Society, New York, September 1963, Abstract 1-C.

21. R. N. Gupta, M. Castillo, D. B. MacLean, I. D. Spenser, and J. T. Wrobel, *J. Am. Chem. Soc.* **90**, 1360 (1968).

22. E. Leete, *J. Am. Chem. Soc.* **86**, 2509 (1964).

23. R. N. Gupta and I. D. Spenser, *Phytochemistry* **8**, 1937 (1969).

24. M. Castillo, R. N. Gupta, D. B. MacLean, and I. D. Spenser, *Can. J. Chem.* **48**, 1893 (1970).

25. M. Castillo, R. N. Gupta, D. B. MacLean, and I. D. Spenser, *Can. J. Chem.* **48**, 2911 (1970).

26. Y. K. Ho, R. N. Gupta, D. B. MacLean, and I. D. Spenser, *Can. J. Chem.* **49**, 3352 (1971).

27. J.-C. Braekman, R. N. Gupta, D. B. MacLean, and I. D. Spenser, *Can. J. Chem.* **50**, 2591 (1972).

28. W. D. Marshall, T. T. Nguyen, D. B. MacLean, and I. D. Spenser, *Can. J. Chem.* **53**, 41 (1975).

29. L. Nyembo, A. Goffin, C. Hootelé, and J.-C. Braekman, *Can. J. Chem.* **56**, 851 (1978).

30. (a) Z.-I. Horii, S.-W. Kim, T. Imanishi, and T. Momose, *Chem. Pharm. Bull.* **18**, 2235 (1970);
     (b) W. A. Ayer, Y. Fukazawa, and P. P. Singer, *Tetrahedron Lett.,* 5045 (1973).

31. Y. Inubushi, H. Ishii, T. Harayama, R. H. Burnell, W. A. Ayer, and B. Altenkirk, *Tetrahedron Lett.,* 1069 (1967).

32. M. Przybylska and L. Marion, *Can. J. Chem.* **35**, 1075 (1957).

33. K. Wiesner and L. Poon, *Tetrahedron Lett.,* 4937 (1967); K. Wiesner, L. Poon, I. Jirkovský, and M. Fishman, *Can. J. Chem.* **47**, 433 (1969).

34. C. A. Grob and J. H. Wilkens, *Helv. Chim. Acta.* **48**, 808 (1965).

35. E. H. W. Böhme, Z. Valenta, and K. Wiesner, *Tetrahedron Lett.,* 2441 (1965).

36. K. Wienser, Z. Valenta, W. A. Ayer, L. R. Fowler, and J. E. Francis, *Tetrahedron* **4**, 87 (1958).

37. Annotinine lactam can be readily prepared from annotinine: D. B. MacLean and H. C. Prime, *Can. J. Chem.* **31**, 543 (1953).

38. K. Wiesner and I. Jirkovský, *Tetrahedron Lett.,* 2077 (1967).

39. E. E. Betts and D. B. MacLean, *Can. J. Chem.* **35**, 211 (1957).

40. L. Mandell, J. U. Piper, and K. P. Singh, *J. Org. Chem.* **28**, 3440 (1963); L. Mandell, B. A. Hall, and K. P. Singh, *J. Org. Chem.* **29**, 3067 (1964).

41. N. J. Leonard and M. J. Middleton, *J. Am. Chem. Soc.* **74**, 5114 (1952).

42. E. Wenkert, K. G. Dave, and R. V. Stevens, *J. Am. Chem. Soc.* **90**, 6177 (1968).

43. E. P. Anderson, J. V. Crawford, and M. L. Sherill, *J. Am. Chem. Soc.* **68**, 1294 (1946).

44. For a review of the synthetic utility of β-acyl pyridinium salts, see E. Wenkert, *Acc. Chem. Res.* **1**, 78 (1968).

45. E. Wenkert, B. Chauncy, K. G. Dave, A. R. Jeffcoat, F. M. Schell, and H. P. Schenk, *J. Am. Chem. Soc.* **95**, 8427 (1973).

46. E. Wenkert, B. Chauncy, and S. H. Wentland, *Synth. Commun.* **3**, 73 (1973).

47. A. Ahond, A. Cavé, C. Kan-Fan, H.-P. Husson, J. de Rostolan, and P. Potier, *J. Am. Chem. Soc.* **90**, 5622 (1968); A. Ahond, A Cavé, C. Kan-Fan, Y. Langlois, and P. Potier, *J. Chem. Soc. Chem. Commun.,* 517 (1970); H.-P. Husson, L. Chevolot, Y. Langlois, C. Thal, and P. Potier, *J. Chem. Soc. Chem. Commun.,* 930 (1972).

48. The corresponding des-acetonyl compound had been previously made in low yield in a preparation of tetrahydrolyconnotine: Z. Valenta, P. Deslongchamps, R. A. Ellison, and K. Wiesner, *J. Am. Chem. Soc.* **86**, 2533 (1964).

49. F. Bohlmann and O. Schmidt, *Chem. Ber.* **97**, 1354 (1964).

50. E. F. Kleinman, Ph.D. dissertation, University of California, Berkeley, 1980.

51. H. Dugas, R. A. Ellison, Z. Valenta, K. Wiesner, and C. M. Wong, *Tetrahedron Lett.,* 1279 (1965).

52. E. Colvin, J. Martin, W. Parker, and R. A. Raphael, *J. Chem. Soc. Chem. Commun.,* 596 (1966).

53. A. C. Cope and M. E. Synerholm, *J. Am. Chem. Soc.* **72**, 5228 (1950); E. Colvin and W. Parker, *J. Chem. Soc.,* 5764 (1965).

54. (a) Z.-I. Horii, T. Imanishi, S.-W. Kim, and I. Ninomiya, *Chem. Pharm. Bull.* **16**, 1918 (1968); (b) Z.-I. Horii, S.-W. Kim, T. Imanishi, and I. Ninomiya, *Chem. Pharm. Bull.* **16**, 2107 (1968).

55. W. A. Ayer, W. R. Bowman, T. C. Joseph, and P. Smith, *J. Am. Chem. Soc.* **90**, 1647 (1968).

56. Compound **110** was also prepared by degradation of lycopodine and was used as a relay compound to complete this project.

57. G. Stork, R. A. Kretchmer, and R. H. Schlessinger, *J. Am. Chem. Soc.* **90**, 1648 (1968).

58. G. Stork, *Pure Appl. chem.* **17**, 383 (1968).

59. H. Dugas, M. E. Hazenberg, Z. Valenta, and K. Wiesner, *Tetrahedron Lett.*, 4931 (1967).

60. W. A. Ayer, T. E. Habgood, V. Deulofeu, and H. R. Juliani, *Tetrahedron* **21**, 2169 (1965).

61. H. Reinshagen, *Annalen* **681**, 84 (1965).

62. K. Wiesner, V. Musil, and K. J. Wiesner, *Tetrahedron Lett.*, 5643 (1968).

63. For reviews of the Mannich condensation, see (a) F. H. Blicke, *Org. React.* **1**, 303 (1942); (b) H. Hellmann and G. Optiz, *Angew. Chem.* **68**, 265 (1956); (c) M. Tramontini, *Synthesis*, 703 (1975).

64. R. A. Kunz, Ph.D. dissertation, University of Wisconsin, Madison, 1974.

65. I. N. Nazarov and S. I. Zavyalov, *Zh. Obshch. Khim.* **23**, 1703 (1953); J. E. Ellis, J. S. Dutcher, and C. H. Heathcock, *Synth. Commun.* **4**, 71 (1974).

66. C. H. Heathcock, E. Kleinman, and E. S. Binkley, *J. Am. Chem. Soc.* **100**, 8036 (1978).

67. (a) C. H. Heathcock, E. F. Kleinman, and E. S. Binkley, *J. Am. Chem. Soc.* **104**, 1054 (1982); (b) C. H. Heathcock, E. F. Kleinman, and E. S. Binkley, *Int. Congr. Ser.—Excepta Med.* **457**, 71 (1979).

68. R. D. Clark and C. H. Heathcock, *J. Org. Chem.* **41**, 636 (1976).

69. A. Hosomi and H. Sakurai, *J. Am. Chem. Soc.* **99**, 1673 (1977); H. Sakurai, *Pure Appl. Chem.* **54**, 1 (1982).

70. T. A. Blumenkopf and C. H. Heathcock, *J. Am. Chem. Soc.* **105**, 2354 (1983).

71. R. B. Woodward, N. L. Wendler, and F. J. Brutschy, *J. Am. Chem. Soc.* **67**, 1425 (1945); H. Rapoport, R. Naumann, F. R. Bissell, and R. M. Borner, *J. Org. Chem.* **15**, 1103 (1950).

72. E. Kleinman and C. H. Heathcock, *Tetrahedron Lett.*, 4125 (1979).

73. D. Schumann, H.-J. Müller, and A. Naumann, *Liebigs Ann. Chem.*, 1700 (1982).

74. E. J. Corey, W. J. Howe, H. W. Orf, D. A. Pensak, and P. Petersson, *J. Am. Chem. Soc.* **97**, 6116 (1975).

75. C. H. Heathcock and E. F. Kleinman, *J. Am. Chem. Soc.* **103**, 222 (1981).

76. S.-W. Kim, Y. Bando, and Z.-I. Horii, *Tetrahedron Lett.*, 2293 (1978).

77. T. Mamose, S. Uchida, T. Imanishi, S. Kim, N. Takahashi, and Z. Horii, *Heterocycles* **6**, 1105 (1977).

78. W. A. Ayer and G. G. Iverach, *Can. J. Chem.* **42**, 2514 (1964).

79. W. L. Scott and D. A. Evans, *J. Am. Chem. Soc.* **94**, 4779 (1972).

80. D. A. Evans, W. L. Scott, and L. K. Truesdale, *Tetrahedron Lett.*, 121 (1972).

81. W. G. Dauben, M. E. Lorber, N. D. Vietmeyer, R. H. Shapiro, J. H. Duncan, and K. Tomer, *J. Am. Chem. Soc.* **90**, 4762 (1968).

82. W. Oppolzer and M. Petrzilka, *J. Am. Chem. Soc.* **98**, 6722 (1976).

83. J. Szychowski and D. B. MacLean, *Can. J. Chem.* **57**, 1631 (1979).

84. K. Nomura, J. Adachi, M. Hanai, S. Nakayama, and K. Mitsuhashi, *Chem. Pharm. Bull.* **22**, 1386 (1974).

85. A. Leniewski, J. Szychowski, and D. B. MacLean, *Can. J. Chem.* **59**, 2479 (1981).

86. W. A. Ayer, J. K. Jenkins, S. Valverde-Lopez, and R. H. Burnell, *Can. J. Chem.* **45**, 433 (1967).

87. W. A. Ayer, J. K. Jenkins, K. Piers, and S. Valverde-Lopez, *Can. J. Chem.* **45**, 445 (1967).

88. W. A. Ayer and K. Piers, *Can. J. Chem.* **45**, 451 (1967).

89. E. Ochiai, S. Okuda, and H. Minato, *J. Pharm. Soc. Japan* **72**, 1481 (1956).

90. For studies of related cyclizations, see W. A. Ayer, R. Dawe, R. A. Eisner, and K. Furuichi, *Can. J. Chem.* **54**, 473 (1976); W. A. Ayer and K. Furuichi, *Can. J. Chem.* **54**, 1494 (1976); R. V. Stevens and A. W. M. Lee, *J. Am. Chem. Soc.* **101**, 7032 (1979).

91. Y. Ban, M. Kimura, and T. Oishi, *Chem. Pharm. Bull.* **24**, 1490 (1976).

92. T. Harayama, M. Ohtani, M. Oki, and Y. Inubushi, *J. Chem. Soc. Chem. Commun.*, 827 (1974); *Chem Pharm. Bull.* **23**, 1511 (1975).

93. (a) T. Harayama, M. Ohtani, M. Oki, and Y. Inubushi, *Chem. Pharm. Bull.* **21**, 25 (1973). (b) 1061 (1973).

94. The methodology employed is based on that developed by Woodward for synthesis of the steroid ring D: R. B. Woodward, F. Sondheimer, D. Taub, K. Heusler, and W. M. McLamore, *J. Am. Chem. Soc* **74**, 4223 (1952); see also reference 93b.

95. For a study of the relative reduction rates of cyclic ketones, see H. C. Brown and K. Ichikawa, *Tetrahedron* **1**, 221 (1957).

96. (a) R. H. Burnell, C. G. Chin, B. S. Mootoo, and D. R. Taylor, *Can. J. Chem.* **41**, 3091 (1963); (b) W. A. Ayer, Y. Fukazawa, P. P. Singer, and B. Altenkirk, *Tetrahedron Lett.*, 5045 (1973).

97. T. Harayama, M. Takatani, and Y. Inubushi, *Tetrahedron Lett.*, 4307 (1979); *Chem. Pharm. Bull.* **28**, 2394 (1980).

98. T. Harayama, H. Cho, and Y. Inubushi, *Chem. Pharm. Bull.* **25**, 2773 (1977); **26**, 1201 (1978).

99. Protection of the amine also resulted in formation of a trifluoroacetate, which was selectively hydrolyzed before oxidation.

100. R. H. Burnell, *Can. J. Chem.* **37**, 3091 (1959).

101. R. H. Burnell, A. Humblet, J. Pelletier, and S. Badripersand, *Can. J. Chem.* **58**, 1243 (1980).

Chapter Six

The Synthesis of Indolizidine and Quinolizidine Alkaloids of *Tylophora, Cryptocarya, Ipomoea, Elaeocarpus,* and Related Species

R. B. Herbert

Department of Organic Chemistry
The University
Leeds LS2 9JT, England

CONTENTS

1. INTRODUCTION

The most typical phenanthroindolizidine alkaloid, (−)-tylophorine (1), is the most widely distributed of this small family of plant bases and has attracted the most attention as a synthetic target.

Phenanthroindolizidine alkaloids have been found in the genus *Tylophora* and the plants *Cyanchum vincetoxicum, Vincetoxicum officinale, Pergularia pallida,* and *Antitoxicum funebre* among the more than 300 genera of the Asclepiadaceae family. Typical of these alkaloids, in addition to (−)-tylophorine (1), are tylophorinine (2) (racemic in *T. asthmatica* but optically active in *P. pallida*), (+)-tylophorinidine (3), (−)-antofine (4), (−)-tylocrebrine (5), and (+)-isotylocrebrine (6) (absolute configuration shown in each case). A single plant of

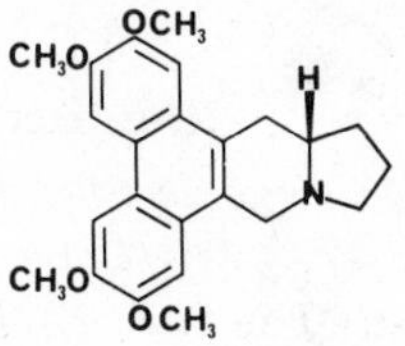

(−)-Tylophorine (1) 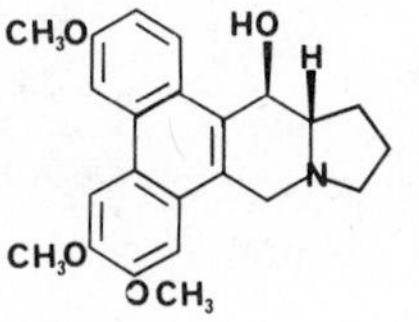   Tylophorinine (2) 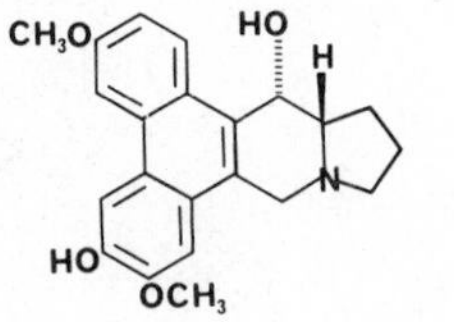   (+)-Tylophorinidine (3)

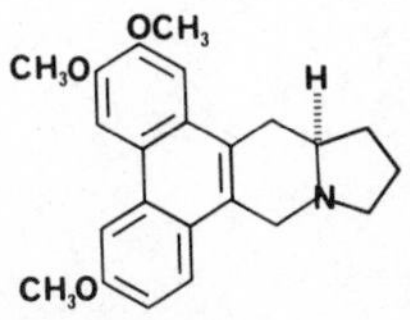

(−)-Antofine (4) 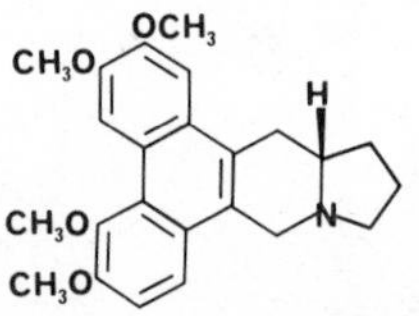   (−)-Tylocrebrine (5) 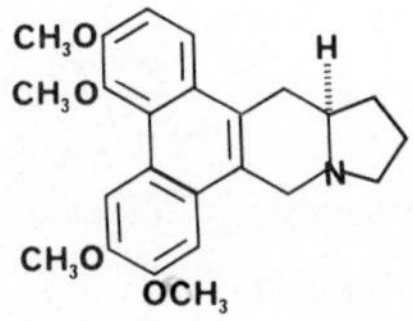   (+)-Isotylocrebrine (6)

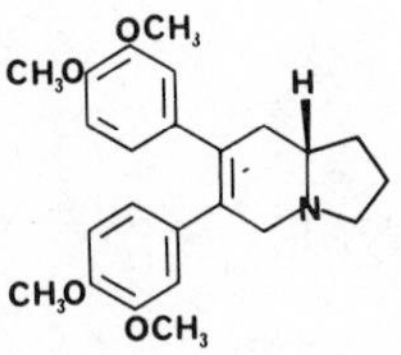

(−)-Septicine (7) 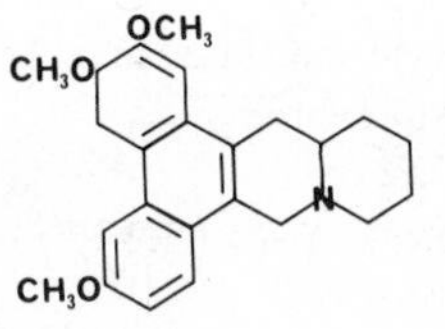   (−)-Cryptopleurine (8) 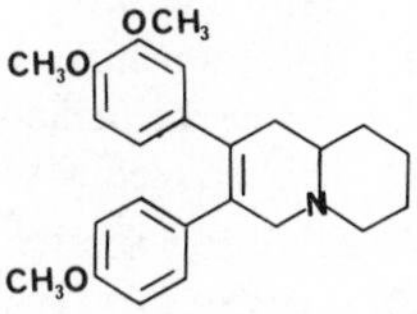   (+)-Julandine (9)

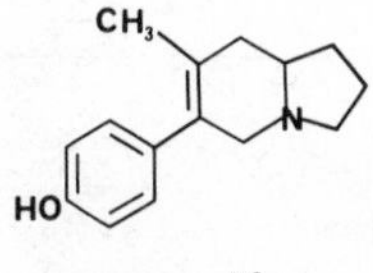

10

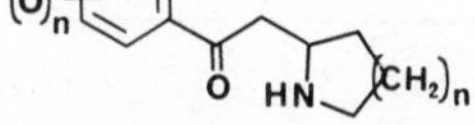

11, n = 1 or 2

the Moraceae family, *Ficus septica,* has been reported to produce alkaloids of this type, namely (−)-tylophorine **(1)**, (+)-tylocrebrine [as **(5)**] and the interesting seco-base, (−)-septicine **(7)**; (+)-septicine [as **(7)**] has been found as a minor alkaloid of *T. asthmatica.* (For reviews on the natural occurrence and structure determination of these alkaloids see refs. 1–4.)

Closely related in structure to the phenanthroindolizidine alkaloids is (−)-cryptopleurine **(8)** found in *Cryptocarya pleurosperma* (Lauraceae) and *Boehmeria platyphylla* [5, 6]. The latter plant also produces the seco-alkaloid (+)-julandine **(9)** [6].

Some syntheses of phenanthroindolizidine and phenanthroquinolizidine alkaloids overlap, and there is overlap also with the synthesis of ipalbidine **(10)**. This alkaloid and its glucoside, ipalbine, have been found in the seeds of *Ipomoea alba* [7].

The biosynthetic routes to alkaloids represented by structures **1–10** are almost certainly similar. Where biosynthetic ideas have been exploited in synthesis a few related, simple, naturally occurring alkaloids of type **11** have been prepared in the course of elaborating the more complex bases. [These simple bases are exemplified by the alkaloids of *Ruspolia hypercrateriformis* (Section 6)].

Syntheses of alkaloids represented by **1–9** are often interwoven with those of the *Elaeocarpus* alkaloids. These alkaloids constitute a group with a unique set of structures, which have been discovered only quite recently [4, 8]. Elaeocarpine **(12)** and elaeokanine C **(13)** may be taken as typical examples of these alkaloids, which are found in *Elaeocarpus* species (Elaeocarpaceae).

There is a line of biosynthetic reasoning for all these alkaloids **(1–13)** that has been successfully tapped in the development of economical syntheses. To appreciate fully these synthetic approaches, understanding of the actual (and likely) biosynthetic pathways is necessary. So, we begin with the topic of biosynthesis and will follow with that of synthesis.

This survey covers the literature up to the end of June 1983.

2. BIOSYNTHESIS

The alkaloids represented by tylophorine **(1)** and cryptopleurine **(8)** are of unique structure, a fact which leads one to expect a unique biosynthesis. This is certainly true for the later stages of biosynthesis, but the origins in primary metabolism and the general features of the early steps are shared with many other alkaloids. Thus tyrosine provides ring B plus C(9) and C(10) in tylophorine **(16)** via dopa **(15)**, and phenylalanine serves as an independent source for ring A plus C(14) and C(15) by way of cinnamic acid **(14)**; ornithine is in all probability the source of ring E and the nitrogen atom [9, 10]. The biosynthesis of cryptopleurine has not been studied, but similar origins and intermediates are expected with lysine replacing ornithine as a precursor.

Important clues to the nature of the intermediates involved in the biosynthesis of tylophorine **(16)** and cryptopleurine **(8)** are provided by the natural occurrence of the seco-bases, septicine (**7**) and julandine (**9**), and alkaloids of type **11**. A reasonable hypothesis for the biosynthesis of these alkaloids is that crypto-pleurine is formed sequentially through intermediates of type **11** and **9**. Similarly the biosynthesis of tylophorine (**16**) and tylophorinine (**2**) was predicted to be through intermediates of type **11** and **7**. This prediction has been verified for the first of these hypothetical intermediates by showing [11] that the phenacylpyr-rolidines **20**, **21**, and **22** are intact precursors for tylophorinine (**2**) in *T. asthmatica*. In addition the oxo-acids **17** and **18**, but not (**23**), were also found to act as precursors. Taken together, these results define the early stages of tylophorinine biosynthesis as that shown in Scheme 1. The incorporation of **18**

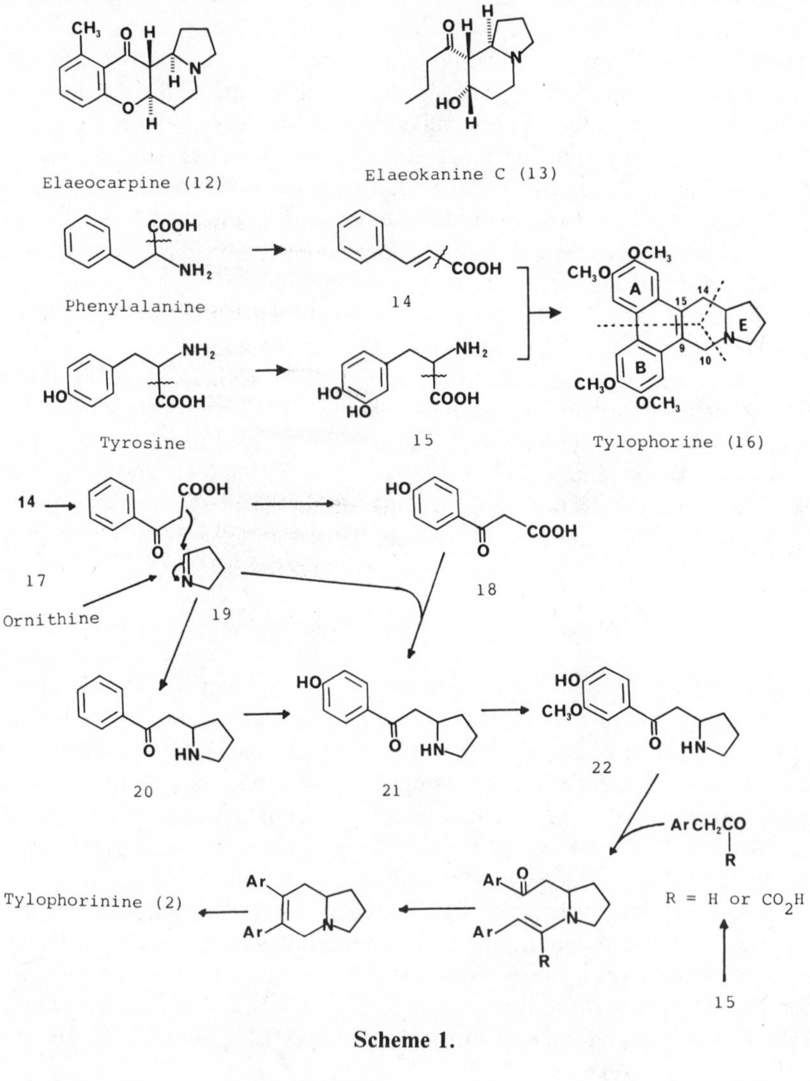

Elaeocarpine (12)

Elaeokanine C (13)

Phenylalanine

14

Tyrosine

15

Tylophorine (16)

17

Ornithine

19

18

20

21

22

Tylophorinine (2)

$R = H$ or CO_2H

15

Scheme 1.

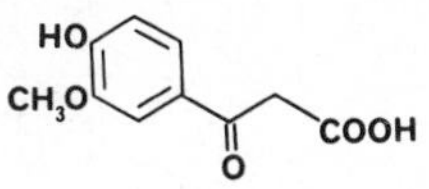

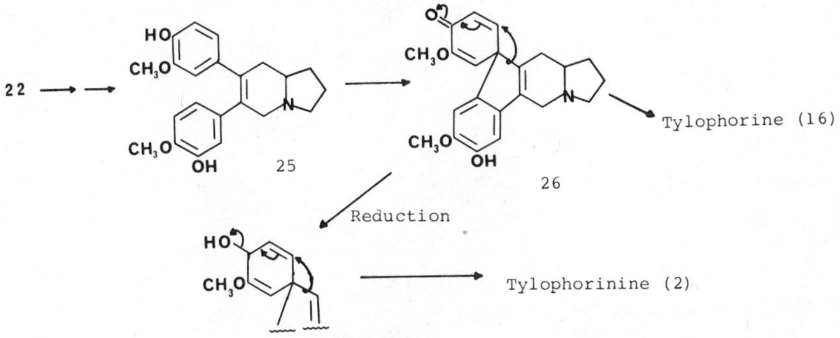

Scheme 2.

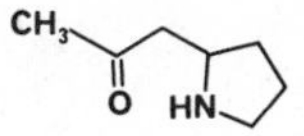

and **20** into **2** indicates that aromatic hydroxylation can occur before and after phenacylpyrrolidine formation [11]. Subsequent to these experiments, **22** was identified as the alkaloid, norruspolinone, found in *R. hypercrateriformis* [12].

The hexahydroindolizine **25** has been clearly identified [10, 13] as a pivotal late intermediate in the biosynthesis of phenanthroindolizidine alkaloids in *T. asthmatica.* It is plausibly derived as outlined in Scheme 1. Phenol oxidative coupling within **25** leads to the dienone **26** from which alternatively by key steps of rearrangement, or reduction and rearrangement, tylophorine **(16)** and tylophorinine **(2),** respectively, may be derived (Scheme 2) [13]. Hexahydro-indolizine intermediates lying earlier on the pathway have been tentatively identified [10, 13] as **27, 28,** and **29.** The latter would have to be derived from

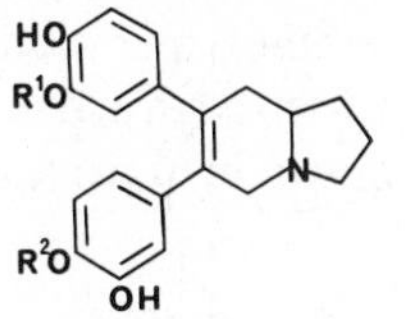

27, R^1 = R^2 = H
28, R^1 = Me, R^2 = H

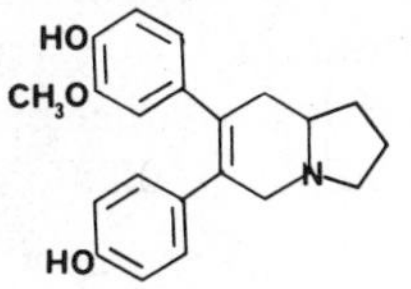

tyrosine without the intermediacy of dopa **(15).** Only minor modification of the pathway to tylophorine **(16)** and tylophorinine **(2),** illustrated in Schemes 1 and 2, is required in order to write schemes for the biosynthesis of other phenanthro-indolizidine alkaloids. Ipalbidine **(10)** is seen to have a similar biogenesis where hygrine **(24)** replaces **22** as an intermediate (cf. Scheme 1).

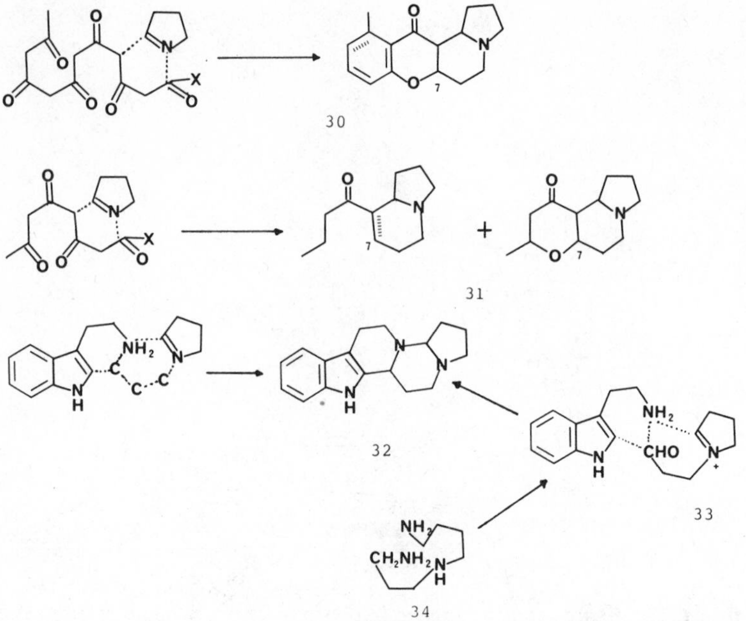

Scheme 3.

1-Pyrroline (19) is placed as a key early intermediate in phenanthroindolizi-dine alkaloid biosynthesis, in part by analogy with the known biosynthesis of a host of pyrrolidine, and also piperidine, alkaloids [14]. It probably is also a key intermediate in the formation of *Elaeocarpus* bases. Here condensation occurs, in a formal sense at least, with a C_{12} or C_8 polyketide leading, respectively, to C_{16}-aromatic and dienone alkaloids 30, and C_{12}-alkaloids 31 (Scheme 3) [15, 16].

Elaeocarpidine (32) could arise similarly from 19, tryptamine and a C_3 unit [17]. A priori this seems less likely than an alternative suggestion [18] that 33 could be the species that condenses with tryptamine; 33 could very reasonably derive from the naturally occurring triamine, spermidine (34). The other *Elaeocarpus* alkaloids could then be biosynthesized in a manner not dissimilar to the phenanthroindolizidine alkaloids, as shown in Scheme 4. In support of this

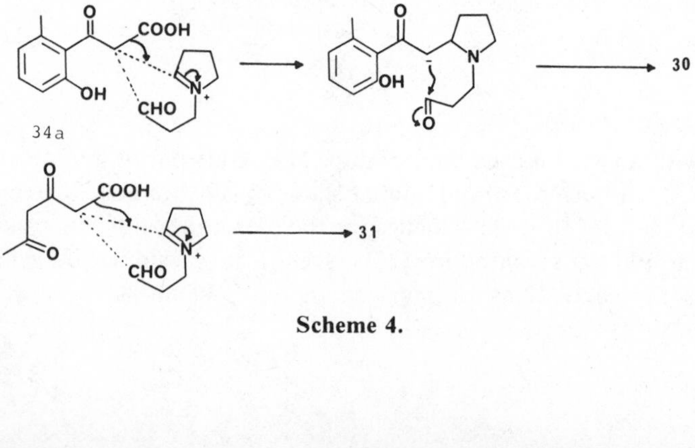

Scheme 4.

idea, 2-hydroxyacetophenone, formed from **34a** by decarboxylation, has been isolated from the same plant as elaeocarpine **(12)**, isoelaeocarpicine, and isoelaeocarpine [19]. One difficulty is associated with this hypothesis: an, a priori, unfavorable reduction of the aromatic ring in **35** is required to give the dienone alkaloids.

3. SYNTHESIS OF PHENANTHROINDOLIZIDINE AND RELATED ALKALOIDS

3.1. Tylophorine

A number of syntheses of ($\pm$)-tylophorine [as **(1)**] have been described. The first allowed confirmation of the structure assigned on the basis of degradative evidence. Later syntheses are illustrations of the art of synthesis. No doubt the most easily accessible phenanthrene substitution pattern seen in **1** makes it the most popular of phenanthroindolizidine targets. One synthesis is modeled on the biosynthetic pathway to tylophorine and related alkaloids and proceeds via septicine **(7)**. Since the latter naturally occurring base may be converted efficiently into tylophorine **(1)** with thallium(III)trifluoroacetate [20], syntheses of septicine **(7)** are formally syntheses of tylophorine also. Septicine **(7)** has also been photocyclized to tylophorine **(1)** [21, 22].

The first synthesis [23] to be described involved the preparation of the phenanthrene **35**. Its condensation with pyrryl magnesium bromide yielded **40,** which was hydrogenated to give the corresponding pyrrolidine. Cyclisation of the *N*-formyl derivative of the latter compound with phosphorus oxychloride followed by reduction gave ($\pm$)-tylophorine [as **(1)**]. Synthetic ($-$)-tylophorine **(1)** obtained by resolution was identical with the natural alkaloid.

In a slightly different approach [20], the acid **36** was treated with methyl

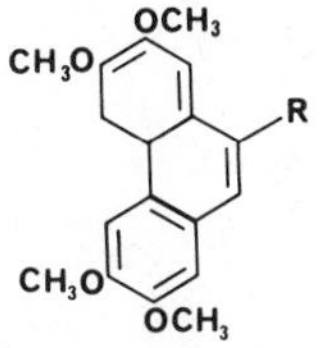

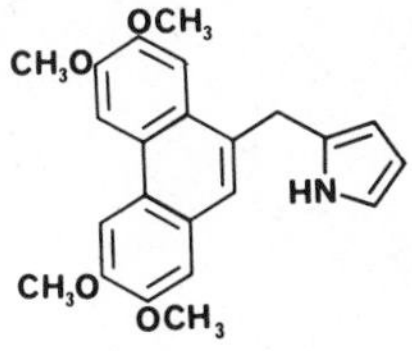

35, R = CH$_2$Cl

36, R = CO$_2$H

37, R = (structure)

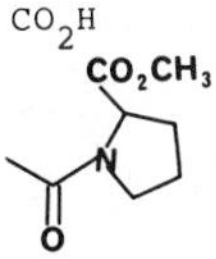

38, R = CO$_2$Me

39, R = (structure)

40

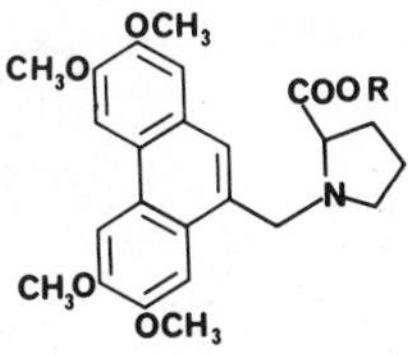

41, R = Me

42, R = H

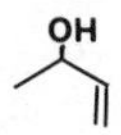

prolinate and dicyclohexylcarbodiimide to give the amide 37. Specific *O*-alkylation of the amide function with triethyloxonium fluoroborate and subsequent reduction with sodium borohydride [cf. 24] yielded the ester 41. The corresponding acid 42 was cyclized with polyphosphoric acid to the ketone 43, which was converted into tylophorine by a Clemmensen reduction. The ketone 43 was also obtained through condensation of 35 with benzyl prolinate, and subsequent hydrolysis and cyclization [25]. Conversion into (±)-tylophorine [as (1)] was effected by sodium borohydride reduction of the tosylhydrazone of 43.

Two novel syntheses of tylophorine (1) complete the review of those published to date, with the exception of those discussed in Section 3.2 in association with the synthesis of septicine (7). The second of these novel syntheses involved an intramolecular imino Diels–Alder reaction on 51 [26]. The first of these syntheses involved the elaboration of 45 and from it the formation in two steps of three further rings leading to tylophorine [27].

Michael addition of the ketoester 44 to 3,4-dimethoxybenzyl cyanide, in the presence of catalytic amounts of anhydrous potassium carbonate gave adduct 45. Hydrogenation of 45 resulted in a double ring closure and the formation of the lactam 46. Vanadyl trifluoride (VOF₃) was shown to be a generally effective reagent for the conversion of stilbenes into phenanthrenes [27]. This reagent effected the cyclization of 46 with simultaneous dehydrogenation to give 47. Reduction of the latter with diborane gave (±)-tylophorine [as (1)] [27].

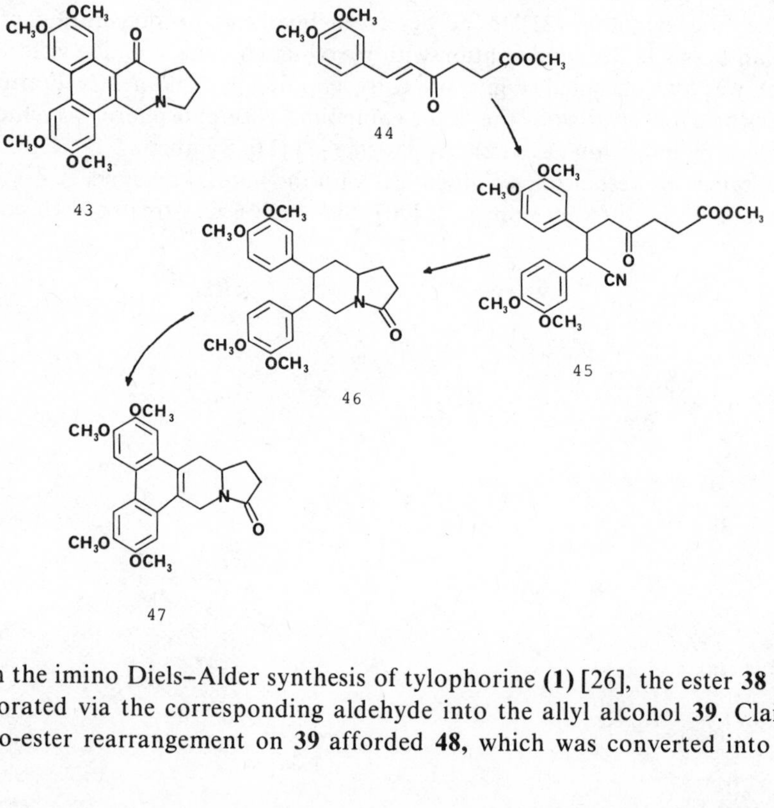

In the imino Diels–Alder synthesis of tylophorine (1) [26], the ester 38 was elaborated via the corresponding aldehyde into the allyl alcohol 39. Claisen ortho-ester rearrangement on 39 afforded 48, which was converted into the

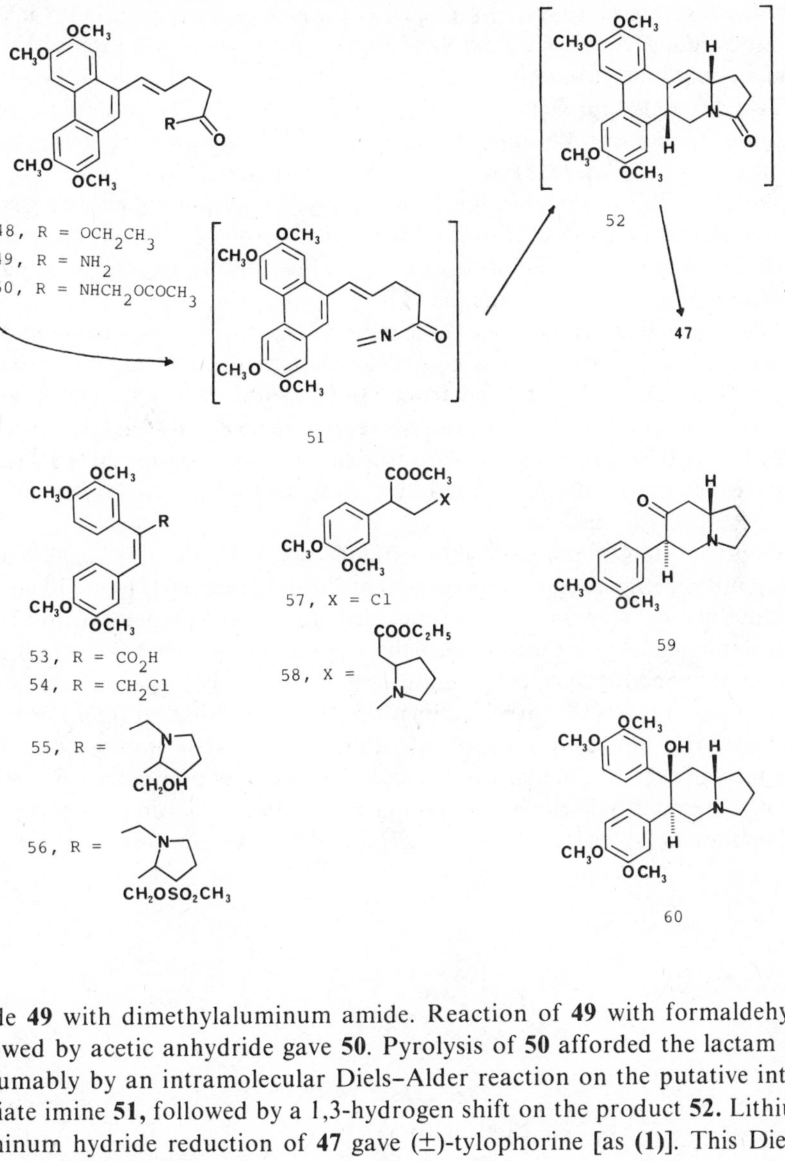

48, R = OCH₂CH₃ — written as $48, R = OCH_2CH_3$

49, R = NH₂ — $49, R = NH_2$

50, R = NHCH₂OCOCH₃ — $50, R = NHCH_2OCOCH_3$

51

52

47

53, R = CO₂H — $53, R = CO_2H$

54, R = CH₂Cl — $54, R = CH_2Cl$

55, R =

56, R =

57, X = Cl

58, X =

59

60

amide **49** with dimethylaluminum amide. Reaction of **49** with formaldehyde followed by acetic anhydride gave **50**. Pyrolysis of **50** afforded the lactam **47**, presumably by an intramolecular Diels–Alder reaction on the putative intermediate imine **51**, followed by a 1,3-hydrogen shift on the product **52**. Lithium aluminum hydride reduction of **47** gave (±)-tylophorine [as (**1**)]. This Diels–Alder reaction has found application in the synthesis of other alkaloids, notably two of the elaeokanines (Section 7).

3.2. Septicine

The first reported synthesis of septicine (**7**) involved the preparation of the stilbene acid **53** and its transformation by conventional means into the chloride **54**. Condensation of the latter with L-prolinol gave the amino-alcohol **55**. Treatment of the corresponding mesylate **56** with sodium hydride in anhydrous dimethylformamide gave (−)-septicine (7), which was identical with material isolated from *F. septica*. Attempts to carry out the cyclization under Friedel–Crafts conditions gave only small yields of the desired product [28].

Two different approaches to the synthesis of septicine (7) required the preparation of the ketone 59; minor modification of these approaches enabled the synthesis of ipalbidine (10) to be carried out (Section 5).

Condensation of the chloride 57 with ethyl 2-pyrrolidinyl acetate gave the diester 58. Dieckmann cyclization yielded the ketone 59, which reacted with 3,4-dimethoxyphenyl lithium to give the carbinol 60. Dehydration of 60 with sulfuric acid gave (±)-septicine [as 7] [29].

The alternative synthesis of the ketone 59 and from it (±)-septicine [as 7] (Scheme 5) [30] involved 64 as a key intermediate formed from 63 and 62, the latter compound being derived from 61. The sulfur-bearing intermediate 65 formed in the key step by acid-catalyzed rearrangement of 64 was conceived of as a relatively stable equivalent of the unstable endocyclic enamine 67 (Scheme 6). The ideas contained in this synthesis have been used for the preparation of other alkaloids [31].

Consideration of the likely biosynthesis of tylophorine (1) suggested that compounds with structures represented by 70 and septicine (7) were likely to be key intermediates. Ideas about the course of biosynthesis (Schemes 1 and 2) have been exploited in a synthesis of septicine (7) which began with the pyrrolidin-2-ylacetophenone 70 [20, 32]. This amine was conveniently prepared by condensation of 1-pyrroline (19) with 3,4-dimethoxybenzoylacetic acid (68). 1-Pyrroline was obtained either by reaction of ornithine with *N*-bromosuccinimide or, routinely in excellent preparative yield, by oxidation of putrescine (69) with pea seedling diamine oxidase; in the latter case an aqueous solution of enzyme, acid 68, and amine 69 was incubated at 27°C and pH7 to give 70.

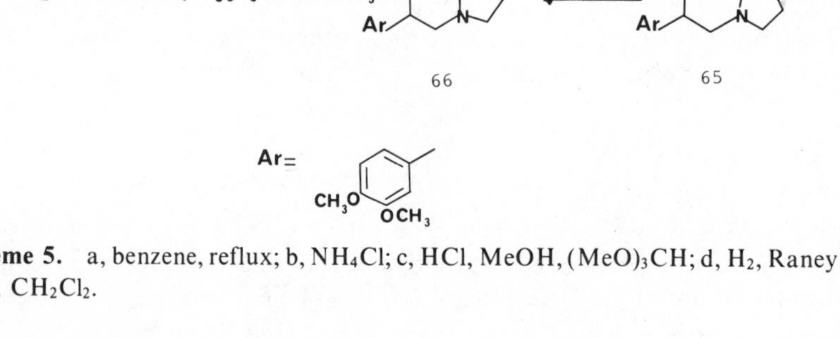

Scheme 5.   a, benzene, reflux; b, NH₄Cl; c, HCl, MeOH, (MeO)₃CH; d, H₂, Raney Ni; e, HCl, CH₂Cl₂.

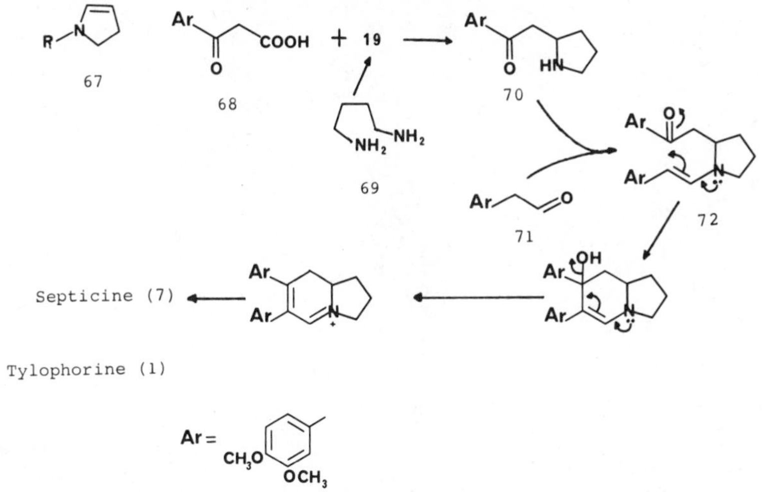

Scheme 6.

The enamine **72** formed spontaneously from **70** and 3,4-dimethyoxyphenyl-acetaldehyde **(71)** in benzene or methanol solution. In the latter solvent cyclization and dehydration also occurred; septicine **(7)** was isolated after treatment with sodium borohydride [20, 32]. The yield for this reaction has been markedly improved over that reported originally by carrying it out in very dilute methanol solution [35]. Cyclization and dehydration of the enamine **(72)** was also catalyzed by titanium(IV)chloride [20] (cf. Section 4).

Reaction of septicine **(7)** with thallium(III)trifluoroacetate, an efficient coupling reagent for the synthesis of biaryls, gave (±)-tylophorine [as **(1)**] in excellent yield as the only coupling product [20].

Adaptation of this biogenetically patterned synthesis to the preparation of other alkaloids is discussed in Sections 4, 5, and 6. This synthesis of septicine readily lends itself to the preparation of isotopically labeled compounds for biosynthetic studies. This route [11, 20, 32] has been applied in the synthesis of **73**. Potassium fericyanide oxidation of **73** and methylation of the product gave tylophorine in 3.8% overall yield [33]. Further, condensation of **70** [20, 32] with 3,4-dimethoxyphenylacetyl chloride gave **83,** which was converted into (±)-septicine [as **(7)**] [34].

1,3-Dipolar cycloaddition of pyrroline oxide **(74)** to olefins, employed so effectively for the synthesis of *Elaeocarpus* alkaloids (Section 7), has found application in two syntheses of septicine **(7)**, the second of which has parallels with the synthesis just discussed.

The butadiene **75** was found to undergo regiospecific [3 + 2] cycloaddition with pyrroline oxide **(74)** to give a 2:1 mixture of the stereoisomers **76** and **77** (Scheme 7). Treatment of **76** with bromine gave **78,** which on reduction with lithium aluminum hydride afforded the epoxide **79** as one of three products. Treatment of **79** with trimethylsilyl iodide gave an intermediate trimethyl-

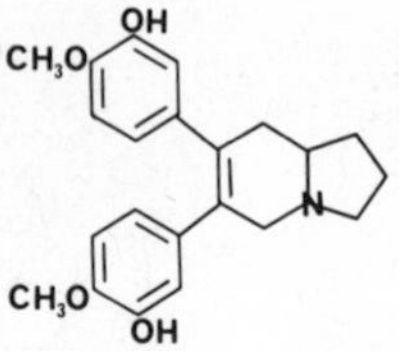

73

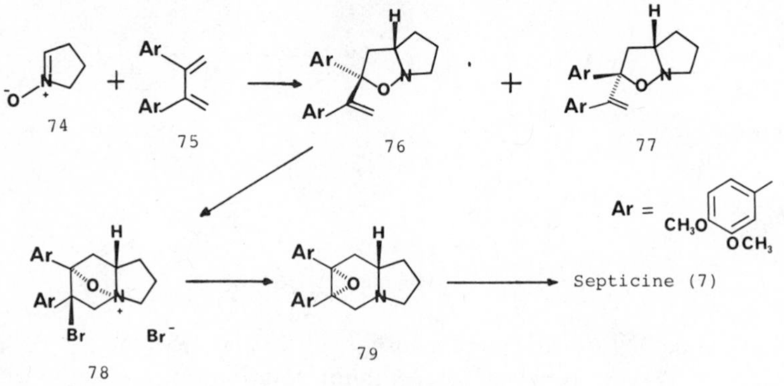

Scheme 7.

silyliodohydrin from which (±)-septicine [as (7)] was obtained by treatment with zinc and acetic acid [36].

[3 + 2] Cycloaddition between pyrroline oxide (74) and 3,4-dimethoxystyrene gave a diastereoisomeric mixture of the two adducts 80 and 81 [22]. Hydrogenolysis gave the alcohol 82 from which the ketoamide 83 could be obtained. Treatment of 83 with alcoholic potassium hydroxide led to the lactam 84.

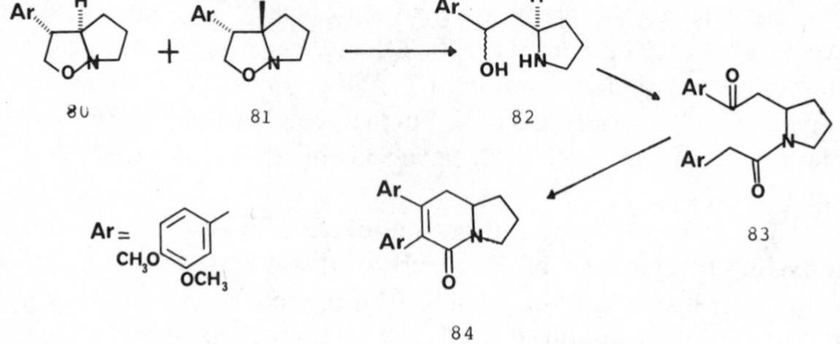

Irradiation of 84 gave a mixture of 9-oxotylophorine (85), 9-oxoisotylocrebrine (86), and 9-oxotylocrebrine (87). Reduction of 85 with diborane or lithium aluminum hydride resulted in recovery of starting material or a low yield of tylophorine. On the other hand, reduction of 84 with the mixed reducing agent prepared from lithium aluminum hydride and aluminum chloride (3:1) gave

(±)-septicine [as (7)] in high yield. Irradiation of the septicine (7) then gave tylophorine (1) as the major product.

3.3. Tylophorinine, Deoxytylophorinine, and Antofine

The presence of the alcohol function in tylophorinine (2) requires particular attention in any synthesis of the alkaloid, and to date only one synthesis has been reported. This synthesis is analogous to one of the early tylophorine syntheses (Section 3.1) and was achieved via the phenanthrene 88 prepared from 9-chloromethyl-3,6,7-trimethoxyphenanthrene and methyl L-prolinate. Poly-phosphoric acid cyclization of the acid corresponding to 88 gave the ketone 89. Tylophorinine (2) was prepared as one of two racemates by sodium borohydride reduction of 89 [37]. Deoxytylophorinine (90) [38] and (±)-antofine [as (4)] [39] have been synthesized by a route similar to one for tylophorine [23] (Section 3.1). Another synthesis of antofine was similar to a different route to tylophorine [25]. One synthesis of deoxytylophorinine (90) involved a biogenetically patterned sequence to 91; the latter gave compound 90 when treated with thallium(III) trifluoroacetate [20] (cf. Scheme 6 for a closely related sequence).

A synthesis of the enantiomer of (−)-antofine (4) has been carried out in which chirality introduced at the beginning of the synthesis was preserved to the end [40]. The absolute configuration assigned [41] to natural (−)-antofine (4) was thereby confirmed. An outline of the route is given in Scheme 8. The Pschorr reaction intended to be used as a means of forming the phenanthrene ring proved unsatisfactory so the nitro-group initially present was removed at a later stage. Cyclization was then achieved by photochemical means; the isomer 93 was formed as a minor product along with required material 92.

An interesting synthesis of antofine (4) involved metallation of the phenan-threne 94 which was assisted by the amide function, and then reaction with N-benzylpyrrol-2-aldehyde. Subsequent manipulation led to (±)-antofine [as (4)] (Scheme 9) [42]. Cryptopleurine (8) has been made in a similar way (Section 4) [42].

3.4. Tylocrebrine and Isotylocrebrine

(±)-Tylocrebrine [as (5)] and (±)-isotylocrebrine [as (6)] have been synthesized [25, 43] through the appropriate chloromethylphenanthrene in ways similar to those discussed for tylophorine [23, 25] (Section 3.1).

3.5. Miscellaneous Bases

The unsubstituted phenanthroindolizidine 95 has been synthesized by several workers [44–47]; the dimethoxy-analog 96 has also been prepared [25]. Syntheses of 98 have also been described [20, 29, 32]. These compounds are not natural products.

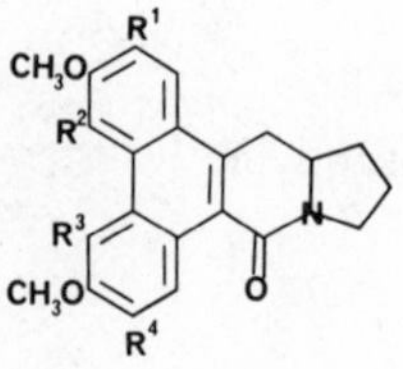

85, $R^1 = R^4 = $ OMe, $R^2 = R^3 = $ H
86, $R^1 = R^3 = $ H, $R^2 = R^4 = $ OMe
87, $R^1 = R^3 = $ OMe, $R^2 = R^4 = $ H

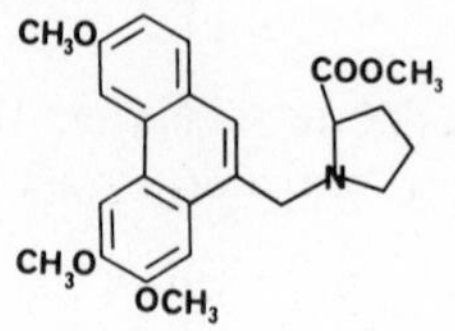

88

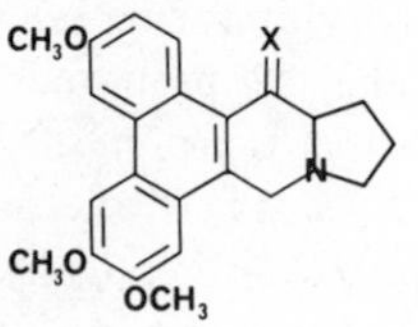

89, X = O
90, X = H$_2$

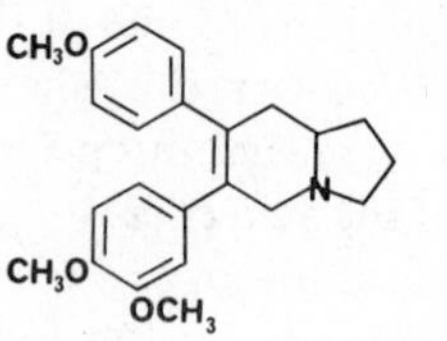

91

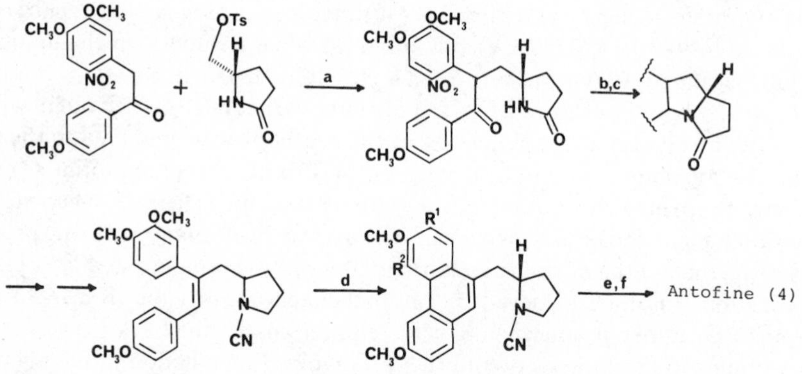

92, $R^1 = $ OMe, $R^2 = $ H
93, $R^1 = $ H, $R^2 = $ OMe

Scheme 8. a, K_2CO_3; b, $NaBH_4$; c, conc. HCl; d, hν; e, $LiAlH_4$; f, HCO_2H.

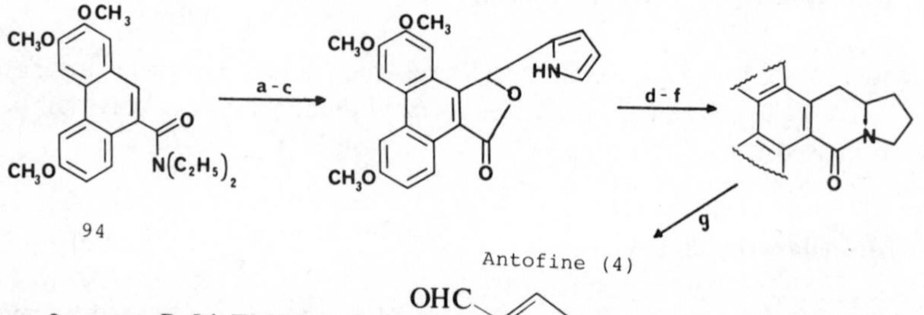

94

Antofine (4)

Scheme 9. a, *sec*-BuLi, TMEDA; b, [structure: OHC— pyrrole —$C_8H_5CH_2N$]; c, silica gel; d, Zn(Cu), KOH, py; e, CH_2N_2; f, H_2, Pt, HOAc; g, $LiAlH_4$.

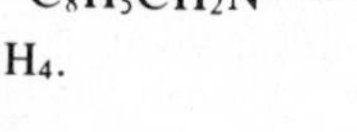

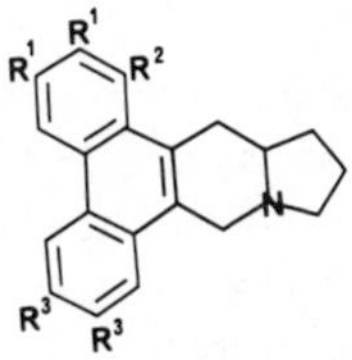

95, $R^1 = R^2 = R^3 = H$
96, $R^1 = R^2 = H$, $R^3 = OMe$
97, $R^1 = R^3 = OMe$, $R^2 = OH$

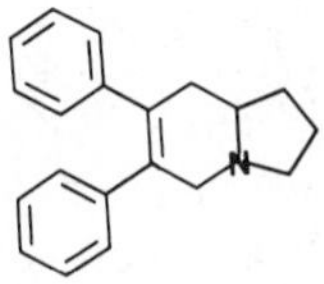

98

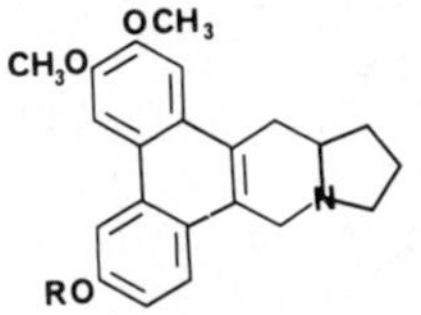

99, R = H
100, R = Me

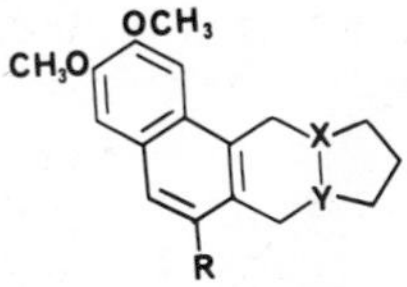

101, R = [structure with OH, CH₃]

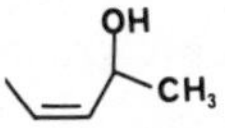

X = CH, Y = N

105, R = Cl
106, R = COOH

107

102, R = H, X = CH, Y = N
103, R = -(CH₂)₃CH₃, X = CH, Y = N
104, R = -(CH₂)₃CH₃, X = N, Y = CH

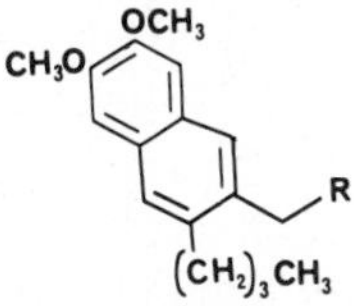

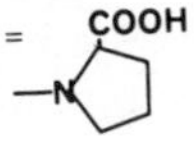

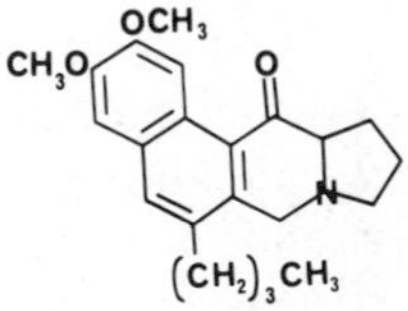

Septicine (7) has been efficiently converted into tylophorine (1) with one equivalent of thallium(III) trifluoroacetate. When two equivalents were used, a different product (namely, 97) was obtained [20].

The structures of two alkaloids (A and C) isolated from *Cyanchum vincetoxicum* were assigned as 99 and 100 on spectroscopic evidence [48]. These assignments have been confirmed through their synthesis [49] by adaptation of a route used for the synthesis of tylophorine (1) (Section 3.1) [23].

Vinceten, a very minor alkaloid of *C. vincetoxicum*, was deduced [50] to have the unusual structure 101. The alkaloid presumably arises by aromatic ring scission of a phenanthroindolizidine precursor.

The structure of vincetin was confirmed by synthesis of dihydrodesoxyvinceten (103). The synthesis was based on an exploratory route to 102 and involved the preparation of the chloride 105. Condensation of 105 with methyl prolinate and hydrolysis afforded the amino acid 106. Cyclization of 106 to give 107 was effected with phosphoric acid–phosphorus oxychloride, chosen as the best reagent from studies in the model series to 102. The keto function in 107 was removed to give 103 by the sequence: reduction with sodium borohydride acetylation, and hydrogenolysis. The racemic dihydrodesoxyvinceten (103) was identical on TLC with material derived from the natural alkaloid; the latter was different from the isomer 104 prepared in a manner similar to that used for 103 [50].

4. SYNTHESIS OF JULANDINE[†] AND CRYPTOPLEURINE

The first synthesis of julandine (9) and an early one of cryptopleurine (8) was modeled along biogenetic lines, and a most attractive synthesis it is [51]. 1-Methylpyridyl lithium reacted with methyl 3,4-dimethoxybenzoate to give the pyridine 108. Hydrogenation of 108 afforded 3′,4′-dimethoxy-2-(2-piperidyl)-acetophenone (109), which has been isolated from plant sources that also contain cryptopleurine [6, 52]. The amide 110 was formed by reaction of 109 with p-methoxyphenylacetyl chloride. Cyclization of 110 to give 111 was effected with potassium t-butoxide. Hydrogenation of 111 afforded the dihydro-derivative 112, which proved a better substrate for the required demethylation in the next step than did 111. The trihydric phenol 113 thus formed was oxidatively cyclized to the dienone 114, and this transformation was best achieved with manganese dioxide in the presence of silica gel in 15–20% yield. The dienone underwent rearrangement with a sulfuric acid–acetic anhydride mixture to give the triacetate of 116, from which (±)-cryptopleurine [as (8)] was simply derived. Reduction of 111 with lithium aluminum hydride gave material with properties corresponding to natural julandine (9) [6], thus confirming the structure tentatively assigned to this alkaloid.

The coupling reaction in this synthesis of cryptopleurine has been improved by electrochemical means [53]. Thus, the quinolizidine 111 gave, on anodic oxidation, the dienone 115 and 117 in 60 and 31% yields, respectively. The dienone 115 was then converted into cryptopleurine by rearrangement (acetic anhydride-sulfuric acid), followed by hydrolysis of the acetate produced, and then methylation (diazomethane) and reduction (lithium aluminum hydride) to give 8. Anodic coupling is an efficient way of forming phenanthrenes which could find useful application elsewhere as an alternative to the use of thallium(III) trifluoroacetate (see below and Section 3.2).

Another route to julandine (9) and cryptopleurine (8) was also modeled on the likely biosynthetic pathway to these alkaloids [54, 55]; it drew on experience gained in a related synthesis of septicine (7) and tylophorine (1) (Section 3.2). 3,4-Dimethoxybenzoylacetic acid was condensed with 3,4,5,6-tetrahydropyridine (119), generated in situ from cadaverine (118) using pea seedling diamine oxidase, to give the amine 109 in excellent preparative yield.

This amine condensed spontaneously with p-methoxyphenylacetaldehyde in benzene solution to give the enamine 120. Unlike the corresponding enamine 72, an intermediate in the synthesis of phenanthroindolizidine alkaloids, 120 would not undergo cyclization and dehydration in methanol solution. However, cyclization and dehydration were effected with various Lewis-acid catalysts, notably titanium (IV) chloride and silicon (IV) chloride; the efficacy of the Lewis-acid catalyst showed a correlation with recorded metal–oxygen bond

---

[†]This alkaloid, which was not given a name when it was first reported, was subsequently whimsically named julandine by one of the workers who synthesized it for the pleasure of his children, Julie and Andrew [54].

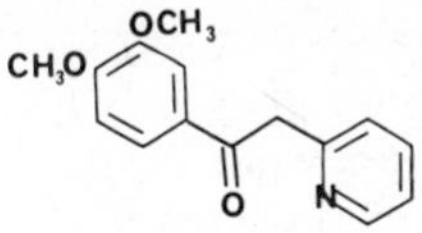

108

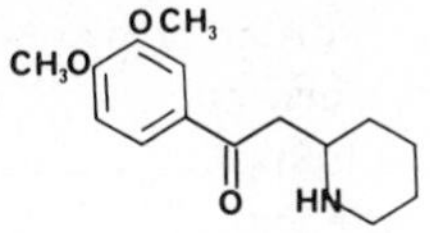

109

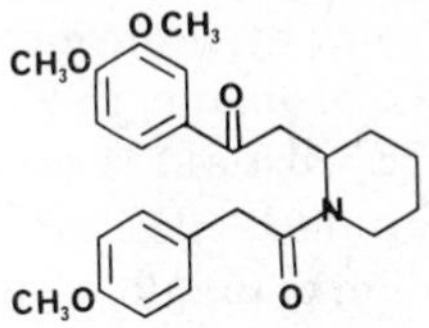

110

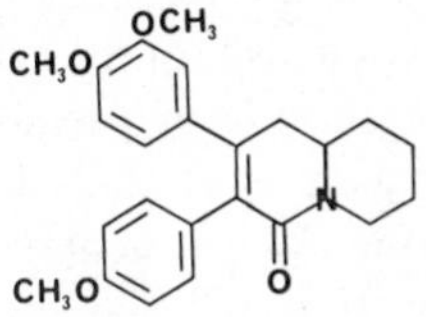

111

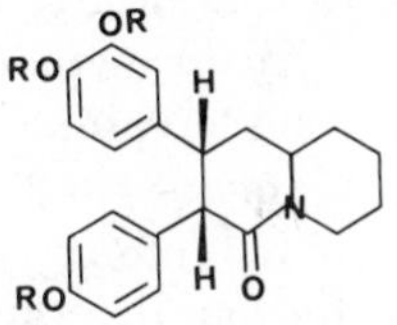

112, R = Me
113, R = H

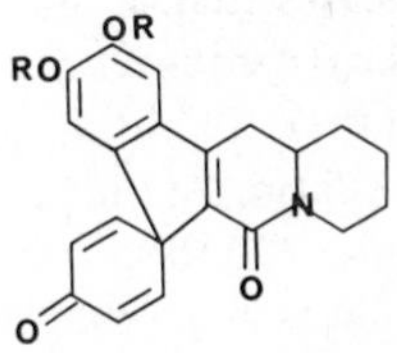

114, R = H
115, R = Me

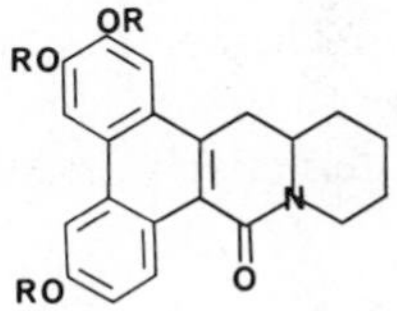

116, R = H
117, R = Me

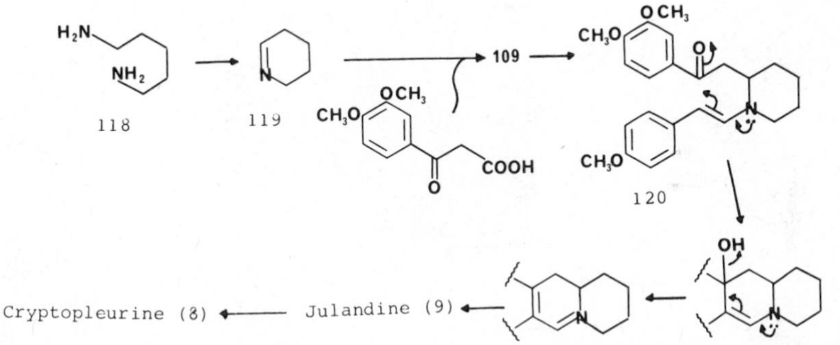

118   119

120

Cryptopleurine (8) ◀—— Julandine (9) ◀——

strengths. By inference the effectiveness of the catalyst was related to its ability to coordinate to the ketonic oxygen of **120**, in preference to the enamine system, thus catalyzing the ring closure of **120** (and subsequent dehydration). Reduction of the cyclized-dehydrated product with sodium borohydride yielded (±)-julandine [as **(9)**], which was oxidatively cyclized to cryptopleurine **(8)** in good yield with thallium(III) trifluoroacetate.

In the reaction catalyzed by thallium(III) trifluoroacetate which converts septicine **(7)** into tylophorine **(1)**, bond formation inevitably occurs *para* to a methoxy-group on each aromatic ring. In the case of julandine **(9)**, coupling was efficient even though bond formation was to one ring bearing a *meta*-methoxy-group. The likely mechanism [56] of the coupling reaction involves transfer of a single electron to thallium from the aromatic ring of lower oxidation potential, followed by electrophilic substitution of the radical-cation thus formed into the other aromatic ring, and further one electron oxidation. This mechanism suggested that **121** could be the first product formed, from which cryptopleurine (**8**) could be derived by rearrangement. In support of this idea a very small amount of the alternative rearrangement product **122** was isolated. However, the 100-fold difference in the yields obtained for the two products, **8** and **122**, indicates that cryptopleurine is formed largely by direct coupling [55].

Two syntheses of cryptopleurine use methods employed also for the synthesis of other alkaloids [42, 57]. Thus cryptopleurine (**8**) has been prepared by metallation of the phenanthrene amide **94** and reaction with pyridine-2-aldehyde to give **123**. Transformation of **123** into cryptopleurine (**8**) was achieved in a similar way to that described for antofine (**4**) [42] (Section 3.3; Scheme 9).

The second of these syntheses [57] involved 1,3-dipolar addition of an unsaturated *N*-oxide to an olefin (cf. Sections 3.2 and 7). Addition of the *N*-oxide **125** to 3,4-dimethoxystyrene **124** occurred regioselectivity to give **126** in high yield. Zinc–acetic acid reduction of **126** gave an amino–alcohol from which the amide **127** could be prepared. This compound was converted simply via the lactam **111** into (±)-julandine [as (**9**)] (Scheme 10). Alternatively, irradiation of

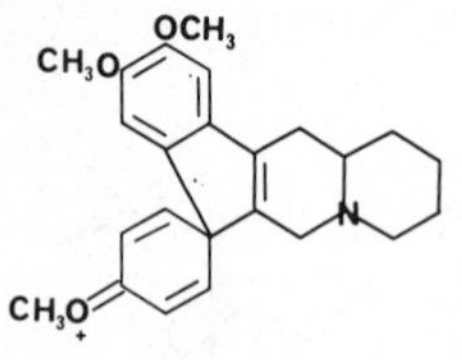

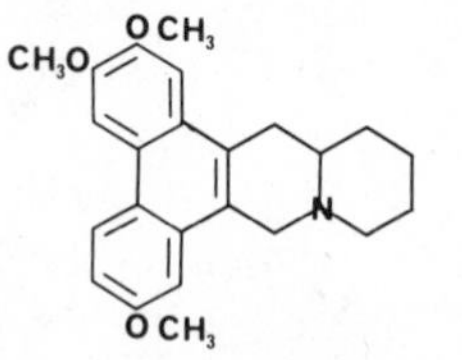

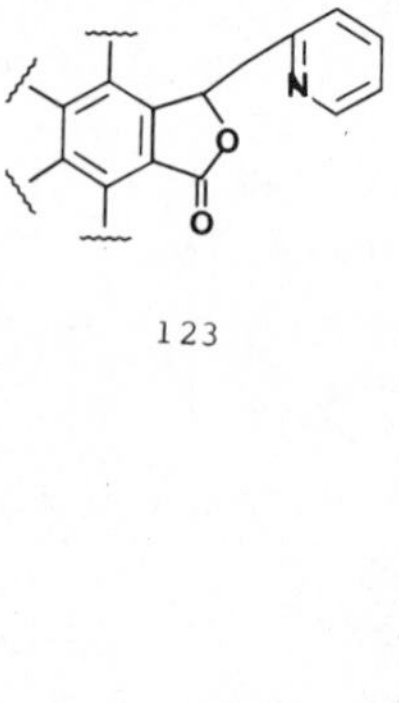

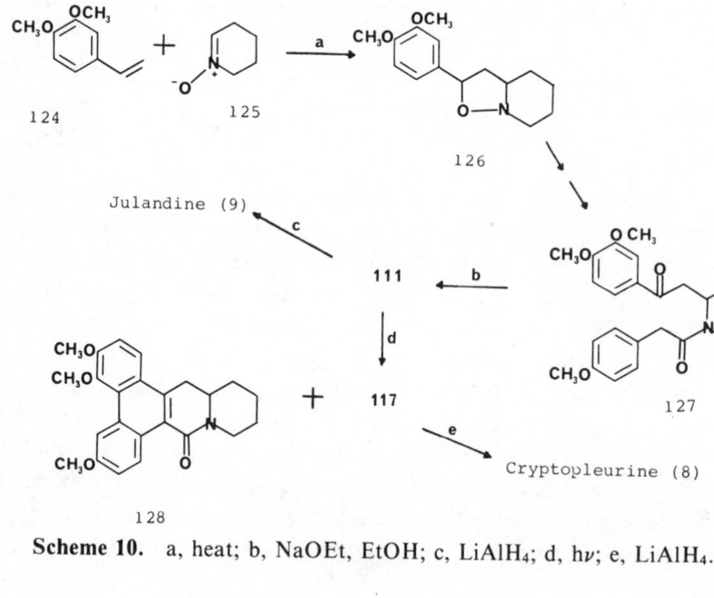

Scheme 10. a, heat; b, NaOEt, EtOH; c, LiAlH₄; d, h*ν*; e, LiAlH₄.

the lactam (111) gave a mixture of products, 117 and 128. ($\pm$)-Cryptopleurine [as (8)] was obtained as the reduction product of the major isomer (117) [57] (cf. Section 3.2 for slightly different results in a strictly analogous synthesis of septicine and tylophorine).

The first synthesis of cryptopleurine to be reported involved preparation of the phenanthrene (129) and its reaction with pyridine-2-aldehyde followed by polyphosphoric acid cyclization to give 131. Hydrogenation of 131 as the chloride afforded material which had properties corresponding to those of natural cryptopleurine (8), thereby affirming the structure assigned to the alkaloid [58].

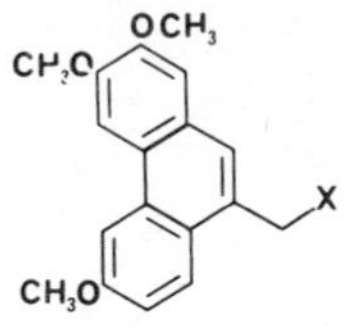

129, X = Br
130, X = Cl

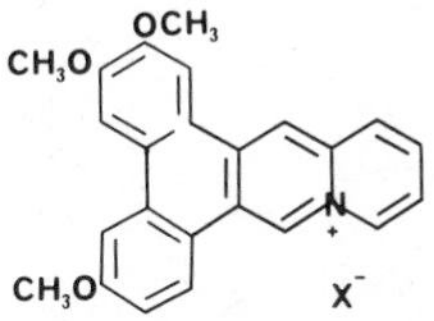

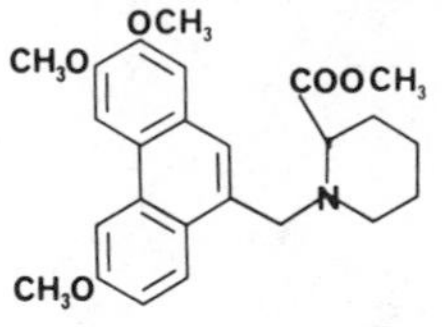

132

A slight variation on this synthesis began with the phenanthrene 130, which was condensed with methyl pipecolate to give 132. The corresponding amino acid was cyclized with polyphosphoric acid to afford 133. This ketone underwent Wolff–Kishner reduction to give a mixture of phenols from which ($\pm$)-cryptopleurine (8) was isolated in low yield after treatment with diazomethane [59]. It may be noted that Wolff–Kishner reduction of the model compound 134 proceeded in high yield, but Clemmensen reduction gave a different compound, which was assigned the structure 135.

This synthesis of cryptopleurine (8) has been improved [60] by the synthesis of the ester 132 in an adaptation of the method used in a synthesis of tylophorine (1) [20, 61]. Conversion of the ketone 133 into cryptopleurine (8) was achieved by a sequence involving lithium aluminum hydride reduction, perchloric acid dehydration, and finally sodium borohydride reduction.

This sequence for removing the keto function in 133 was employed earlier in the synthesis of cryptopleurine analogs [62]. These analogs were prepared otherwise following the earlier synthesis [59] of cryptopleurine.

5. SYNTHESIS OF IPALBIDINE

This alkaloid has been the subject of five quite different synthetic approaches. In some of these syntheses parallel preparations of septicine (7) were reported. The ketone 140 is common to three approaches; the analog 59 gives septicine.

The first reported synthesis of ipalbidine (10) includes a resolution of the synthetic alkaloid and preparation also of (+)-ipalbine (138) through the reaction

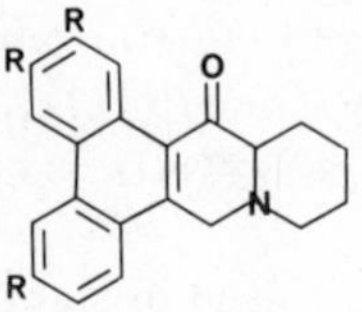

133, R = OMe
134, R = H

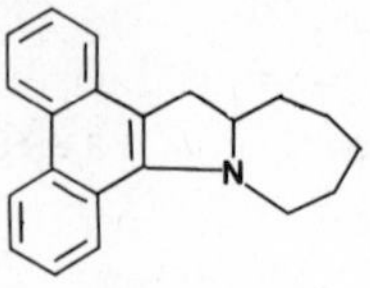

135

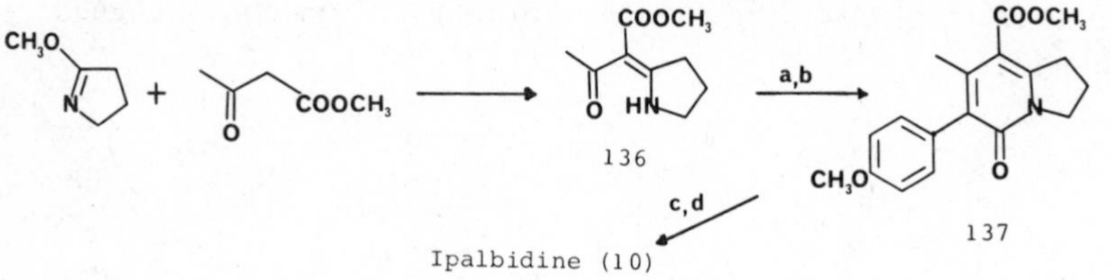

136

137

Ipalbidine (10)

Scheme 11. a, NaH; b, 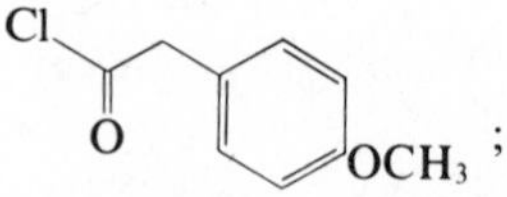 ; c, HBr; d, AlH₃, THF.

of (+)-ipalbidine with tetraacetyl-α-D-bromo-glucose. The synthesis of ipalbidine involved construction of the pyridone (137) and is illustrated in Scheme 11. It may be noted that (a) the first reaction proceded without base catalysis; catalytic amounts of pyridine or triethylamine were without effect on the reaction rate; (b) acylation of 136 with p-methoxyphenylacetyl chloride was effected via the sodium salt of 136 but not using triethylamine as base [63].

It seems probable that the biosynthetic route to ipalbidine proceeds via hygrine (24), which upon condensation with p-hydroxyphenylacetaldehyde (or equivalent) would lead to ipalbidine. A simple synthesis modeled on this hypothesis and drawing on experience gained in a related synthesis of septicine (7) (cf. Scheme 6) [20], has yielded ipalbidine in good yield (Scheme 12) [64]. These two ipalbidine syntheses (Schemes 11 and 12) are related in involving similar C—C and C—N bond making steps but at different oxidation levels.

The three other syntheses of ipalbidine depend on different preparations of the ketone 140. Two of the syntheses [30, 65] involved exactly parallel sequences of reactions to those described for 59 in the syntheses of septicine (7) discussed above. The ketone 140 was converted by standard methods into ipalbidine [30, 65].

A third synthesis [66] of the ketone 140 involved the vinylogous urethane 143. The properties of 143 as a stabilized enamine were exploited in an intramolecular cyclization on the mixed anhydride of 143 to give 144; conventional procedures gave 140 (Scheme 13). The key compound 142 was prepared from 141 by the Eschenmoser "sulfide contraction" method [67] which proceeds particularly well on the tertiary thioamide 141. Different application of the synthetic ideas embodied in this work has led to the synthesis of lupinine [68] and *Elaeocarpus* alkaloids (Section 7).

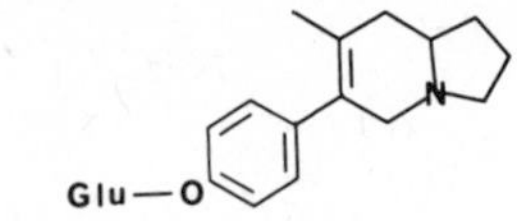

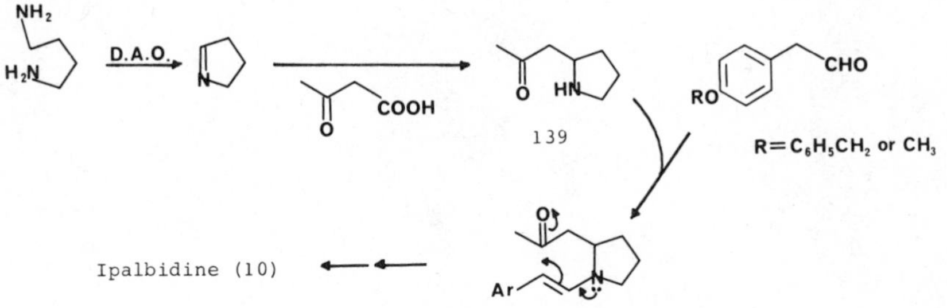

Scheme 12.

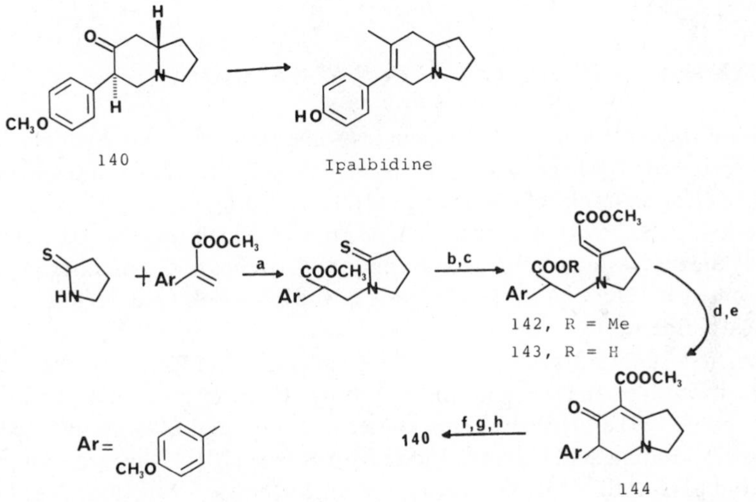

Scheme 13. a, NaOH; b, BrCH₂CO₂Me; c, PPh₃, NEt₃; d, NaOH, H₂O; e, Bu₄NI, ClCO₂Me; f, KOH, H₂O; g, HCl, H₂O, H, LiAlH₄.

6. SYNTHESIS OF *RUSPOLIA HYPERCRATERIFORMIS* ALKALOIDS

The synthesis of ruspolinone **(70)** has been described above (Section 3.2) where this alkaloid served as an intermediate in the synthesis of septicine **(7)** [20, 32]. Norruspolinone **(22)** has been synthesized in a similar way through the benzyl-protected compound **145** [11]. Reduction of **145** gave the alcohol **146**, which on treatment with titanium(IV) chloride satisfyingly underwent debenzylation and

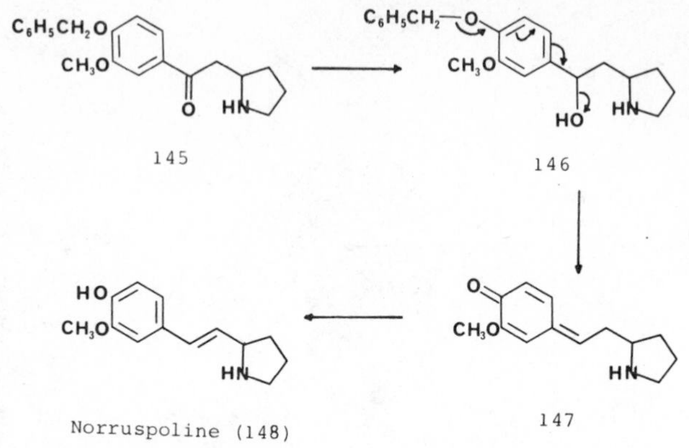

145

146

Norruspoline (148)

147

concomitant dehydration to give a third *R. hypercrateriformis* alkaloid, norruspoline (148) in good yield. The reaction proceeds from 146 via 147 as shown [69].

7. SYNTHESIS OF *ELAEOCARPUS* ALKALOIDS

Work on these alkaloids, which includes some syntheses, has been authoritatively reviewed [8]. Discussion here will be first on the syntheses of C_{16}-aromatic alkaloids, for example, elaeocarpine (12), followed by the syntheses of the C_{12} alkaloids, for example, elaeokanine C (13), with which there is some overlap. Finally the discussion will turn to synthetic approaches to C_{16}-dienone alkaloids represented by 212, and to the synthesis of the maverick *Elaeocarpus* alkaloid, elaeocarpidine (221).

The first reported synthesis was of elaeocarpine (12) and isoelaeocarpine (159). It involved the preparation of 150 by the reaction of pyrrole with the diazoketone 149 catalyzed by copper powder; compound 149 was obtained from the corresponding acid chloride and diazomethane [70]. Hydrogenation of 150 gave the pyrrolidine 151 without reduction of the keto function, presumably because of steric hindrance. Reaction of the pyrrolidine 151 with ethyl acrylate gave the amino-ester 152, which was cyclized using sodium hydride as the base to give the diketoindolizidine 153. Demethylation of 153 with boron tribromide was accompanied by spontaneous cyclization to give 154. In a parallel sequence 155 was prepared, but it could not be satisfactorily hydrogenated or demethylated.

Dehydration of 154 afforded the chromone 156. Attempts to reduce the double bond in 156 directly were unsuccessful, but reduction with sodium borohydride gave the two isomeric alcohols 157 and 158 from which (±)-elaeocarpine (12) and (±)-isoelaeocarpine (159), respectively, could be obtained by use of chromium trioxide in acetic acid [70].

The ketone 153 has also been prepared by application of regiospecific *C*-alkylation of an acyl cyanide with a lithium enolate. Reaction of the acyl

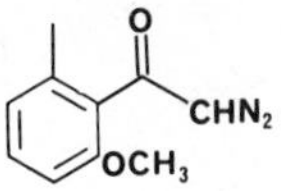

149

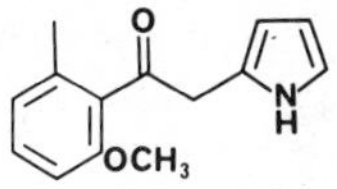

150

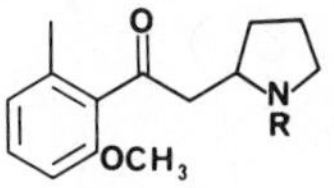

151, R = H
152, R = $CH_2CH_2CO_2Et$

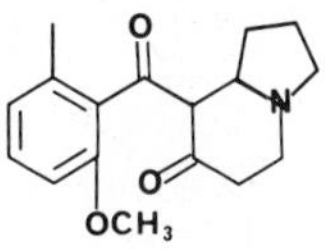

153

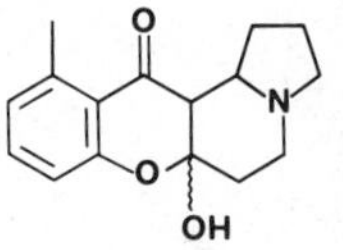

154

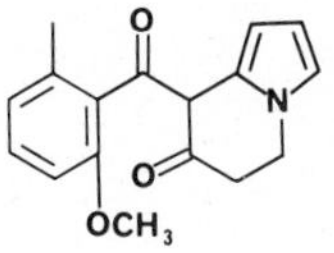

155

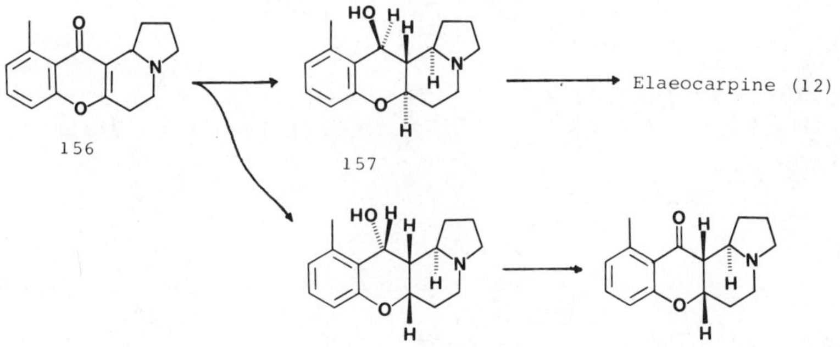

156

157

⟶ Elaeocarpine (12)

158

Isoelaeocarpine (159)

cyanide **166** with **165** gave the diketone **153** in 73% yield [71]. The enolate **165** was prepared as shown in Scheme 14 [72]. The thione **160** was converted into the vinylogous urethane **161** by a sulfide contraction procedure [67]; subsequent steps were as shown. The lithium enolate **165** reacted [72] with *n*-propionyl cyanide to give **167**, which has been converted into elaeokanine A **(168)** [73]. It may be noted that vinylogous urethanes have been used in a synthesis of ipalbidine (Section 5) and of lupinine [68].

A different and ingenious approach to the synthesis of elaeocarpine **(12)** and isoelaeocarpine **(159)** involved condensation of 6-methylsalicylaldehyde **(169)** with the dienamine **170** prepared *in situ* by lithium aluminum hydride reduction of 2,3-dihydro-1*H*-indolizinium bromide [18]. The product **171**, obtained in 18% yield, then gave a mixture of **12** and **159** on Jones oxidation.

A generally consistent structural unit associated with the alkaloids covered in this review is an amine of type **172**. These amines may be seen clearly as a part of the alkaloid structure or appear in masked form (cf. **1–10, 12** and **13**); they are intermediates of critical importance in the biosynthesis of all these alkaloids with the possible exception of the *Elaeocarpus* alkaloids (see Section 2). Moreover, they often serve as key intermediates in the synthesis of these alkaloids. There are

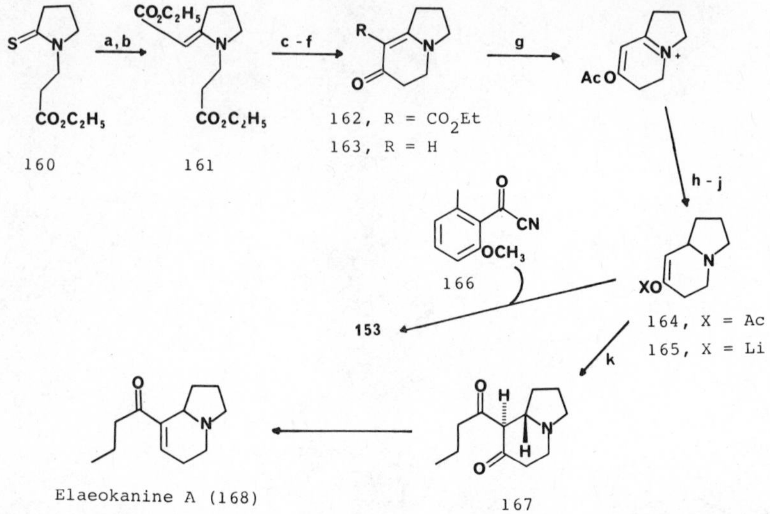
a, b
c - f
g
h - j
k
$CO_2C_2H_5$
$CO_2C_2H_5$
$CO_2C_2H_5$
160
161
162, R = CO_2Et
163, R = H
AcO
166
153
164, X = Ac
165, X = Li
Elaeokanine A (168)
167

Scheme 14. a, $BrCH_2CO_2Et$; b, Ph_3P, Et_3N; c, NaOH, H_2O; d, Ac_2O; e, KOH, H_2O; f, H^+, H_2O; g, CH_3COCl, $AgClO_4$; h, $NaBH_4$; i, HOAc; j, MeLi; k, n-C_3H_7COCN.

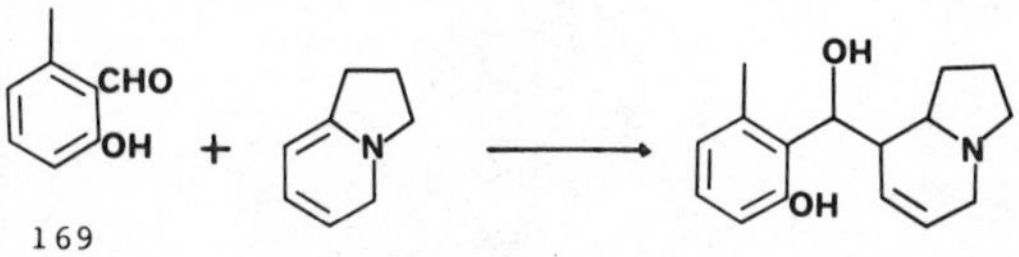

169 170 171

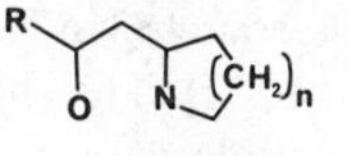

172, n = 1 or 2

several ways of generating **172**. One is modeled along biosynthetic lines (Sections 3.2, 4 and 5), and another involves [2,3]-cycloaddition of a nitrone to an olefin (Sections 3.2 and 4. For a general review of applications of these methods in alkaloid synthesis, see ref. 74).

A synthesis of elaeocarpine **(12)** and isoelaeocarpine **(159)** using [2, 3]-cycloaddition was as follows. Reaction of pyrroline oxide **(74)** and the styrene **173** proceeded smoothly to give **174** with high regioselectivity and stereoselectivity. Of three routes used for the further elaboration of **174** to give **12** and **159,** that which involved alkylation with 3-bromo-1-propanol to give **175** was the best. Treatment of **175** in benzene with potassium *tert*-butylate and benzophenone at reflux afforded **176**. Involved in this one-flask operation was base-induced oxazolidine ring opening, modified Oppenauer oxidation of alcohol to aldehyde, and finally aldol ring closure.

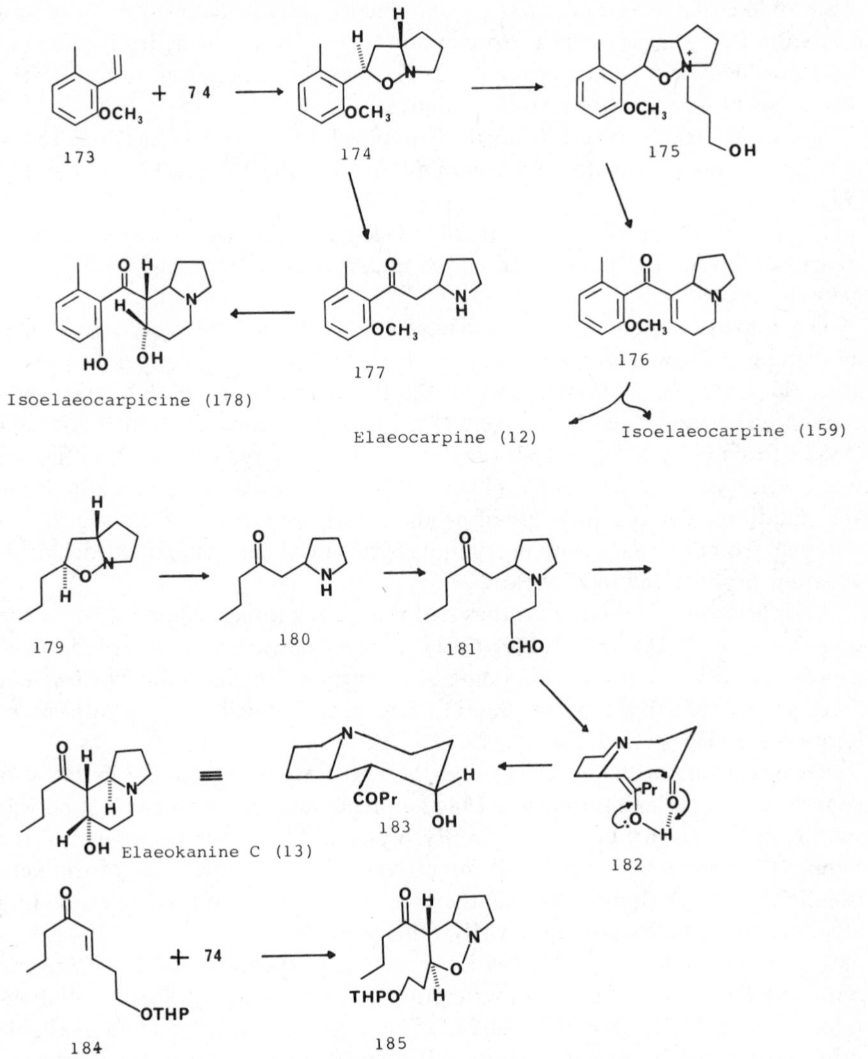

173   + 74 ⟶   174   ⟶   175

Isoelaeocarpicine (178)

177

176

Elaeocarpine (12)   Isoelaeocarpine (159)

179   ⟶   180   ⟶   181   CHO   ⟶

Elaeokanine C (13)   ≡   COPr   183   OH   182

184   + 74   ⟶   185

The ketone **176** was converted smoothly (boron tribromide, then aqueous sodium hydroxide) into a mixture of elaeocarpine **(12)** and isoelaeocarpine **(159)**. The amino-ketone **177** derived from **174** was elaborated to isoelaeocarpicine **(178)** by condensation with acrolein and treatment with acid rather than base (which caused dehydration), and finally demethylation (boron tribromide) [75].

In an exactly similar way [76] elaeokanines A **(168)** and C **(13)** have been prepared starting with the cycloaddition of pyrroline oxide to pent-1-ene which gave **179**. The ketone **180**, derived from **179**, underwent reaction with acrolein under base-catalysis to yield elaeokanine A **(168)**. Treatment of **180** with acrolein, followed by acid afforded elaeokanine C **(13)**. None of the C(7) epimer of 13 could be detected, and it is suggested that the intermediate 181 cyclizes via a transition state that involves hydrogen bonding of the enolic hydroxy-function with the aldehyde carbonyl group (cf. 182). A similar rationale accounts for the formation of isoelaeocarpicine (178) above.

A sensible extension of this approach involved the addition of an enone 184 to pyrroline oxide (74). Subsequent manipulation of the 185 produced gave 167 [77].

The first synthesis [73] of elaeokanine C (13), which was used to confirm its structure, followed a route [70] used for the synthesis of elaeocarpine (12) (see below).

The diazoketone obtained by reaction of n-butyryl chloride with diazo-methane was allowed to react with pyrrole in the presence of copper powder to give 186. Catalytic hydrogenation of 186 in acetic acid gave 187, which was condensed with ethyl acrylate to yield 188. Dieckmann condensation of this ester (188) afforded the diketone 167, which was hydrogenated to 13 with difficulty: the reaction gave elaeokanine C (13) in 30% yield in ethanol solution; in acetic acid solution a complex mixture of products was obtained. The major product obtained from hydrogenation in ethanol solution was surprisingly the dehydro-genation product 189 (60% yield) [73].

An ingenious, if lengthy, synthesis of the elaeokanines, exemplified by the preparation of A (168) and B (200)/(201), involved construction of the indolizine ring system present in these alkaloids by an intramolecular imino-Diels–Alder reaction on 195; this type of reaction has been applied to the synthesis of tylophorine (1) (Section 3.1) [26, 78].

Because of its probable instability, the necessary diene moiety in 195 was constructed in masked form using 194 as a precursor. The latter compound was synthesized as shown in Scheme 15. Pyrolysis of 193 gave a poor yield of the required lactam which it was felt might result from the presence of the keto function. The compound 192 was therefore converted into the corresponding alcohol and then into the trimethylsilyl ether 194. Careful pyrolysis of 194 gave a 68% yield of the lactams 196/198 from which the alcohols 197 and 199 were obtained. Reduction of the lactam functionality was achieved with diisobutyla-luminum hydride to give 200 and 201. The authors concluded from available information that natural elaeokanine B might be a mixture of the diastereo-isomers 200 and 201. Oxidation of either of these alcohols (trifluoroacetic anhydride–dimethylsulfoxide) gave elaeokanine A (168).

Inspection [26] of molecular models suggested two favorable transition states, 202 and 203, for the cycloaddition reaction which was presumed to go via 195. There appears to be very little difference in energy between 202 and 203 which explains the nearly equal amounts of the diastereoisomeric cyclization products 196 and 198 obtained.

The ester 162, whose synthesis and use in the elaboration of *Elaeocarpus* alkaloids was discussed above, has been prepared independently and used for the synthesis of elaeokanines [79, 80].

Reaction of 2-ethoxy-1-pyrroline with ethyl 3-oxo-pentenoate by adaptation

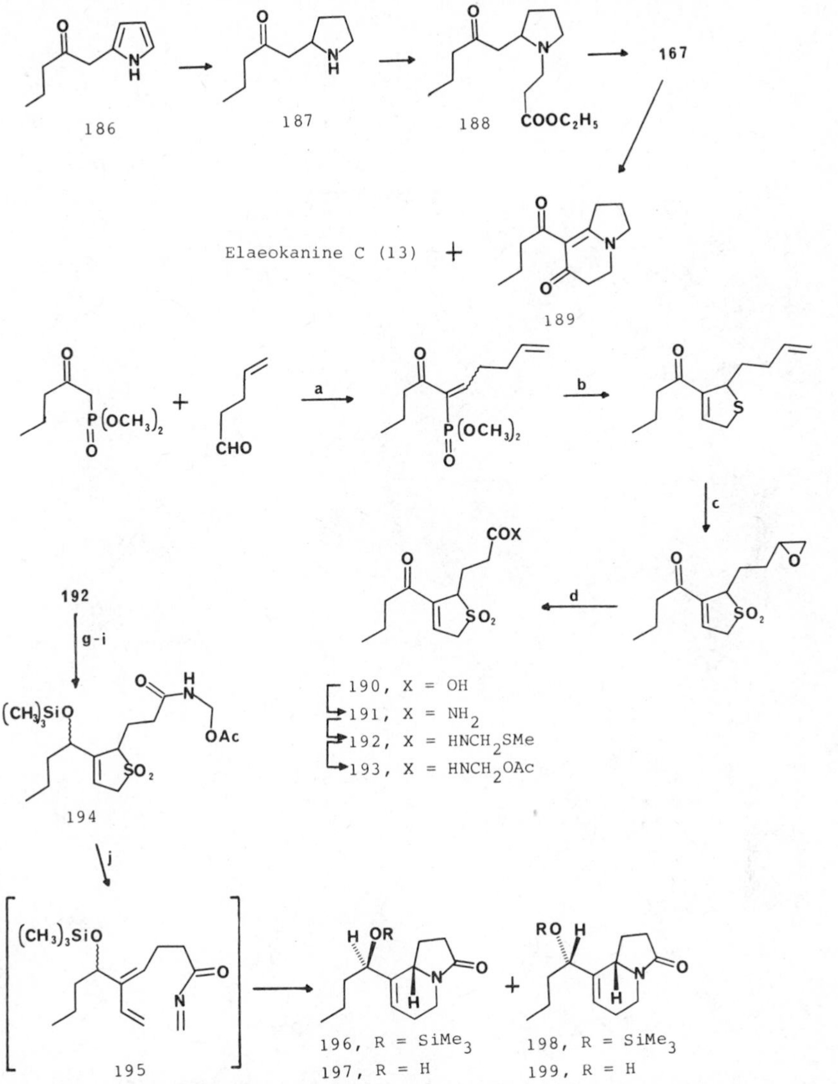

Scheme 15. a, piperidine, HOAc; b, HSCH₂CHO, Et₃N; c, *m*-CPBA; d, H₅IO₆, CrO₃; e, ClCH₂SMe, TFA; f, Hg(OAc)₂, HOAc; g, NaBH₄, CeCl₃; h, Hg(OAc)₂, HOAc; i, hexamethyldisilazane, Me₃SiCl; j, 370°.

of a general literature procedure [81] gave **162** in good yield [79]; this method appears to be the most economical way of making what is a versatile synthetic intermediate for *Elaeocarpus* alkaloids. This compound was converted by standard methods into **204**. Reaction of **204** with *n*-propyl magnesium bromide gave an alcohol that was transformed smoothly into **167**; this ketone had previously been converted into elaeokanine C **(13)** [73].

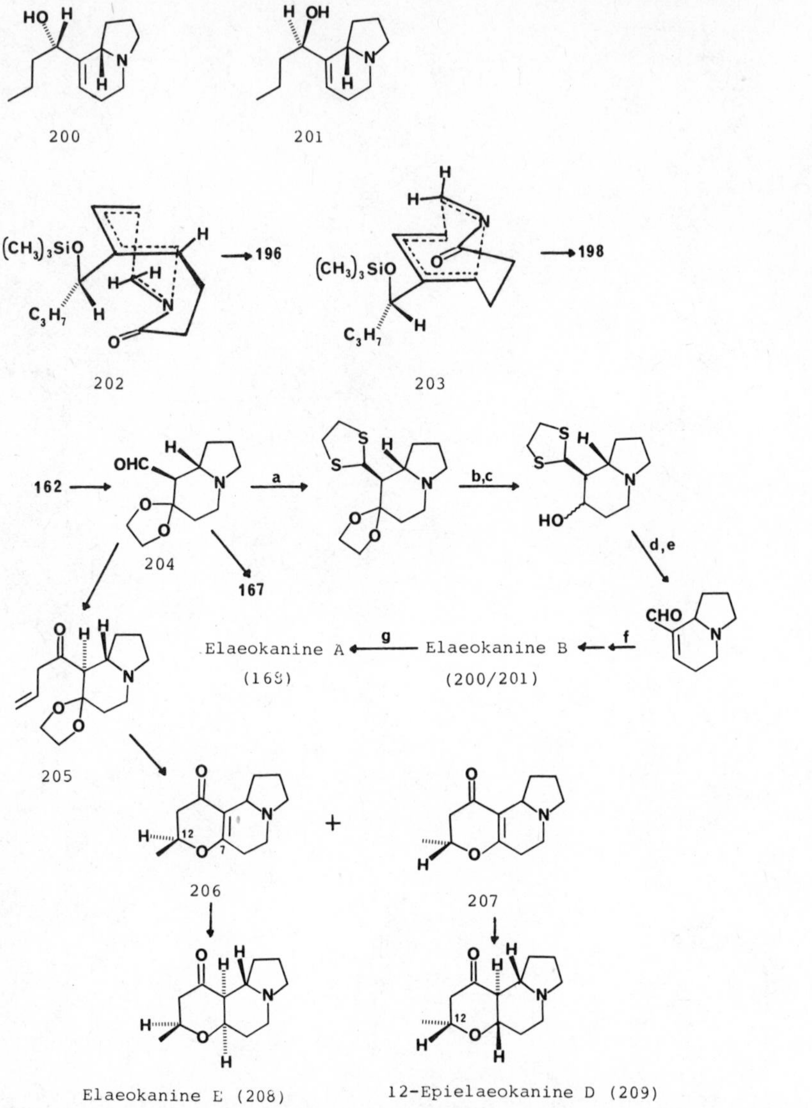

Scheme 16.   a, $(CH_2SH)_2$, BF_3-Et_2O, HOAc; b, HBr, H_2O; c, $LiAlH_4$; d, MeI; e, MeCN, HCl; f, n-C_3H_7MgBr; g, Jones oxidation.

The aldehyde **204** has also been converted into elaeokanines A **(168)** and B **(200)/(201)** (Scheme 16) [79]. On the other hand [80], reaction of **204** with allylmagnesium bromide, followed by Jones oxidation, gave **205** which underwent deketalization and cyclization with aqueous acid to give **206** and **207** in a ratio of 3.5 : 1. Elaeokanine E **(208)** was obtained as the sole product on reduction of **206** either with lithium in liquid ammonia–ethanol or by hydrogenation. Lithium-ammonia reduction of **207** gave 12-epielaeokanine D **(209)** [80].

Elaeokanine B **(200)/(201)** has been synthesized by a route which involved as a key step the acid-catalyzed (HCl/MeOH) addition of the double bond in **210** to the α-acyliminum ion which was formed by dehydration of the hydroxylactam function in **210** [82]. The product **211** was converted by standard means into elaeokanine B.

The presence of the labile dienone moiety in alkaloids such as elaeocarpiline **(212)** makes them a much more difficult synthetic target than those so far discussed; so far no complete synthesis has been described. The model compounds **215** and **216** have been made, however [83].

Reaction of arecaidine **(213),** as its hydrochloride, with oxalyl chloride followed by hydrogen sulfide in pyridine gave a product that was condensed with

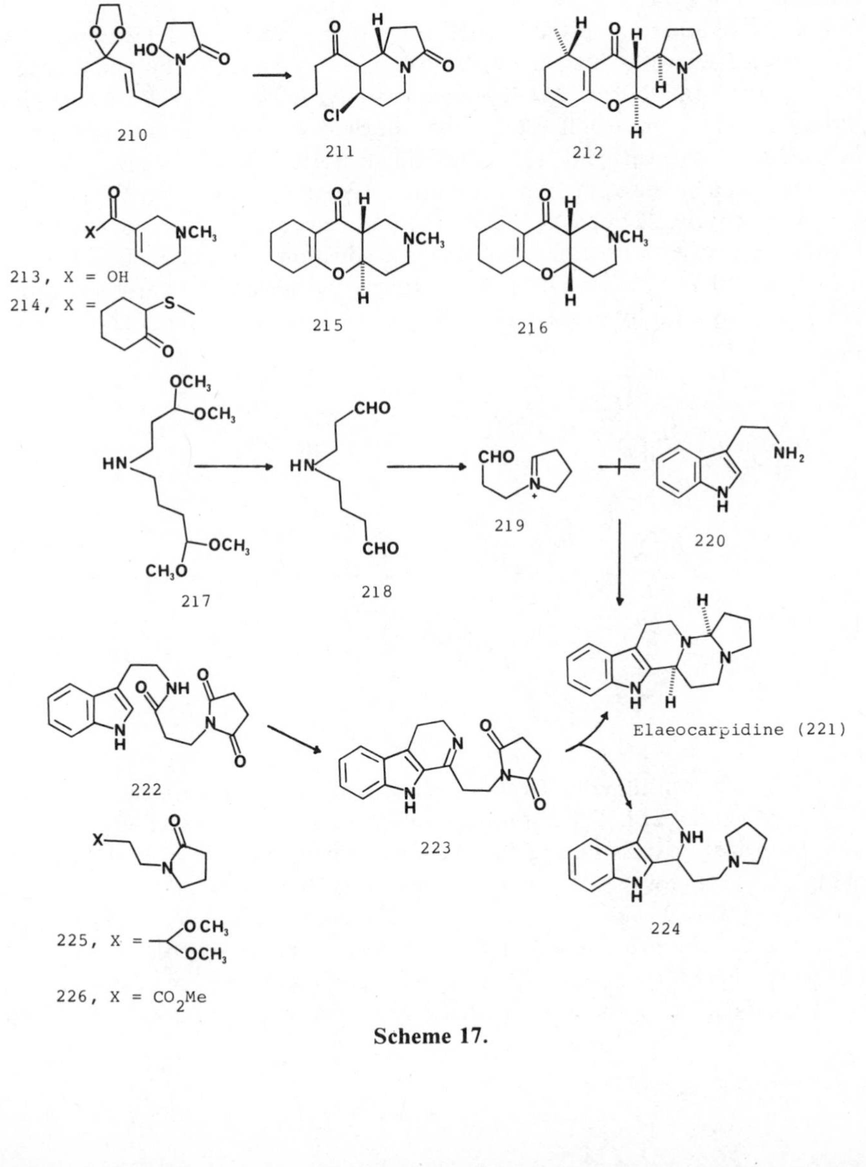

Scheme 17.

2-bromocyclohexanone to yield **214**. Reaction with triethylamine, lithium perchlorate, and triphenylphosphine (cf. ref. 67) presumably gave the expected β-diketone, which subsequently underwent ring closure to give a mixture of **215** and **216** as the products isolated [83].

There are three reported syntheses of elaeocarpidine (**221**). One, which is biogenetically modeled (cf. Section 2), is the most interesting [84]. It began with a synthesis of the bisacetal **217**, which reacted at reflux with tryptamine (**220**), first in buffer solution at pH 5.5 then at pH 1.5, to give elaeocarpidine (**221**) as the more stable of two possible products (Scheme 17). As expected, the dialdehyde **218** had undergone preferential cyclization to **219** rather than alternative cyclization to the four-membered ring immonium ion or intermolecular reaction with tryptamine; once formed, **219** then reacted with tryptamine to give elaeocarpidine (**221**).

Another synthesis involved amides rather than imines. Condensation [85] of tryptamine (**220**) with a mixed anhydride of *N*-succinimido-3-propionic acid and ethyl chloroformate gave **222**. Phosphorus oxychloride catalyzed cyclization afforded the dihydrocarboline **223** which upon reduction with lithium aluminum hydride gave elaeocarpidine (**221**) and dihydroelaeocarpidine (**224**).

A variation on this approach was to treat tryptamine with either **225** or **226** to give, respectively, **227** or **228**; the latter could be converted into the former by polyphosphate–ester treatment followed by reduction with sodium borohydride [86]. Reduction of **227** with lithium aluminum hydride afforded elaeocarpidine (**221**) stereospecifically and also **224**. It seems clear that reduction of **227** proceeds

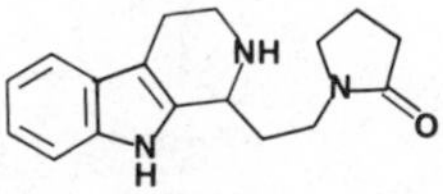

227

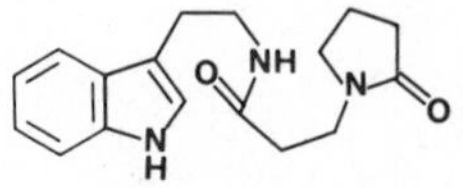

228

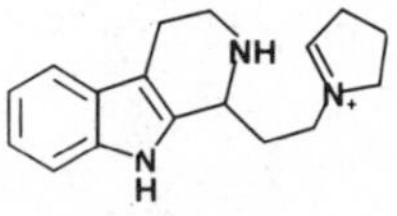

229

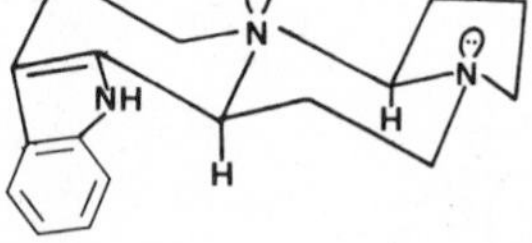

230

via **229**, which then either undergoes cyclization to give **221** or further reduction to give unwanted **224**. The author thought that an external secondary amine present in large excess during the reduction might serve as a more efficient (temporary) trap for the immonium functionality than the tryptamine nitrogen, and that by this means overreduction to **224** could be avoided. And indeed, reduction by lithium aluminum hydride in the presence of pyrrolidine afforded elaeocarpidine (**221**) in improved yield and, significantly, no **224** was produced.

An independent synthesis of dihydroelaeocarpidine (**224**) has been carried out and evidence was presented to show that elaeocarpidine has the *trans*-anti-*trans*-conformation **(230)** [86].

REFERENCES

1. T. R. Govindachari, in *The Alkaloids*, Vol. 9, R. H. F. Manske, Ed., Academic, New York, 1967, p. 517; *J. Ind. Chem. Soc.* **50**, 1 (1973).

2. T. R. Govindachari and N. Viswanathan, *Heterocycles* **11**, 587 (1978).

3. E. Gellert, *J. Nat. Prod.* **45**, 50 (1982).

4. J. E. Saxton, Ed., *The Alkaloids*, (Specialist Periodical Reports), Vols. 1–5, The Chemical Society, London, 1971–1975; M. F. Grundon, Ed., Vols. 6–12, 1976–1982.

5. E. Gellert and N. V. Riggs, *Aust. J. Chem.* **7**, 113 (1954); J. Fridrichsons and A. McL. Mathieson, *Nature* **173**, 732 (1954); *Acta Crystallogr.* **8**, 761 (1955); E. Gellert, R. Rudzats, J.C. Craig, S. K. Roy, and R. W. Woodward, *Aust. J. Chem.* **31**, 2095 (1978).

6. N. K. Hart, S. R. Johns, and J. A. Lamberton, *Aust. J. Chem.* **21**, 2579 (1968).

7. J. M. Gourley, R. A. Heacock, A. G. McInnes, B. Nikolin, and D. G. Smith, *J. Chem. Soc., Chem. Commun.*, 709 (1969).

8. S. R. Johns and J. A. Lamberton, in *The Alkaloids*, Vol. 14, R. H. F. Manske, Ed., Academic, New York, 1973, p. 325; J. E. Saxton, in ref. 4, Vol. 1, 1971, p. 76.

9. N. B. Mulchandani, S. S. Iyer, and L. P. Badheka, *Phytochemistry* **8**, 1931 (1969); **10**, 1047 (1971); **15**, 1697 (1976).

10. D. S. Bhakuni and V. K. Mangla, *Tetrahedron* **37**, 401 (1981).

11. R. B. Herbert, F. B. Jackson, and I. T. Nicolson, *J. Chem. Soc., Chem. Commun.*, 865 (1976); *J. Chem. Soc., Perkin Trans.* **1**, 825 (1984).

12. F. Roessler, D. Ganzinger, S. Johne, E. Schöpp, and M. Hesse, *Helv. Chim. Acta* **61**, 1200 (1978).

13. R. B. Herbert and F. B. Jackson, *J. Chem. Soc., Chem. Commun.*, 955 (1977).

14. R. B. Herbert, *The Biosynthesis of Secondary Metabolites*, Chapman-Hall, London, 1981; in *Rodds's Chemistry of Carbon Compounds*, Vol. 4, Part L., S. Coffey, Ed., Elsevier, Amsterdam, 1980, p. 291.

15. S. R. Johns, J. A. Lamberton, and A. A. Sioumis, *Aust. J. Chem.* **22**, 793 (1969).

16. N. K. Hart, S. R. Johns, and J. A. Lamberton, *Aust. J. Chem.* **25**, 817 (1972).

17. S. R. Johns, J. A. Lamberton, and A. A. Sioumis, *Aust. J. Chem.* **22**, 801 (1969).

18. T. Onaka, *Tetrahedron Lett.* **12**, 4395 (1971).

19. S. R. Johns, J. A. Lamberton, A. A. Sioumis, and R. I. Willing, *Aust. J. Chem.* **22**, 775 (1969).

20. J. E. Cragg, R. B. Herbert, F. B. Jackson, C. J. Moody, and I. T. Nicolson, *J. Chem. Soc., Perkin Trans.* **1**, 2477 (1982).

21. J. H. Russel, *Naturwiss* **50**, 443 (1963).

22. H. Iida, M. Tanaka, C. Kibayashi, *J. Chem. Soc., Chem. Commun.*, 271 (1983).

23. T. R. Govindachari, M. V. Lakshmikantham, and S. Rajadurai, *Tetrahedron* **14**, 284 (1961).

24. R. F. Borch, *Tetrahedron Lett.* **9**, 61 (1968).

25. B. Chauncy and E. Gellert, *Aust. J. Chem.* **23**, 2503 (1970).

26. N. A. Khatri, H. F. Schmitthenner, J. Shringarpure, and S. M. Weinreb, *J. Am. Chem. Soc.* **103**, 6387 (1981).

27. A. J. Liepa and R. E. Summons, *J. Chem. Soc., Chem. Commun.*, 826 (1977).

28. J. H. Russel and H. Hunziker, *Tetrahedron Lett.* **10**, 4035 (1969).

29. T. R. Govindachari and N. Viswanathan, *Tetrahedron* **26**, 715 (1970).

30. R. V. Stevens and Y. Luh, *Tetrahedron Lett.* **18**, 979 (1977).

31. R. V. Stevens, Y. Luh, and J.-T. Sheu, *Tetrahedron Lett.* **17**, 3799 (1976).

32. R. B. Herbert, F. B. Jackson, and I. T. Nicolson, *J. Chem. Soc., Chem. Commun.*, 450 (1976).

33. V. K. Mangla and D. S. Bhakuni, *Tetrahedron* **36**, 2489 (1980).

34. V. K. Mangla and D. S. Bhakuni, *Ind. J. Chem.* **19B**, 748 (1980).

35. R. B. Herbert, C. J. Noble, and P. C. Wormald, unpublished results.

36. T. Iwashita, M. Suzuki, T. Kusumi, and H. Kakisawa, *Chem. Lett.*, 383 (1980).

37. T. R. Govindachari, B. R. Pai, S. Prabhakar, and T. S. Savitri, *Tetrahedron* **21**, 2573 (1965).

38. T. R. Govindachari, B. R. Pai, I. S. Ragade, S. Rajappa, and N. Viswanathan, *Tetrahedron* **14**, 288 (1961).

39. T. R. Govindachari, I. S. Ragade, and N. Viswanathan, *J. Chem. Soc.*, 1357 (1962).

40. L. Faber and W. Wiegrebe, *Helv. Chim. Acta* **59**, 2201 (1976); **56**, 2882 (1973).

41. W. Wiegrebe, L. Faber, and Th. Breyhan, *Arch. Pharm.* **304**, 188 (1971).

42. M. Iwao, K. K. Mahalanabis, M. Watanabe, S. O. de Silva, and V. Snieckus, *Tetrahedron* **39**, 1955 (1983); M. Iwao, M. Watanabe, S. O. de Silva, and V. Snieckus, *Tetrahedron Lett.* **22**, 2349 (1981).

43. E. Gellert, T. R. Govindachari, M. V. Lakshmikantham, I. S. Ragade, R. Rudzats, and N. Viswanathan, *J. Chem. Soc.*, 1008 (1962).

44. T. R. Govindachari, M. V. Lakshmikantham, K. Nagarajan, and B. R. Pai, *Tetrahedron* **4**, 311 (1958).

45. B. Chauncy, E. Gellert, and K. N. Trivedi, *Aust. J. Chem.* **22**, 427 (1969).

46. S. Takano, K. Yuta, and K. Ogasawara, *Heterocycles* **4**, 947 (1976).

47. D. O. Shah and K. N. Trivedi, *Ind. J. Chem.* **15B**, 599 (1977).

48. W. Wiegrebe, L. Faber, H. Brockmann, Jr., H. Budzikiewicz, and U. Krüger, *Liebigs Annalen der Chemie* **721**, 154 (1969).

49. W. Wiegrebe, L. Faber, and H. Budzikiewicz, *Liebigs Annalen der Chemie* **733**, 125 (1970).

50. H. Budzikiewicz, L. Faber, E.-G. Hermann, F. F. Perrollaz, U. P. Schlunegger, and W. Wiegrebe, *Liebigs Annalen der Chemie*, 1212 (1979).

51. J. M. Paton, P. L. Pauson, and T. S. Stevens, *J. Chem. Soc.* (C), 1309 (1969).

52. N. R. Farnsworth, N. K. Hart, S. R. Johns, J. A. Lamberton, and W. Messmer, *Aust. J. Chem.* **22**, 1805 (1969); N. K. Hart, S. R. Johns, and J. A. Lamberton, **21**, 1397 (1968).

53. E. Kotani, M. Kitazawa, and S. Tobinga, *Tetrahedron* **30**, 3027 (1974).

54. R. B. Herbert, *J. Chem. Soc., Chem. Commun.*, 794 (1978).

55. J. E. Cragg and R. B. Herbert, *J. Chem. Soc., Perkin Trans.*, 2487 (1982).

56. A. McKillop, A. G. Turrell, and E. C. Taylor, *J. Org. Chem.* **42**, 764 (1977); A. McKillop, A. G. Turrell, D. W. Young, and E. C. Taylor, *J. Am. Chem. Soc.* **102**, 6504 (1980); E. C. Taylor, J. G. Andrade, and A. McKillop, *J. Chem. Soc., Chem. Commun.*, 538 (1977); E. C. Taylor, J. G. Andrade, G. J. H. Rall, and A. McKillop, *J. Am. Chem. Soc.* **102**, 6513 (1980).

57. H. Ida and C. Kibayashi, *Tetrahedron Lett.* **22**, 1913 (1981).

58. C. K. Bradsher and H. Berger, *J. Am. Chem. Soc.* **80**, 930 (1958).

59. P. Marcini and B. Belleau, *Can. J. Chem.* **36**, 581 (1958).

60. G. C. Trigo, E. Gálvez and M. M. Söllhuber, *J. Heterocyclic Chem.* **17**, 69 (1980).

61. R. B. Herbert and C. J. Moody, *J. Chem. Soc., Chem. Commun.*, 121 (1970).

62. S. Foldeak, *Tetrahedron* **27**, 3465 (1971).

63. A. E. Wick, P. A. Bartlett, and D. Dolphin, *Helv. Chim. Acta* **54**, 513 (1971).

64. S. H. Hedges and R. B. Herbert, *J. Chem. Research* (S), 1 (1979); J. E. Cragg, S. H. Hedges, and R. B. Herbert, *Tetrahedron Lett.* **22**, 2127 (1981).

65. T. R. Govindachari, A. R. Sidhaye, and N. Viswanathan, *Tetrahedron* **54**, 3829 (1970).

66. A. S. Howard, G. C. Gerrans, and J. P. Michael, *J. Org. Chem.* **45**, 1713 (1980).

67. M. Roth, P. Dubs, E. Götschi, and A. Eschenmoser, *Helv. Chim. Acta* **54**, 710 (1971).

68. G. C. Gerrans, A. S. Howard, and B. S. Orlek, *Tetrahedron Lett.* **16**, 4171 (1975).

69. R. B. Herbert and P. C. Wormald, *J. Chem. Res.* (S), 299 (1982).

70. T. Tanaka and I. Iijima, *Tetrahedron* **29**, 1285 (1973); *Tetrahedron Lett.* **11**, 3963 (1970).

71. A. S. Howard, C. A. Meerholz, and J. P. Michael, *Tetrahedron Lett.* **20**, 1339 (1979).

72. A. S. Howard, G. C. Gerrans, C. A. Meerholz, *Tetrahedron Lett.* **21**, 1373 (1980).

73. N. K. Hart, S. R. Johns, and J. A. Lamberton, *Aust. J. Chem.* **25**, 817 (1972).

74. J. J. Tufariello, *Acc. Chem. Res.* **12**, 396 (1979).

75. J. J. Tufariello and S. A. Ali, *J. Am. Chem. Soc.* **101**, 7114 (1979).

76. J. J. Tufariello and S. A. Ali, *Tetrahedron Lett.* **20**, 4445 (1979).

77. H. Otomasu, N. Takatsu, T. Honda, and T. Kametani, *Heterocycles* **19**, 511 (1982); *Tetrahedron* **38**, 2627 (1982).

78. H. F. Schmitthenner and S. M. Weinreb, *J. Org. Chem.* **45**, 3372 (1980).

79. T. Watanabe, Y. Nakashita, S. Katayama, and M. Yamauchi, *Heterocycles* **14**, 1433 (1980).

80. T. Watanabe, Y. Nakashita, S. Katayama, and M. Yamauchi, *Heterocycles* **16**, 39 (1981).

81. B. M. Trost and R. A. Kunz, *J. Am. Chem. Soc.* **97**, 7152 (1975).

82. B. P. Wijnberg and W. N. Speckamp, *Tetrahedron Lett.* **22**, 5079 (1981).

83. T. H. Jones and P. J. Kropp, *Tetrahedron Lett.* **15**, 3503 (1974).

84. R. W. Gribble and R. M. Soll, *J. Org. Chem.* **46**, 2433 (1981).

85. J. Harley-Mason and C. G. Taylor, *J. Chem. Soc., Chem. Commun.*, 281 (1969).

86. G. W. Gribble, *J. Org. Chem.* **35**, 1944 (1970).

Chapter Seven

Recent Advances in the Total Synthesis of Pentacyclic *Aspidosperma* Alkaloids

Larry E. Overman
Department of Chemistry
University of California
Irvine, California 92717

Michael Sworin
Department of Chemistry
University of Missouri
St. Louis, Missouri 61321

CONTENTS

1. INTRODUCTION

The *Aspidosperma* alkaloids are the largest group of indole alkaloids [1]. The pentacyclic subgroup, or Aspidospermine family, is the largest class, numbering nearly 140 members. Simple pentacyclic *Aspidosperma* alkaloids such as (+)-1,2-dehydroaspidospermidine, (+)-vincadifformine, and (−)-tabersonine are widely distributed and have been isolated from a variety of plant species [1]. The most complex of the pentacyclic *Aspidosperma* alkaloids is vindoline. Vindoline is one component of the dimeric *Catharanthus* alkaloids vinblastine and vincristine, which are clinically used oncolytic agents [2].

The first total synthesis of a pentacyclic *Aspidosperma* alkaloid was Stork and Dolfini's pioneering preparation of *dl*-aspidospermine, communicated in 1963 [3]. The interest of synthetic chemists in this area has remained intense since that time. The most recent comprehensive review in this area is due to Cordell [1] and surveyed the literature from 1967–1977. The pinnacle of synthetic accomplishments during that period was the synthesis of *dl*-vindoline accomplished by Büchi, Ando, and Ohnuma [4a].

An impressive number of fundamentally new methods for assembling pentacyclic *Aspidosperma* alkaloids have been developed during the past 6 years. In particular, several of the new approaches, for the first time, achieve truly practical entries to this complex ring system.

We have attempted in this review to survey all total syntheses of pentacyclic *Aspidosperma* alkaloids published since 1977, which entail some new chemistry. To highlight new synthetic strategies, we have chosen to organize this review along chemical lines, rather than by target structure. For the most part, we have not included total syntheses of tetracyclic *Aspidosperma* alkaloids such as quebrachamine, even though they constitute [1] formal total syntheses of pentacyclic *Aspidosperma* alkaloids. We have chosen to employ the systematic numbering system for the pentacyclic *Aspidosperma* skeleton that is utilized in *Chemical Abstracts*, rather than the more common biogenetic system of LeMen and Taylor [5].

2. BIOMIMETIC TOTAL SYNTHESES

Considerable data now exists for inferring that the biosynthesis of *Aspidosperma* alkaloids occurs from *Strychnos* family alkaloids such as preakuammicine (1) via so-called dehydrosecodine intermediates, as outlined in Scheme 1 [6]. A biomimetic approach to tabersonine (3) and pseudotabersonine (5) would require the formation of dehydrosecodines 2 and 4. The inherent instability of these intermediates as well as problems associated with controlling alternate modes of their cyclization makes the synthetic challenge of a biomimetic approach large indeed. These problems are greatly simplified if the tricyclic intermediate is a so-called secodine (or "isosecodine") [1b] and contains a tetrahydropyridine

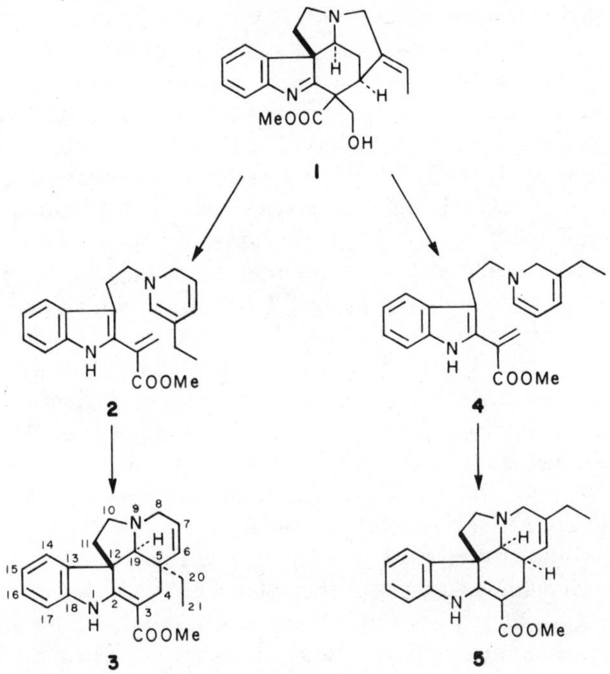

Scheme 1. Proposed biosynthetic pathway to pentacyclic *Aspidosperma* alkaloids.

rather than a dihydropyridine ring. It is not surprising that the most successful efforts in the biomimetic area involve syntheses based on secodine intermediates.

2.1. Via Secodine Intermediates

In 1978, Kuehne and co-workers reported the first biomimetic total synthesis of a pentacyclic *Aspidosperma* alkaloid, which is believed to proceed via a secodine intermediate [7]. Preliminary work [8] on the reaction of chloroindolenine **6** with

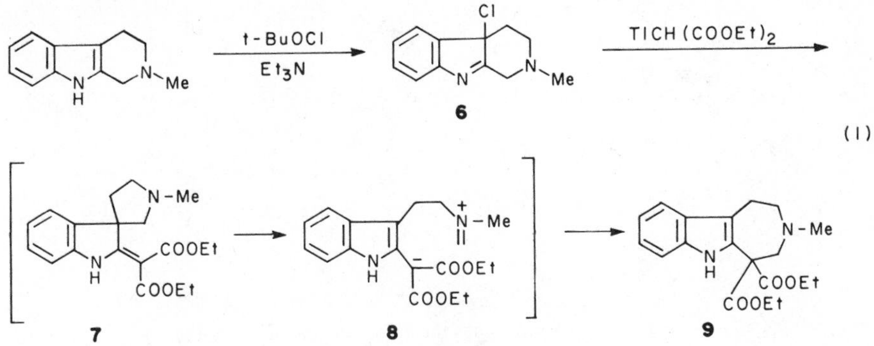

(1)

thallium diethyl malonate did not provide the expected spiro system **7** [8], but indoloazepine **9**. This facile reaction, presumably proceeding via **8,** provided a convenient conceptual base for attempts at generating secodines.

The Kuehne synthesis of *dl*-vincadifformine [7] is outlined in Scheme 2 and begins with N_b-benzyltetrahydro-β-carboline **10,** which is available in three steps from tryptamine hydrochloride [7]. Chlorination of **10,** followed by addition of thallium *tert*-butyl methyl malonate gave the rearranged indoloazepine **11** in 63–85% overall yield. Cleavage of the *tert*-butyl ester with concomitant decarboxylation was achieved with trifluoroacetic acid in refluxing 1,2-dichloroethane. Catalytic debenzylation using 5% Pd on carbon in acetic acid then provided indoloazepine **12** in 78% yield. The required bromoaldehyde **14** was prepared from methyl 4-formylhexanoate **(13)** [9] in 72% overall yield using a straightforward sequence. The biomimetic synthesis of *dl*-vincadifformine **(15)** was concluded by mixing equimolar amounts of indoloazepine **12** and bromoaldehyde **14** in methanol. Subsequent addition of triethylamine led directly to *dl*-vincadifformine in 70% yield [7]. This remarkable one-pot conversion was originally suggested [see Eq. (2)] to proceed via the intermediacy of the spiroenammonium salt **16,** which then underwent fragmentation to the secodine intermediate **17** upon reaction with triethylamine.

Since the original synthesis relied on nonsubstantiated mechanistic considerations, Kuehne explored alternative transformations to characterize the

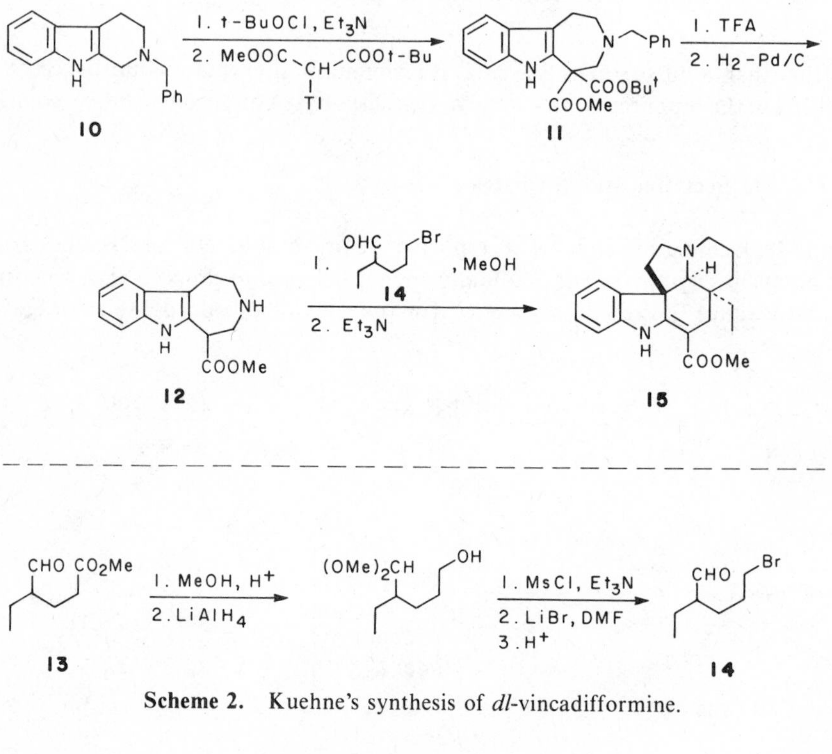

Scheme 2.   Kuehne's synthesis of *dl*-vincadifformine.

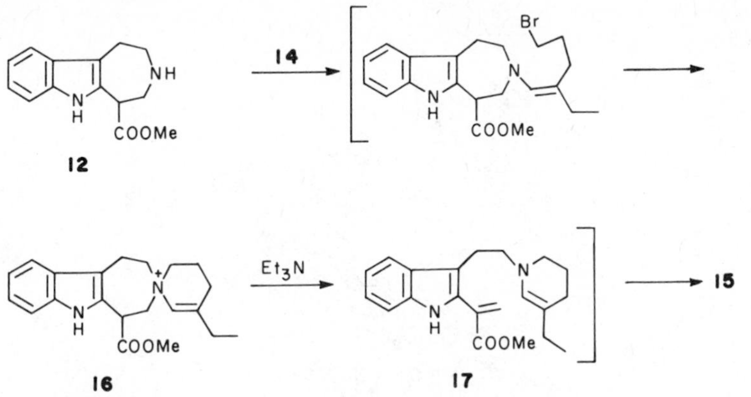

(2)

reactive intermediates and to extend this powerful new *Aspidosperma* entry. The Fischer indole synthesis provided a convenient route to substituted tetrahydro-γ-carbolines **19** from *N*-benzyl-4-piperidone **(18)** as outlined in Scheme 3 [10]. Chlorination of **19** followed by the addition of thallium dimethyl malonate and subsequent catalytic hydrogenation provided indoloazepines **20**. Since this sequence with either a tetrahydro-β- or γ-carboline starting material provided the desired indoloazepines, the common intermediacy of spirocycle **7** and zwitterion **8** [see Eq. (1)] was indicated. Indeed, the 3-spiropyrrolidinylindoline **22** was isolated in 79% yield when the malonate addition reaction was conducted at room temperature. Heating **22** at reflux in tetrahydrofuran then gave the corresponding indoloazepine in good yield.

The final conversion to the *Aspidosperma* skeleton could be done directly from the indoloazepine diesters **20** by reaction with bromoaldehyde **14** and then triethylamine. The spiroenammonium ion **23** formed in the first step could be isolated [10] and was readily converted to *dl*-vincadifformine when treated in

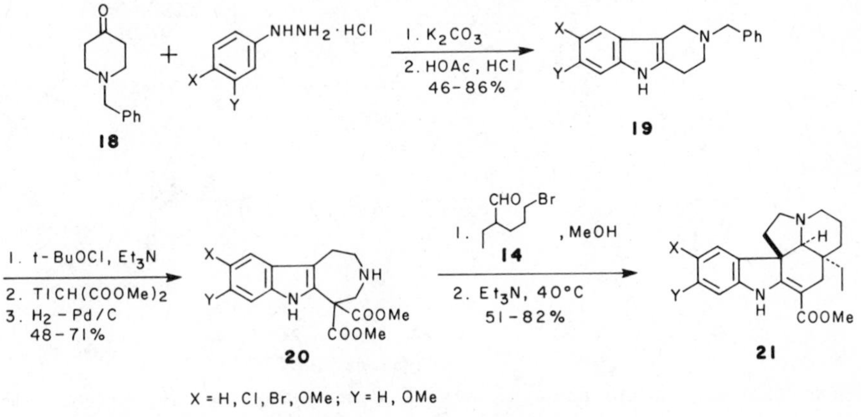

X = H, Cl, Br, OMe; Y = H, OMe

Scheme 3. Kuehne's synthesis of *dl*-ervinceine and related compounds.

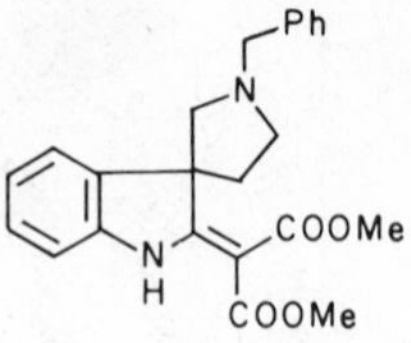

22

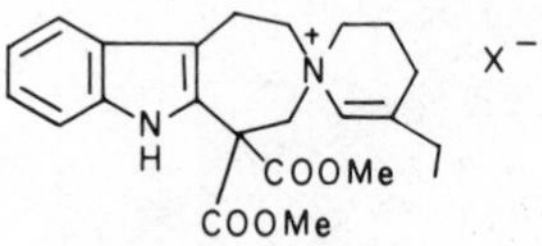

23 (X⁻ = Br⁻, BPh₄⁻)

methanol with triethylamine at 20° C. This unusually facile decarboalkoxylation of diester **23** upon treatment with triethylamine implies that the fragmentation and decarboalkoxylation steps likely occur in concert.

If the required tryptamine is available, tetrahydro-β-carboline esters **24** are readily assembled by condensation with methyl pyruvate (see Scheme 4) [11]. Acid-catalyzed reaction of **24** with chloroaldehyde **25** [10], followed by base-initiated fragmentation provided a second route to *dl*-vincadifformine (**15**) and *dl*-ervinceine (**26**), as well as *dl*-minovine (**27**). This alternate biomimetic approach provided *dl*-vincadifformine in *three-steps and 60% overall yield* from tryptamine hydrochloride [11].

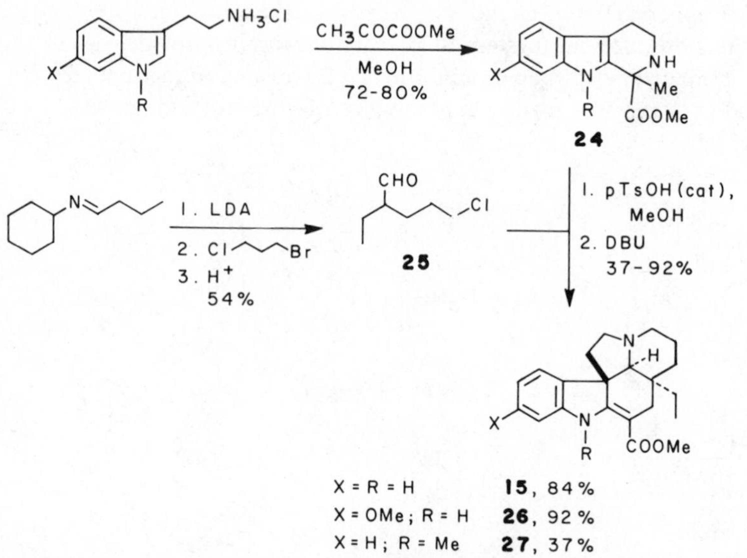

X = R = H   **15**, 84 %
X = OMe; R = H   **26**, 92 %
X = H; R = Me   **27**, 37 %

Scheme 4. Kuehne's alternate synthesis of *dl*-vincadifformine, *dl*-ervinceine and *dl*-minovine.

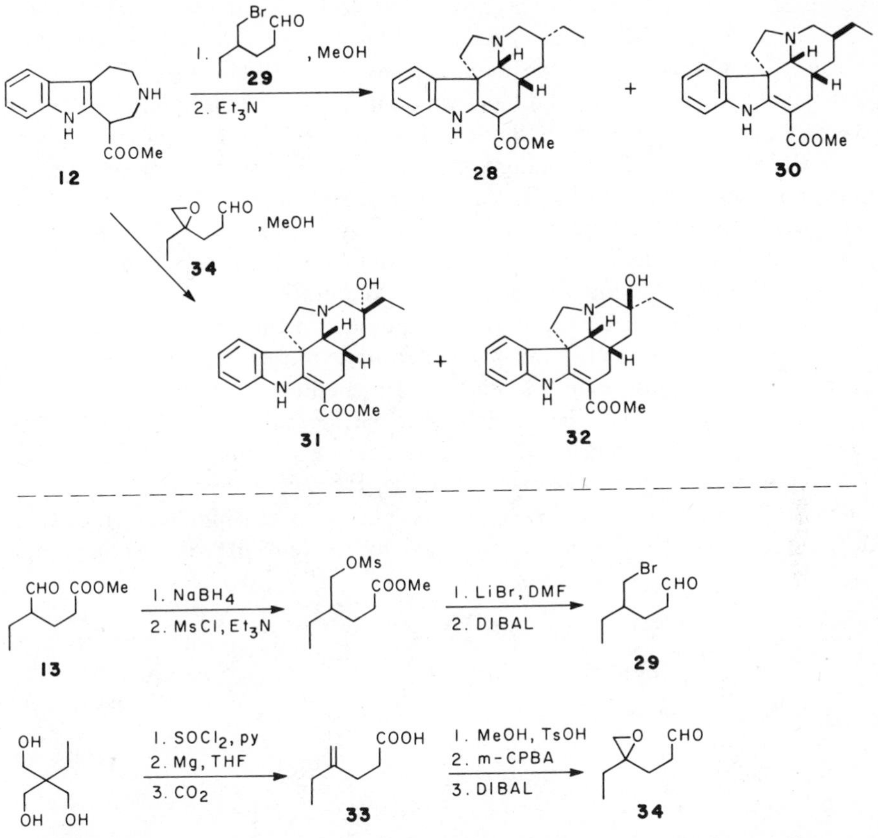

Scheme 5. Kuehne's synthesis of the *dl*-8-vincadifformines and *dl*-pandolines.

Biomimetic synthesis of pentacyclic *Aspidosperma* alkaloids with the "pseudo"-*Aspidosperma* ring system requires the generation of isomeric dehydrosecodines containing a 1,2-dihydro-3-ethylpyridine ring (**4**, Scheme 1), or secodines containing a 3-ethyl-1,2,3,4-tetrahydropyridine ring. Only minor modification of Kuehne's approach is required to access the latter intermediates [12]. The Kuehne synthesis of *dl*-pseudo-vincadifformine **28** starts with methyl 4-formyl-hexanoate (**13**) [9], which was readily converted to bromoaldehyde **29** in 50% overall yield as outlined in Scheme 5 [12]. Condensation of **29** with indoloazepine **12** (conveniently formed from the N_b-benzyl-dimethyl ester **20** (X=Y=H) by mono-decarbomethoxylation with lithium chloride in DMF) [12], followed by addition of triethylamine, provided two isomers in 41% yield. The 4:1 ratio of *dl*-pseudo-vincadifformine (**28**) and *dl*-7-epi-pseudo-vincadifformine **30** thus formed was identical with the ratio of natural pseudo-vincadifformines isolated from plant sources [12]. Crystallographic analysis [12] of both synthetic products established, for the first time, the stereostructures of these two alkaloids.

The synthesis of the 7-hydroxylated alkaloids *dl*-pandoline (**31**) and *dl*-7-epipandoline **(32)** required unstable epoxyaldehyde **34**, which was prepared in 31% overall yield (and 67% purity) from 4-ethyl-4-pentenoic acid **(33)** [13], see Scheme 5. Condensation of indoloazepine **12** with an excess of impure epoxyaldehyde **34** resulted in direct formation of the racemic pandolines **31** and **32** in a 1:1 ratio and 63% yield, but it may be noted that the epimeric ratio can be varied (2:1 versus 1:2) by changes in solvents [15]. The selective, presumably intramolecular, opening of an epoxide intermediate at only the primary carbon was crucial to the success of this synthesis. As anticipated [14, 15] no such selectivity was seen in cyclization reactions of related epoxides lacking the ethyl group, since here the intrinsic stereoelectronic preference [14] for forming 5-membered rings in such intramolecular epoxide openings becomes dominant.

In more recent work [16], the Kuehne group explored further the reaction of aldehydes with indoloazepines, and the fragmentation of these products to secodine intermediates. The original Kuehne mechanism [7] [Eq. (2)] drew support from the previously cited isolation of spiroenammonium salt **23**, [10] which contains the geminal diester grouping. However, when simple aldehydes or methyl 4-formylhexanoate **(13)** [9] were condensed with indoloazepine **12**, the bridged indoloazepines **35** and **36** were isolated (Scheme 6) [16]. Although these

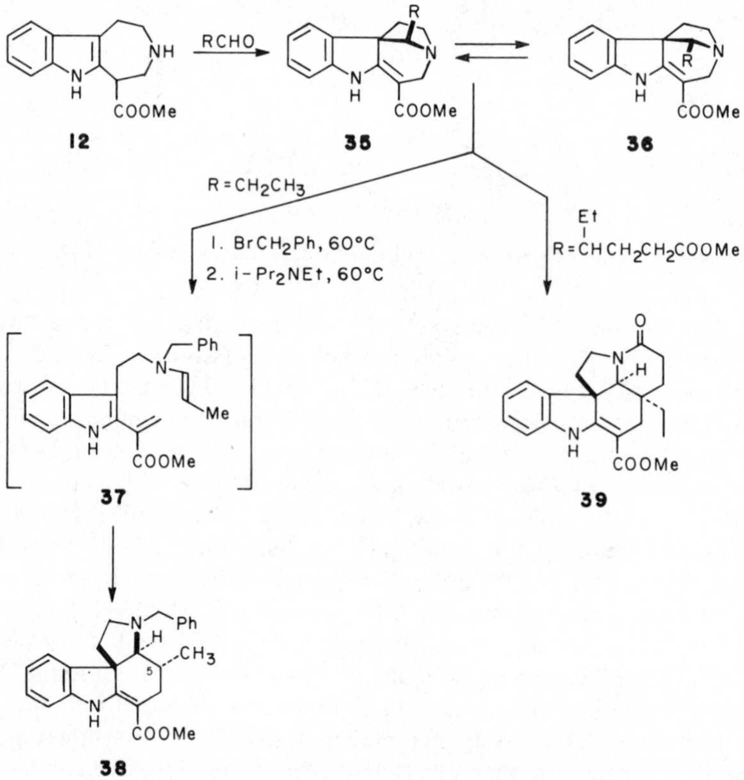

Scheme 6. Kuehne's synthesis of *dl*-8-oxovincadifformine and *Aspidosperma* analogs lacking ring D.

epimers could be separated by rapid chromatography, they interconverted readily and the mixture was employed in most subsequent reactions. Reaction of the **35/36** (R=CH₃) mixture with benzyl bromide occurred at the tertiary nitrogen, and subsequent treatment of the derived ammonium salt with diisopropylethylamine at 60°C for 2 days gave the tetracyclic *Aspidosperma* alkaloid analog **38** in 84% yield [16]. Interestingly, only a single stereoisomer having an α-methyl group at C(5) was obtained. This result indicates either preferential generation or preferential reaction of intermediate **37** with the *E*-configuration of the enamine grouping. The initial quaternization step was crucial, since **35** and **36** gave only traces of tetracyclic products when heated directly at 100°C for 20 h [16]. Activation of **35/36** could be achieved also by intramolecular acylation. Thus, heating the mixture of **35** and **36** formed from **12** and methyl 4-formylhexanoate in refluxing toluene provided *dl*-8-oxovincadifformine **(39)**, in 85% yield [16]. Taken together, these experiments suggest that the two-step conversion of indoloazepine monoester **12** to the pentacyclic *Aspidosperma* nucleus occurs by the sequence of steps outlined for the formation of *dl*-vincadifformine in Eq. (3). It should be stressed that, although this sequence is quite reasonable, the presumed secodine intermediate in this reaction has not been isolated or detected. However, in two important papers [60] published after this review was written, Kuehne has described the total synthesis of *dl*-minovincine from an isolated and thoroughly characterized oxosecodine intermediate.

The true alkaloid secodine, which is isolated from *Rhazya* species, contains a 1,2,5,6-tetrahydropyridine ring, and not the enamine grouping. It is reasonably stable and has been synthesized by several groups [17].

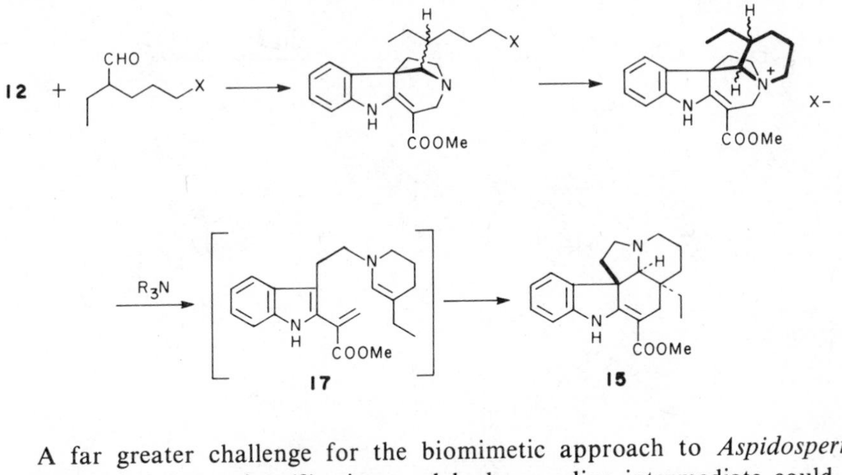

$$(3)$$

A far greater challenge for the biomimetic approach to *Aspidosperma* alkaloids is tabersonine **(3)**, since a dehydrosecodine intermediate could, in principle, cyclize to provide either the *Aspidosperma* or *Iboga* skeletons, see Eq. (4) [6]. In the Kuehne synthesis of *dl*-tabersonine [15], this issue was avoided by introducing the D-ring unsaturation after the pentacyclic skeleton was established (Scheme 7). The synthesis starts with butanoic acid, which was allylated, chlorolactonized, and reduced to give lactol **40** in 68% yield, as a mixture of

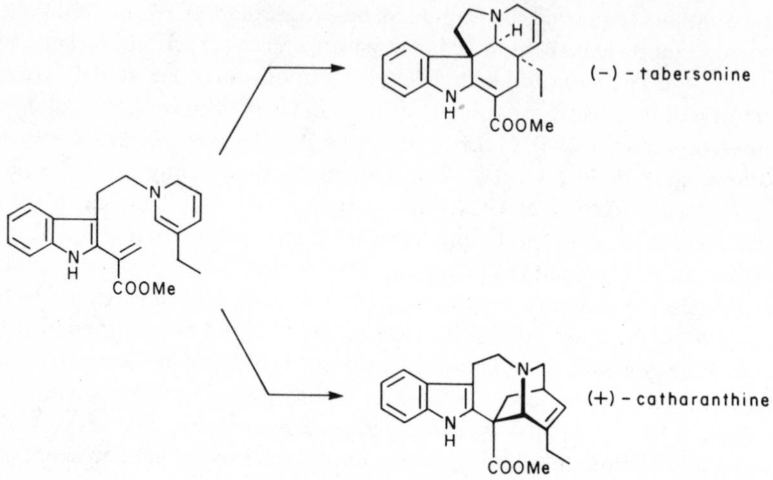

(−) − tabersonine

(4)

(+) − catharanthine

stereoisomers. Condensation of **40** with indoloazepine **12** provided an epimeric mixture of 7-hydroxyvincadifformines **41** and **42** in up to 83% yield. The final conversion to tabersonine presented a major obstacle, since only the β-hydroxy isomer **41** would undergo the desired elimination reaction. To optimize the formation of isomer **41**, the reaction had to be conducted in methanol, providing only a 43% yield of **41** and 11% of epimer **42**. Dehydration of **41** to give *dl*-tabersonine (**3**) was accomplished in 68% yield by treatment with triphenyl-phosphine and CCl₄, followed by addition of triethylamine. Attempts to

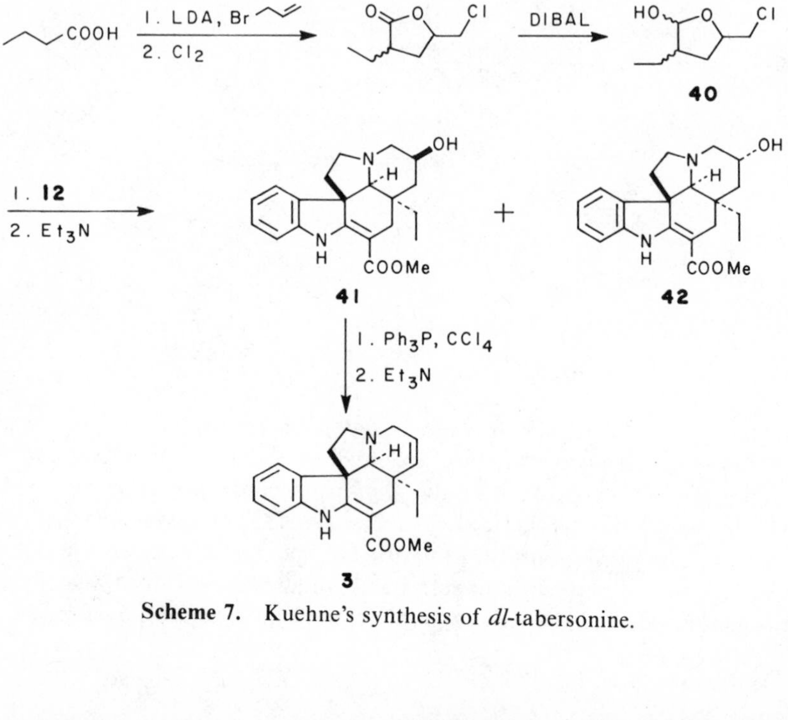

Scheme 7.   Kuehne's synthesis of *dl*-tabersonine.

dehydrate the 7α-hydroxy isomer **42** were frustrated by the formation of 7α-(chloromethyl)-D-norvincadifformine **(44)**, which presumably arose from trans-annular displacement of the equatorial leaving group to generate the aziridium intermediate **43** [15]. The Kuehne synthesis of *dl*-tabersonine **(3)** was accomplished in eight steps with an overall yield of 13% from *N*-benzyl-4-piperidone **(18)**.

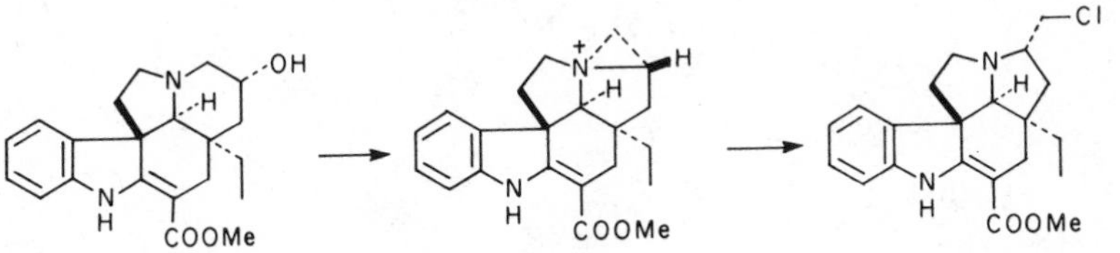

Application of the Kuehne [11] procedure to prepare 21-methylenevincadifformine **(46)** was reported by Hajicek and Trojanek, and is outlined in Scheme 8 [18]. Chloroaldehyde **45** was prepared from 4-pentenal. Reaction of **45** with β-carboline **24**, followed by treatment with DBU provided a 50% yield of racemate **46**, which was resolved with D-tartaric acid to provide (−)-21-methylenevincadifformine.

Kuehne has also used his biomimetic approach to prepare the recently isolated alkaloids ibophyllidine **(53)** and 7-epiibophyllidine **(50)** in stereoselective fashions, see Scheme 9 [19]. Reaction of 4-chlorohexanal **(47)**, available in 28% yield from 4-ethylbutyrolactone, with indoloazepine **12** in refluxing THF provided the quaternary ammonium salt **48**. Treatment of **48** with triethylamine then gave *dl*-7-epiibophyllidine **(50)** in 82% overall yield from **12**. The epialkaloid was the only isomer isolated from this sequence, and this result indicates that the 2-ethyl-2,3-dihydropyrrole ring of intermediate **49** reacted with the vinyl ester completely from the face of the five-membered ring opposite the ethyl group. It should be noted that no such selectivity [12] was observed with analogous secodine intermediates (see Scheme 5).

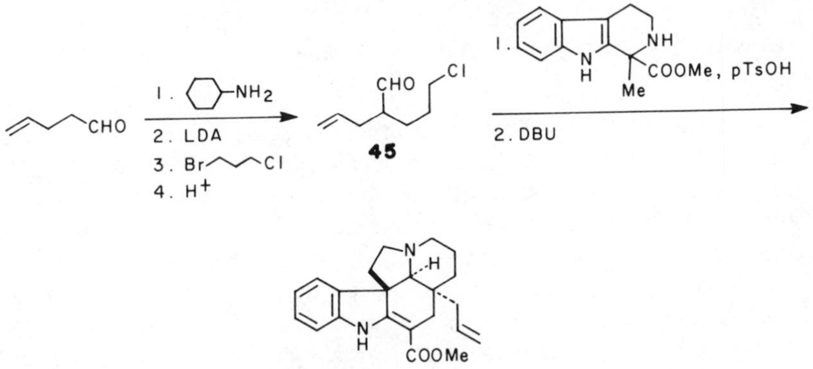

Scheme 8. Hajicek and Trojanek's synthesis of (−)-21-methylenevincadifformine.

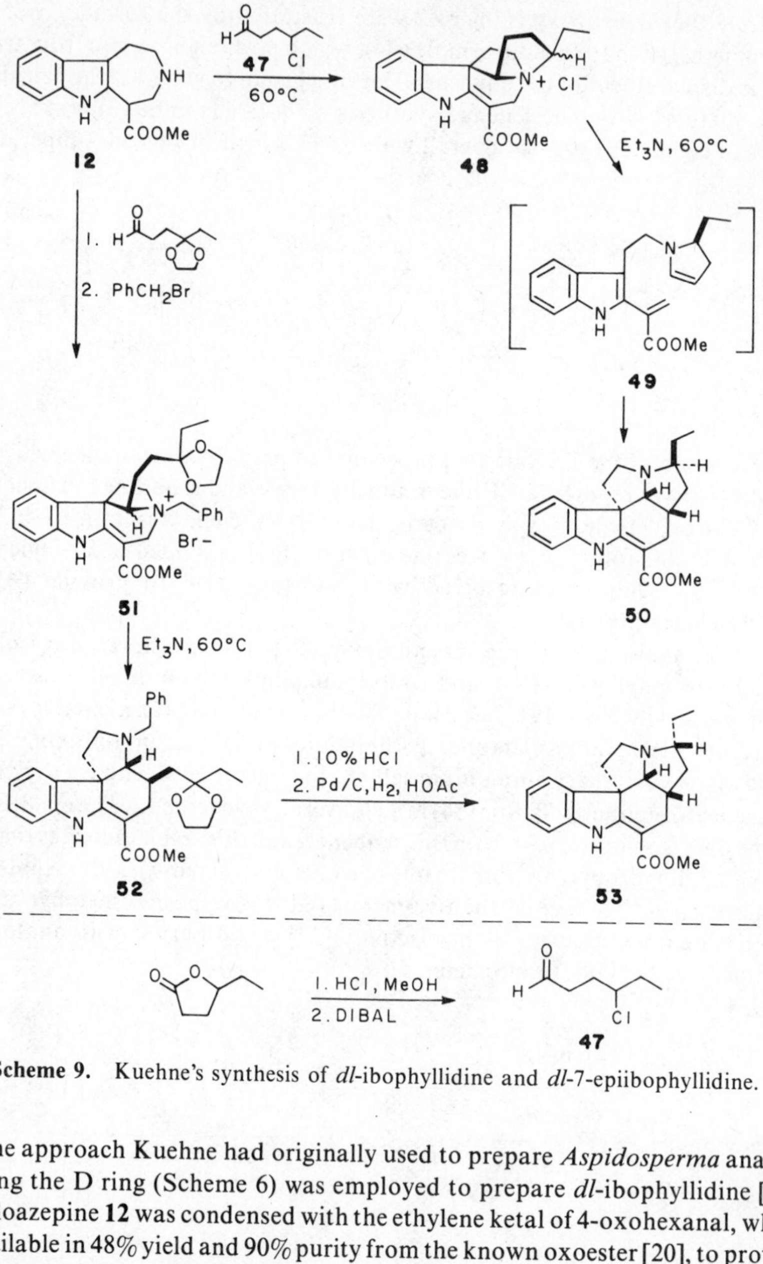

Scheme 9. Kuehne's synthesis of *dl*-ibophyllidine and *dl*-7-epiibophyllidine.

The approach Kuehne had originally used to prepare *Aspidosperma* analogs lacking the D ring (Scheme 6) was employed to prepare *dl*-ibophyllidine [19]. Indoloazepine **12** was condensed with the ethylene ketal of 4-oxohexanal, which is available in 48% yield and 90% purity from the known oxoester [20], to provide the corresponding bridged indoloazepines. These intermediates were separated and individually quaternized to give **51**. Independent treatment of these isomers with triethylamine in refluxing methanol gave the tetracyclic intermediate **52** in 82–85% yield. Hydrolysis of the ketal, followed by hydrogenation under acidic conditions provided *dl*-ibophyllidine **53** in 62% yield.

2.2. Dehydrosecodine Intermediates

There is to date but a single report [21] of constructing the pentacyclic *Aspidosperma* alkaloid skeleton in a reaction that is believed to proceed via a biosynthetically postulated dehydrosecodine intermediate. The Kutney [21] plan was to mask the dienamine portion of a reactive dehydrosecodine intermediate as a tricarbonylchromium(0) complex. This approach (see Scheme 10) began with

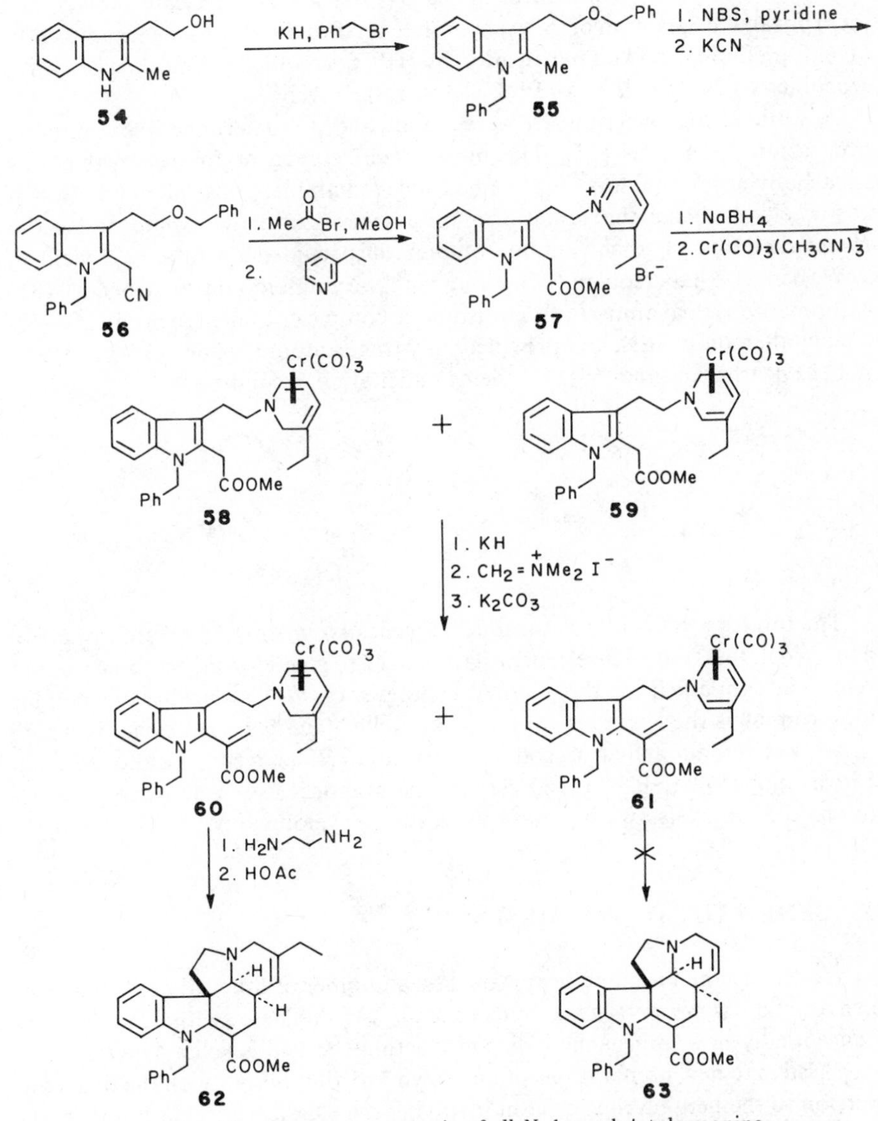

Scheme 10. Kutney's synthesis of *dl*-N_a-benzyl-ψ-tabersonine.

2-methyltryptophol (**54**) [22], which was converted to the dibenzyl derivative **55**. Bromination, pyridinium salt formation, and displacement with cyanide gave nitrile **56**. The addition of acetylbromide to a methanol solution of **56** cleaved the benzyl ether and converted the nitrile to a methyl ester. Chromatographic purification and reaction of the derived bromide with 3-ethylpyridine gave pyridinium salt **57** in 11% overall yield from indole **54**.

Phase transfer borohydride reduction of **57** and complexation of the resulting product with *tris*(acetonitrile)tricarbonylchromium(0) provided chromium complexes **58** and **59** in a 9:1 ratio and a 54% yield. The acrylate moiety was incorporated by reaction of the ester enolate of **58** or **59** with Eschenmoser's salt, and after elimination and purification by HPLC provided the masked dehydrosecodines **60** and **61** in 27 and 8% yields, respectively.

A number of experiments were conducted to determine the optimum cyclization conditions [21]. The most effective reagent for removal of the tricarbonylchromium was found to be ethylenediamine. Subsequent addition of acetic acid resulted in the formation of products with both the *Aspidosperma* and *Iboga* skeletons. The chromium complex **60** provided up to a 22% yield of *dl*-N_a-benzyl-ψ-tabersonine (**62**), along with varying amount of *N*-benzyl-18β-carbomethoxycleavamine (**64**). The isomeric complex **61** failed to produce *dl*-N_a-benzyltabersonine (**63**), and provided only trace amounts of **64** (11% yield) and *N*-benzylcatharanthine (**65**) (1% yield) under similar conditions.

6 4   **65**

The total synthesis of a natural pentacyclic *Aspidosperma* alkaloid was not achieved in this study, since debenzylation of **62** to give *dl*-ψ-tabersonine was not reported. Nonetheless, the Kutney synthesis of *dl*-N_a-benzyl-ψ-tabersonine (**62**) from **60** is the first entry to the pentacyclic *Aspidosperma* skeleton which may proceed via a dehydrosecodine intermediate. The overall yield of **62** from **54** was quite low (~0.3%), and clearly, the practical use of dehydrosecodines in organic synthesis awaits significant further developments.

3. OTHER TOTAL SYNTHESES

In addition to the biomimetic approaches, a number of other total syntheses of pentacyclic *Aspidosperma* alkaloids have appeared. These syntheses vary from conceptually new approaches for constructing the pentacyclic *Aspidosperma* ring system to new preparations of unknown hydrolilolidines with the final conversion to the pentacyclic skeleton involving the classical Fischer indole cyclization, which was originally employed by Stork [3].

3.1. Tandem Aza-Cope–Mannich Approach

The stereocontrolled synthesis of alkaloids using "Mannich-directed" cationic aza-Cope (2-azonia-[3,3]-sigmatropic) rearrangements has been extensively developed in recent years by Overman and co-workers [23]. In 1981, Overman and Sworin reported [24] a stereoselective aza-Cope–Mannich approach to *Aspidosperma* alkaloids, which proceeded via the intermediacy of 9a-arylhydro-lilolidines. Aryl-substituted hydrolilolidines of this type (e.g., **73** in Scheme 11) have considerable advantage over the simpler hydrolilolidines employed in the early Stork and Ban syntheses [1, 3] of pentacyclic *Aspidosperma* alkaloids, since the quaternary aryl grouping need not be incorporated at a late synthetic stage by the typically low-yielding Fischer indole procedure. Application of this new strategy for the total synthesis of *dl*-16-methoxytabersonine (**74**) is outlined in Scheme 11 [25].

The convergent synthesis of the aza-Cope precursors **70** and **71** required the attachment of a 1-arylvinyl group stereoselectively to the convex α-face of the *cis*-hexahydro-7H-1-pyrindin-7-one intermediate **68**. Two syntheses of this ketone were reported, the most direct and efficient of which is summarized in Scheme 11 [25]. Zinc bromide catalyzed reaction of the thermodynamic silyl enol ether of 2-ethylcyclopentanone with 1,3-dichloro-1-(phenylthio)-propane provided chloroketone **77**, which was readily converted to bicyclic enecarbamate **78**. Selective oxidation of **78** at $-40°$ C provided the unstable epoxide sulfoxide **79** as a complex mixture of stereoisomers, which was directly transformed to bicyclic ketone **68** upon heating at $165°$ C in the presence of $CaCO_3$ [26]. A single bicyclic ketone was isolated, presumably reflecting a large thermodynamic bias for a *cis*-ring fusion in this bicyclic ring system. This sequence provided access to bicyclic ketone **68** in six steps (two isolated and purified intermediates) and 25% overall yield from silyl enol ether **76**.

The required aromatic A-ring fragment **67** was prepared in a one-pot reaction from the known arene **66** [27] in 63% yield. The critical coupling reaction was accomplished by converting aldehyde **67** to the silyl cyanohydrin, formation of the dianion at $-78°$ C with butyllithium, followed by the addition of hydropyrin-dinone **68**. Treatment of the reaction product with LiOH provided tetracyclic carbamate **69** in 80% yield. The formation of a single stereoisomer at this stage followed from addition of the silycyanohydrin anion from only the convex side of the *cis*-fused ketone **68**. Methylenation of the ketone and basic hydrolysis using 40% methanolic KOH provided aniline **70** in 93% yield. Selective hydrolysis using 20% methanolic KOH gave a 7 : 3 mixture of pivalamide **71** and aniline **70**, from which pivalamide **71** could be isolated in 49% yield.

Tandem aza-Cope–Mannich rearrangement of pivalamide **71** was effected by treatment with paraformaldehyde and a catalytic amount of camphorsulfonic acid in refluxing benzene to provide hydrolilolidine **73** in 96% yield. The stereoselectivity of the aza-Cope–Mannich rearrangement follows directly [23, 24] from the *trans* relationship of the vinyl and amine groups on the cyclopentane ring in precursor **71**, since iminium ion intermediate **80** can undergo rearrange-

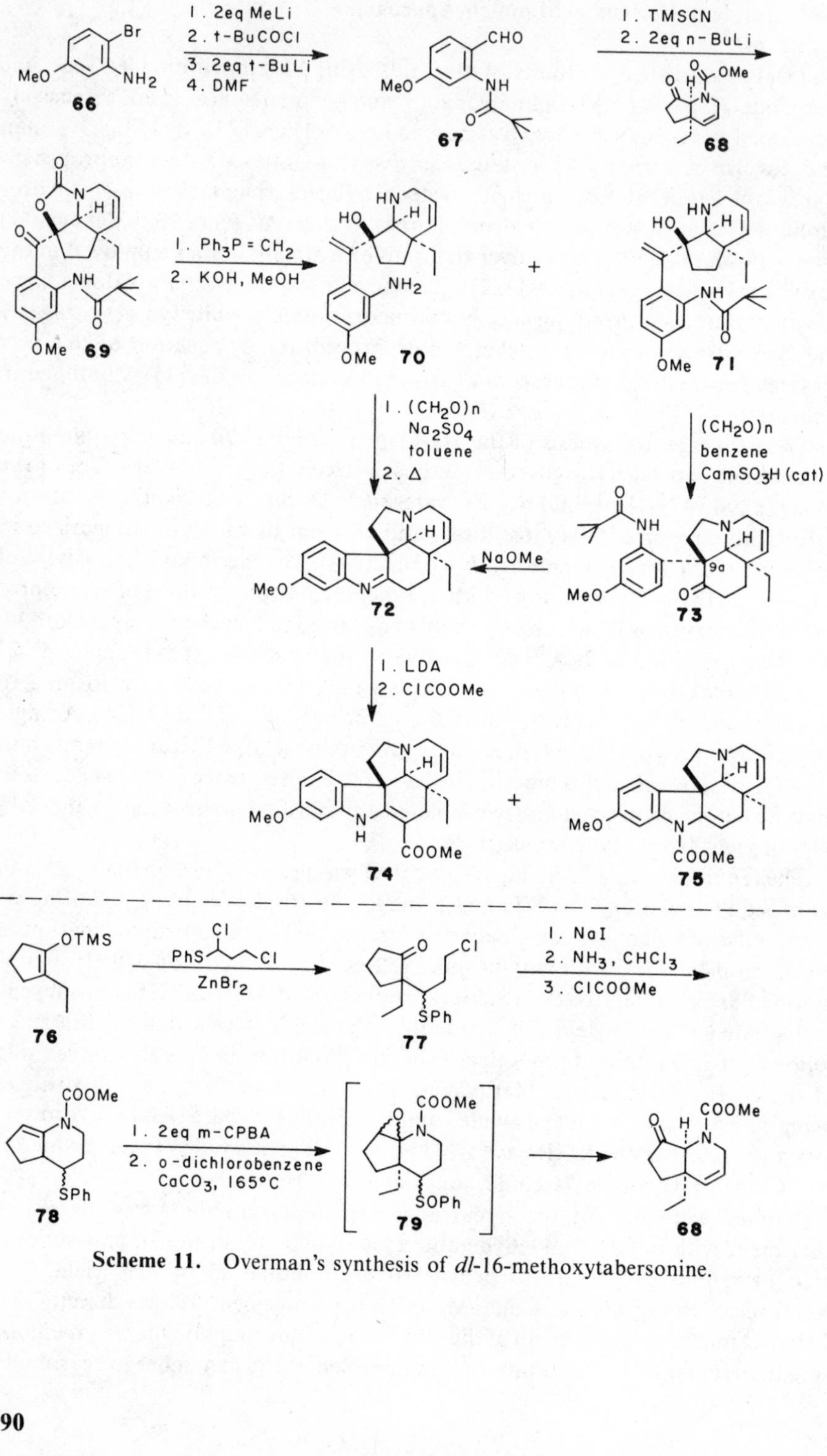

Scheme 11. Overman's synthesis of *dl*-16-methoxytabersonine.

ment via only a single "chairlike" transition state [23, 24]. Intramolecular Mannich closure of the bridged *trans, trans*-1,5-azacyclononadiene **81** thus produced would yield the all *cis*-hydrolilolidine **73**.

Conversion to the pentacyclic ring system **72** was accomplished in 67% yield from hydrolilolidine **73** by treatment with 25% sodium methoxide in refluxing methanol. A more concise entry involved direct aza-Cope–Mannich rearrangement of aniline **70**. This was accomplished by treatment of **70** with one equivalent of paraformaldehyde to form the corresponding tetracyclic oxazolidine, which was directly rearranged in refluxing toluene to the pentacyclic imine **72** in yields of 70–90%. This notable step elaborates three of the five rings of the desired pentacyclic skeleton with complete stereocontrol.

Acylation of the lithium salt of imine **72** provided *dl*-16-methoxytabersonine (**74**) in 32% yield from aniline **70,** along with a 22% yield of the *N*-acylated product **75** [25]. The total synthesis of *dl*-16-methoxytabersonine (**74**) was accomplished in 11 steps and 6% overall yield from silyl enol ether **76**.

3.2. Via Photoisomerization of 1-Acylindoles

Ban had previously observed that 1-acylindoles photorearrange to a variety of products including 3-acylindolenines [28]. Application of an intramolecular version of this photoisomerization for the total synthesis of the pentacyclic *Aspidosperma* alkaloid *dl*-1-acetylaspidospermidine (**87**) is outlined in Scheme 12 [29].

The synthesis of tryptamine **83** was accomplished in 86% overall yield from indole **82** via a Curtius rearrangement sequence. Irradiation of **83** with a 300-W high-pressure mercury lamp afforded lactam **84** in 80% yield. This novel ring expansion likely results from a photochemical [1, 3]-acyl migration to afford the labile carbazolenine **88**, which is subsequently converted, via hemiaminal **89**, to the stable macrolactam **84**. Efficient intramolecular trapping of the unstable carbazolenine is undoubtedly responsible for the excellent yield observed in this complex transformation.

The conversion of lactam **84** to tetracyclic lactam **85** involved a series of standard transformations and proceeded in 35% overall yield. Partial reduction to the hemiaminal followed by acid-catalyzed cleavage of the tetrahydropyranyl group and iminium ion cyclization provided pentacyclic imine **86** in 48% yield. The structure of imine **86** was established by reduction with LiAlH₄ and subsequent acetylation to provide *dl*-1-acetylaspidospermidine (**87**) in 64% yield.

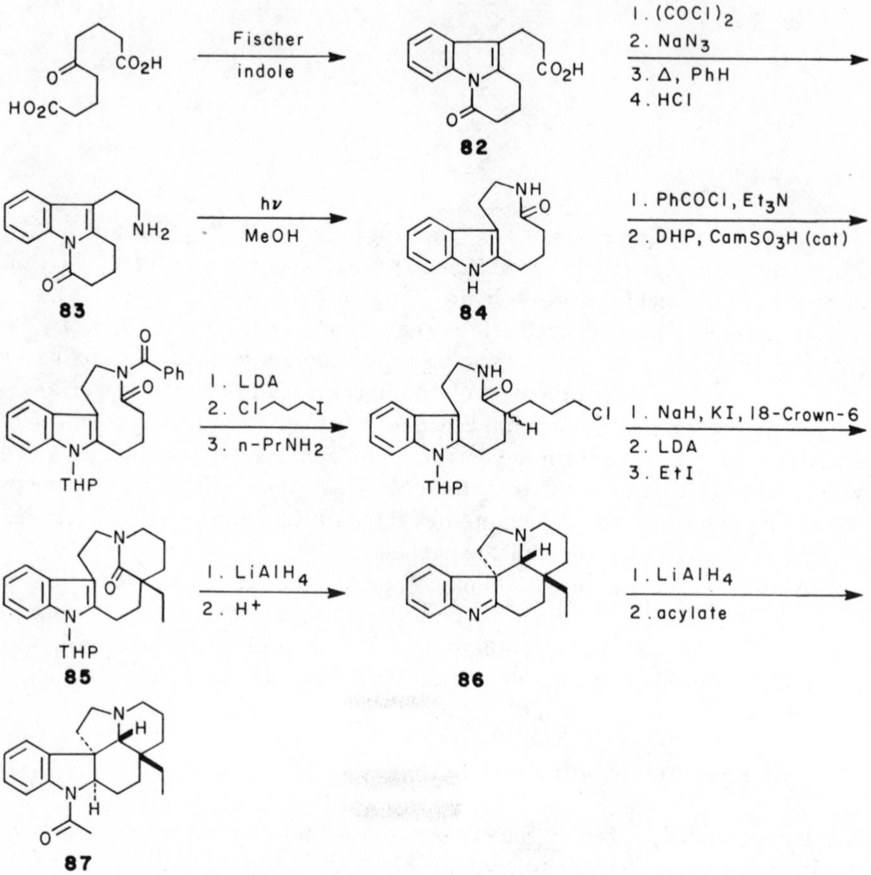

Scheme 12.   Ban's synthesis of *dl*-1-acetylaspidospermidine.

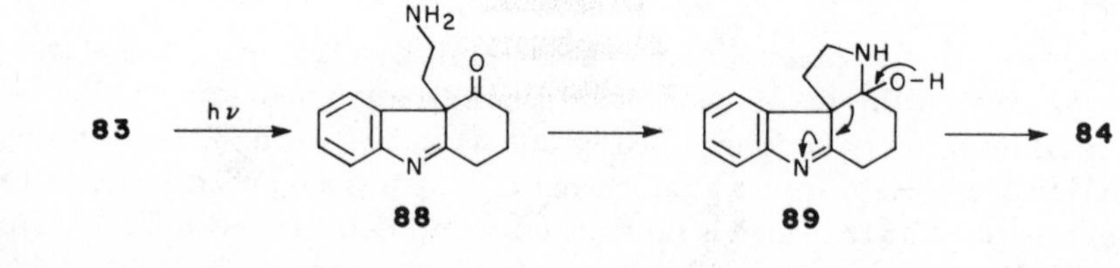

Lactam **84** has also been employed by Ban to achieve formal total syntheses of the *Strychnos* alkaloids *dl*-tubifoline and *dl*-condyfoline [29].

The total synthesis of *dl*-1-acetylaspidospermidine (**87**) was accomplished in 15 steps and 7.5% overall yield from indole **82** [29].

3.3.   Intramolecular Diels–Alder Approach

The synthesis of indole alkaloids via indole-2,3-quinodimethanes has been extensively developed in recent years by Magnus and co-workers [30–32].

Although formation of these intermediates by fluoride anion-induced fragmentation of 2-[(trimethylsily)methyl]indoles was unsuccessful due to competing protodesilylation [30], these intermediates can be generated by direct acylation of imine derivatives of 3-formyl-2-methylindoles [31]. Two syntheses of *dl*-aspidospermidine **(96)** by this approach are outlined in Scheme 13 [32].

The synthesis of indole-2,3-quinodimethane precursor **92** was readily ac-

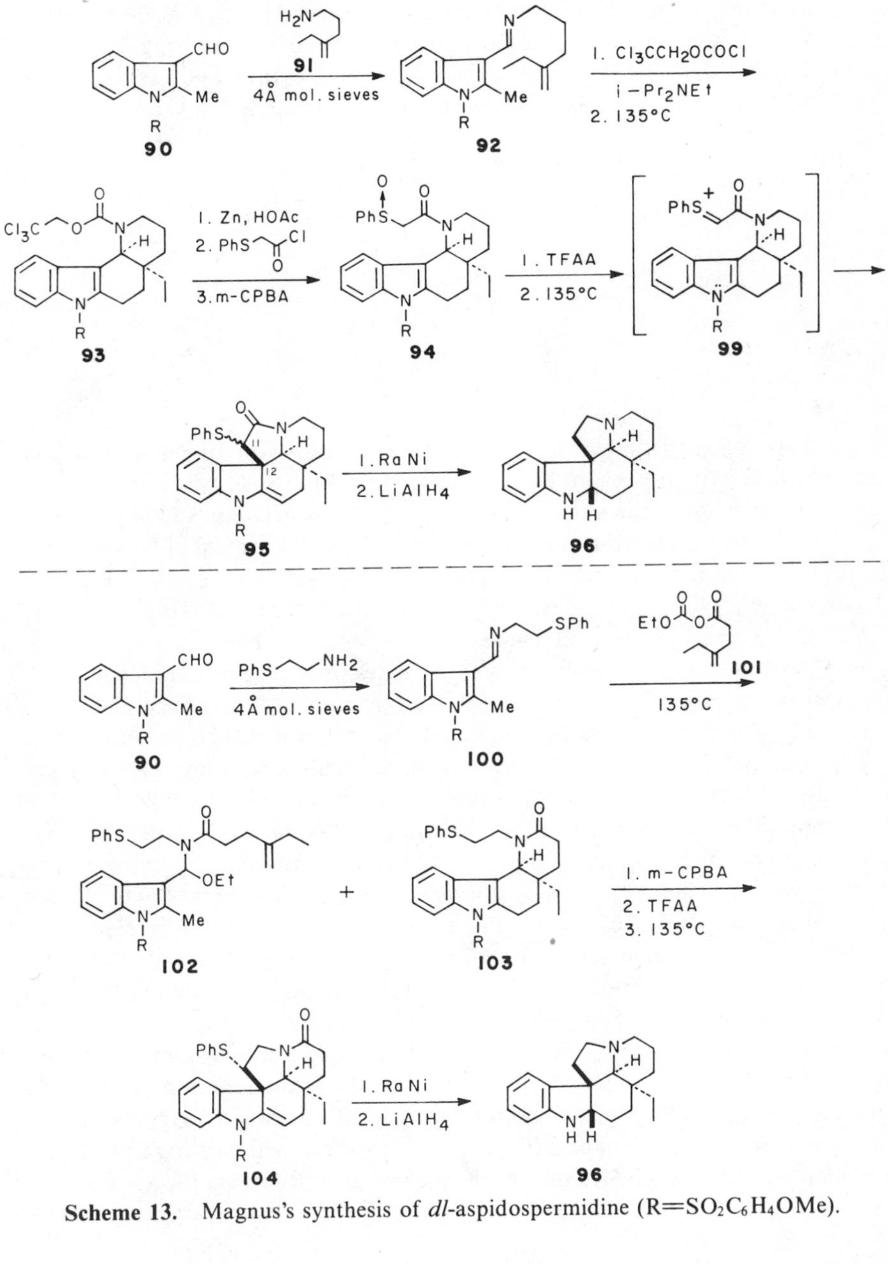

Scheme 13.   Magnus's synthesis of *dl*-aspidospermidine (R=SO₂C₆H₄OMe).

complished in 95% yield from the reaction of the previously reported 3-formylindole **90** [30] and amine **91** (obtained by reduction of the amide of 4-ethyl-4-pentenoic acid with LiAlH₄) [13]. When imine **92** was treated with 2,2,2-trichloroethyl chloroformate in the presence of diisopropylethylamine [31] followed by heating at 135°C, the *cis*-tetracyclic carbamate **93** was isolated in 46% yield. The *cis*-ring junction is believed to arise from preferential Diels–Alder cyclization of the (*E*)-indole-2,3-quinodimethane intermediate **98** in the exo sense. Exo cyclization of the corresponding (*Z*)-isomer **97** should be destabilized by steric interactions between the carbamate and the benzenoid ring. Although not specifically discussed by Magnus, cyclization of **97** in an endo sense (which would also lead to **93**) is less likely, since the amide and diene π-systems would be highly twisted.

97   **98**

Cleavage of carbamate **93,** subsequent acylation with (phenylthio)acetyl chloride, and oxidation to the sulfoxide provided **94** in 36% yield from **93**. The critical conversion to the pentacyclic ring system was accomplished by cyclization of an α-acylsulfenium ion intermediate **99**. Thus, treatment of **94** under the Potier–Pummerer conditions provided **95** in 91% yield, as a 3:1 mixture of diastereomers at C(11). The high yield obtained from this method of forming the C(11)–C(12) bond of the pentacyclic *Aspidosperma* skeleton, should be contrasted with the earlier low-yielding approaches involving intramolecular S$_N$2 displacements [32]. The conversion of the sulfide mixture **95** to *dl*-aspidospermidine **(96)** was achieved in 43% overall yield by stepwise reduction with Raney nickel and LiAlH₄. This total synthesis of *dl*-aspidospermidine **(96)** was accomplished in eight steps and 6% overall yield from formylindole **90** [32].

A somewhat shorter approach [32], in which the two carbon fragment which is destined to be the E ring is directly incorporated into the starting imine, is also summarized in Scheme 13. This synthesis begins with the reaction of imine **100** with mixed anhydride **101** to give the desired *cis*-tetracycle **103** in 33% yield, together with a 21% yield of the ethanol adduct **102** [31]. The formation of **102** is best rationalized by ethanol trapping of acyliminium ion intermediate **105,** which occurs apparently competitively with deprotonation to the desired diene carbamate. Unfortunately, the use of different mixed anhydrides gave comparable amounts of alcohol-trapped products, while the reaction of **100** with other acylating agents provided no tetracyclic products at all. Carrying out the acylation in the presence of base in the hope of facilitating deprotonation of **105,** was also unsuccessful and led only to the formation of β-lactam **106** as the major

105   106

by-product, presumably arising from ketene-imine cycloaddition [31]. Oxidation of sulfide **103** followed by Pummerer rearrangement and cyclization provided a single diastereomeric sulfide **104** in 81% yield. The previously employed reduction sequence completed an alternative total synthesis of *dl*-aspidospermidine. This second total synthesis of *dl*-aspidospermidine (**96**) was accomplished in six steps and 11% overall yield from formylindole **90**.

This intramolecular Diels–Alder approach has also been utilized by Magnus to prepare complex heptacyclic *Aspidosperma* alkaloids [33].

3.4. Via Organoiron Intermediates

Tricarbonylcyclohexadienyliumiron(0) complexes are synthetic equivalents of γ-cyclohexenone cations [34]. The utilization of these complexes in the total synthesis of *Aspidosperma* alkaloids was recently reported by Pearson and Rees [35]. The synthesis of *dl*-limaspermine (**116**), a C(21) oxygenated pentacyclic *Aspidosperma* alkaloid, is outlined in Scheme 14.

The initial tricarbonyliron complex **108** was prepared in 50% overall yield from *p*-methoxycinnamic acid (**107**) by acid-catalyzed isomerization of the Birch reduction product, followed by addition of iron pentacarbonyl. Conversion of **108** to the phthalamide **109**, followed by regioselective hydride abstraction with trityl tetrafluoroborate provided **110** in 77% yield [35, 36]. Reaction of **110** with potassium dimethyl malonate provided a 4.6:1 mixture of regioisomeric-alkylated products. Crystallization provided iron complex **111** in 68% yield. Oxidative removal of the tricarbonyliron group with trimethylamine oxide gave the corresponding dienyl ether in 85% yield. Conversion to *cis*-hydroquinolone **112** was accomplished in a reported 99% yield by regenerating the primary amine, hydrolysis of the enol ether, and base-promoted cyclization.

Transformation of **112** to the α-chloroamide **113** required eight relatively uneventful transformations and was accomplished in 34% overall yield. Conversion of **113** to *dl*-*O*-methylcylindrocarpinol (**115**) was accomplished in 13% overall yield via the hydrolilolidine intermediate **114**, using procedures worked out by Saxton [37] and Stork [3] in closely related systems. The synthesis of *dl*-limaspermine (**116**) was completed in 24% yield by transforming **115** to the propionamide, followed by cleavage of both methyl ethers with trimethylsilyl iodide. This synthesis of *dl*-limaspermine (**116**) required 30 steps and proceeded in 0.24% overall yield from **107**.

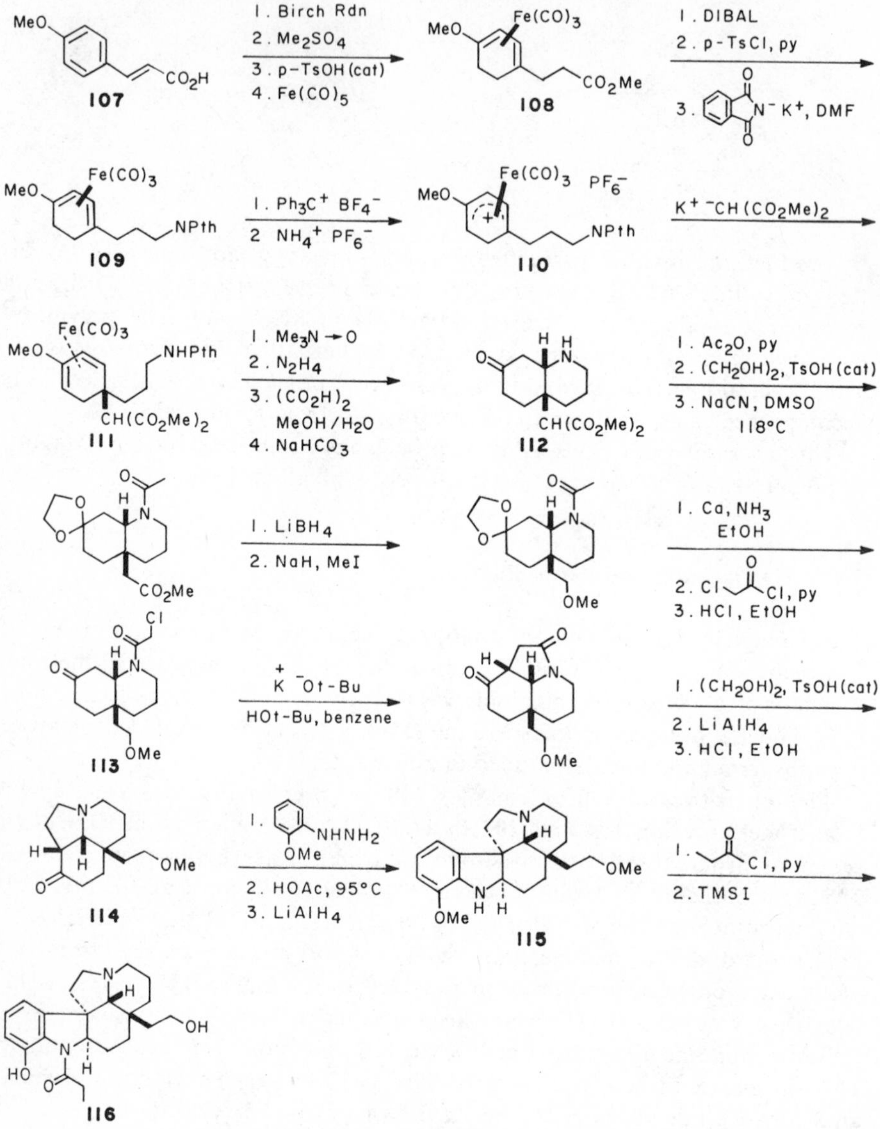

Scheme 14.   Pearson's synthesis of *dl*-limaspermine.

Since the first route to *dl*-limaspermine was extremely long, Pearson developed a somewhat more direct synthesis of chloroamide **113**, which is outlined in Scheme 15 [38]. The required arene **117** was prepared from *p*-hydroxyphenylacetic acid in 57% overall yield. The conversion of **117** to iron complex **119** paralleled the previous route [35] and proceeded in 28% overall yield. Transformation of the dimethyl malonate moiety in **119** to nitrile **120** was accomplished by cyanide displacement of a tosylate intermediate. Removal of the metal and

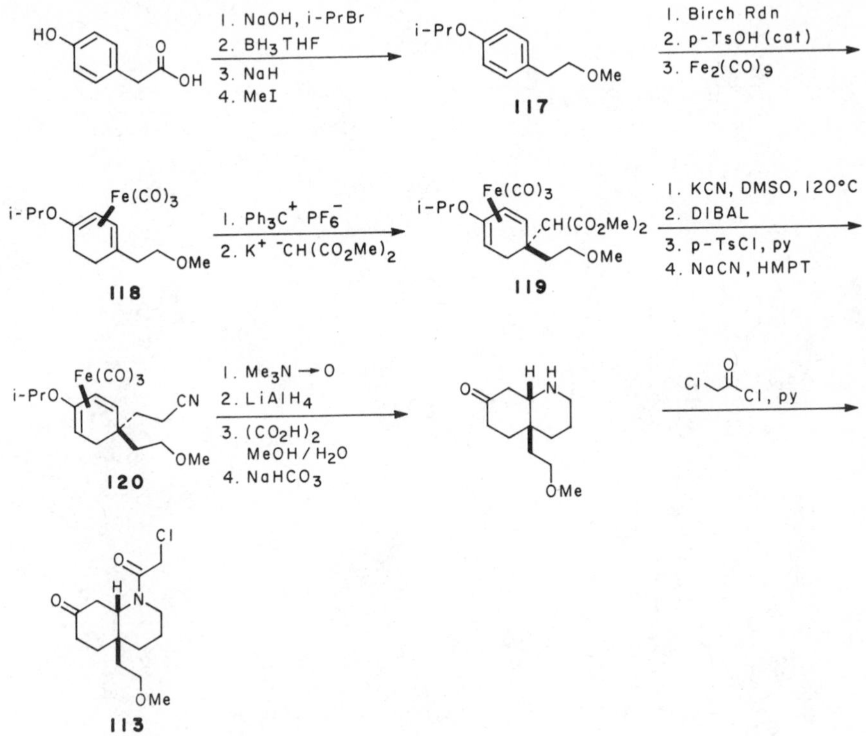

Scheme 15.   Pearson's formal synthesis of *dl*-limaspermine.

reduction of the nitrile gave the primary amine, which was converted to chloro-amide **113** by a sequence which paralleled the previous synthesis.

Unfortunately, this synthesis of **113**, although shorter, proceeded in similar yield. Amide **113** was prepared in 14 steps and 6.8% overall yield from arene **117** (Scheme 15), versus 21 steps and 7.5% overall yield from **107** (Scheme 14).

3.5.   Via Rearrangement of Oxindoles

Levy and co-workers had previously prepared [39] *dl*-8-oxovincadifformine **(39)** from 2-hydroxytryptamine **(121)** in 12% overall yield. This intermediate is also available by the Kuehne biomimetic procedure (Scheme 6) [16]. Levy has reported the conversion of (−)-8-oxovincadifformine (prepared from degrada-tion of (−)-tabersonine) to (−)-tabersonine by the sequence summarized in Scheme 16 [40].

Treatment of (−)-8-oxovincadifformine with lithium diisopropylamide and a large excess of phenylselenyl chloride gave the diselenyl compound **122**. Deselenylation with thiophenoxide afforded an epimeric mixture of the mono-selenides in 85% yield. Oxidative elimination provided (−)-8-oxotabersonine

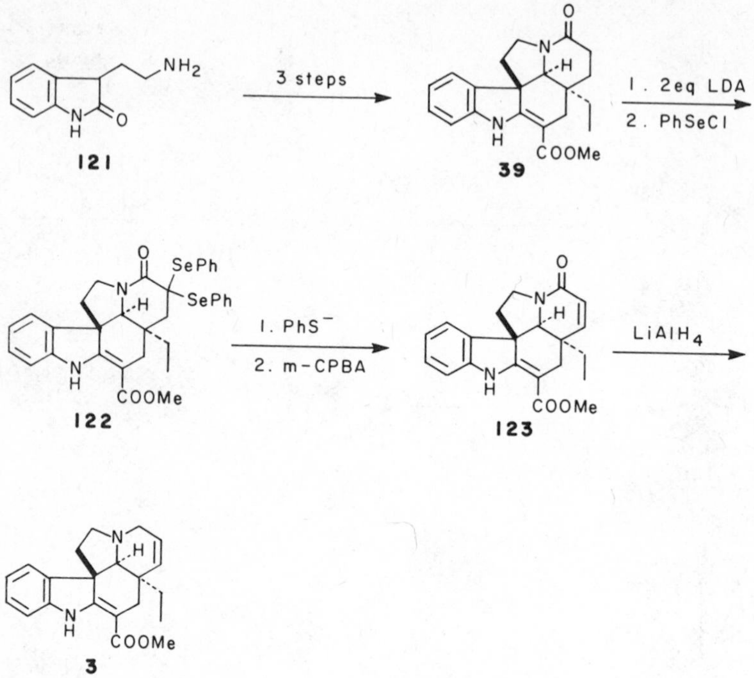

Scheme 16. Levy's relay synthesis of (−)-tabersonine.

(123). Reduction of the amide function with LiAlH₄ at 0° C gave (−)-tabersonine in 38% yield together with 55% of recovered amide **123** [40].

4. PARTIAL TOTAL SYNTHESES

4.1. Alternate Preparations of Intermediates in the Büchi Syntheses of *dl*-Vindorosine and *dl*-Vindoline

Tetracyclic intermediates **124** and **125** are fairly advanced intermediates in Büchi's pioneering syntheses of the complex pentacyclic *Aspidosperma* alkaloids *dl*-vindorosine (**126**) and *dl*-vindoline (**127**) [4]. As such, they have attracted

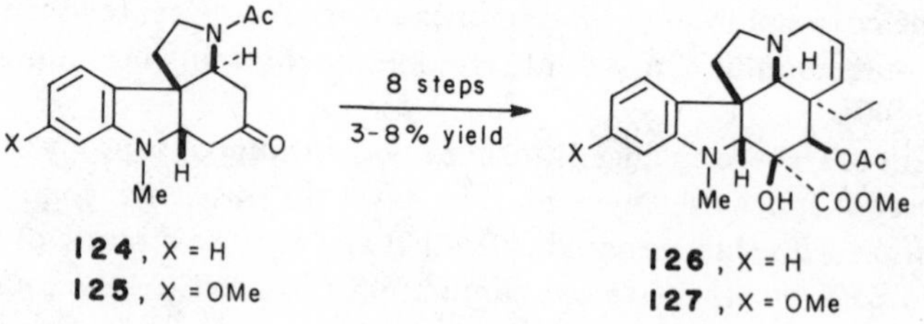

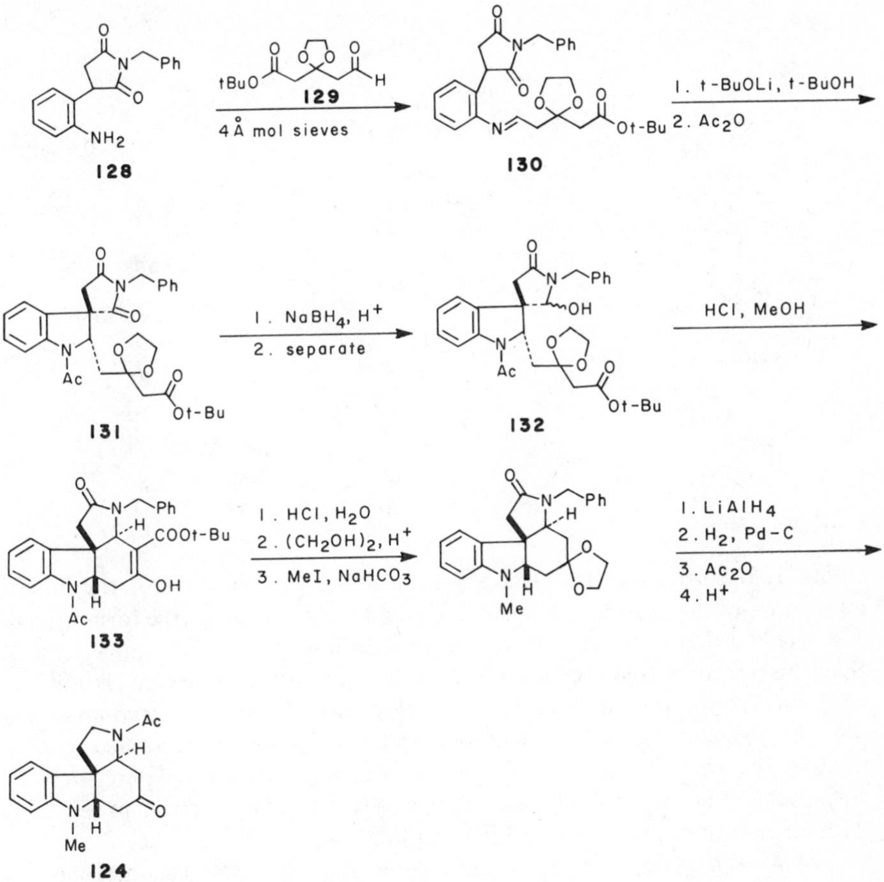

Scheme 17. Speckamp's formal synthesis of *dl*-vindorosine.

considerable attention in recent years and several improved preparations of these precursors have now been developed. Efficient preparations of **124** by Takano [47] and Ban [42] were discussed in the Appendix of the Cordell review [1], and will not be included in this report.

A new stereocontrolled synthesis of 2,3-dihydroindoles was reported by Speckamp in 1981 [43], and the application of this chemistry for the preparation of *dl*-vindorosine intermediate **124** is summarized in Scheme 17 [44]. The key transformation was the electrocyclic ring closure of the conjugate base of **130** to give **131** in 84% yield. Imine **130** was assembled from amine **128** and aldehyde **129** [44].

The stereoselectivity of this cyclization reaction was dramatically affected by reaction conditions [43]. When model imines **134** (R = alkyl or aryl) were cyclized with sodium ethoxide in ethanol, only isomer **136** was obtained, while similar treatment with sodium *tert*-butoxide in *tert*-butanol provided only isomer **138**. Since 6-electron electrocyclization is favored in a disrotatory mode, cyclization of the (*E*)-imine isomer **135** would yield **136,** while the formation of **138** would require disrotatory closure of the less favored (*Z*)-imine isomer. Speckamp rationalized the involvement of the (*Z*)-imine isomer in *tert*-butanol as arising from steric interactions in an imine–alcohol association complex **137** [46]. Experimental evidence to support this conclusion was obtained by conducting the ring closure in the presence of enantiomerically pure alcohols (menthol and borneol) which resulted in the formation of the chiral product **138** with an enantiomeric excess of 17–31% [46].

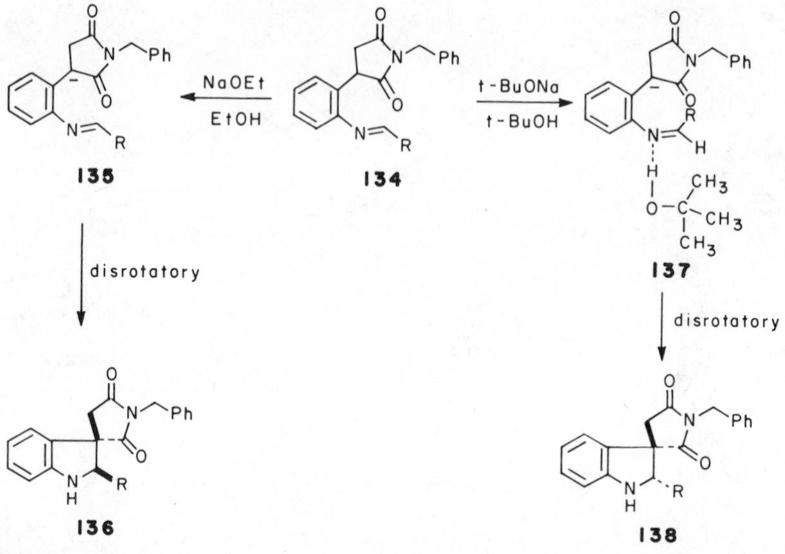

Completion of the formal *dl*-vindorosine synthesis, was accomplished by reduction [47] of imide **131**, separation of the unwanted isomer, followed by acid-promoted acyliminium ion cyclization [44, 45] of **132** to afford the tetracycle **133**. The cyclization step was reported to occur in 70% yield; however the regioselectivity and yield of the reduction step were not reported in this preliminary communication [44]. Elaboration of **133** to *dl*-vindorosine precursor **124** was accomplished via standard transformations; yields were not reported [44].

An extremely efficient synthesis of the more advanced Büchi vindorosine intermediate **144** has been developed by Langlois [48] and is summarized in Scheme 18. Reaction of readily available imine **140** with methyl 2,4-pentadienoate afforded a mixture of three cycloadducts in 71% yield. Treatment of this mixture with lithium diisopropylamide and ethyl iodide gave a *single* diastereomeric alkylation product **141** in 98% yield. Reaction of **141** with dimsyllithium provided the β-ketosulfoxide **142**, which rearranged upon treatment with *p*-toluenesulfonic acid to give pentacycle **143** in 59% overall yield from **141**. Reduction of the vinylogous amide grouping in **143** followed by desulfuriza-

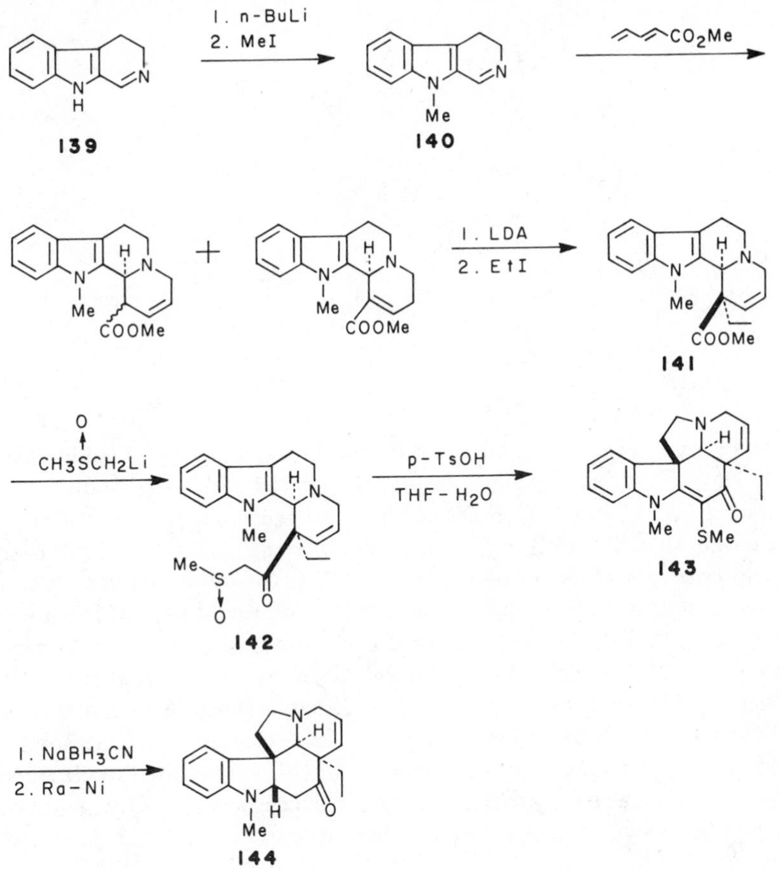

I . n–BuLi
2. MeI

139

140

Me

CO$_2$Me

+

Me
COOMe

Me
COOMe

I . LDA
2. EtI

Me
COOMe

141

O
↑
CH$_3$SCH$_2$Li

Me

Me–S
O
O

142

p–TsOH

THF–H$_2$O

Me SMe
O

143

I . NaBH$_3$CN
2. Ra–Ni

N
.H

Me H
O

144

Scheme 18. Langlois' formal synthesis of *dl*-vindorosine.

tion provided **144** in a remarkable 41% overall yield from dihydro-β-carboline **139**.

The crucial rearrangement step of the Langlois synthesis presumably occurs by initial cyclization [49] of the α-ketosulfenium ion **145** to the pentacyclic iminium ion **146** (either directly or via rearrangement of a spiroindolenine). Fragmentation of **146** to **147** followed by a biomimetic-type ring closure, or direct rearrangement of **146,** would give the pentacycle **148**. This basic rearrangement strategy was first employed by Harley–Mason [50] in an early synthesis of *dl*-aspidospermidine [1].

In his studies of models for folate coenzymes, Pandit [51] has developed new methods for transferring oxidized carbon fragments to imines to form vinylogous amides. Use of this chemistry to prepare a stereoisomer of a Büchi tetracyclic vindorosine intermediate is outlined in Scheme 19 [51]. Addition of

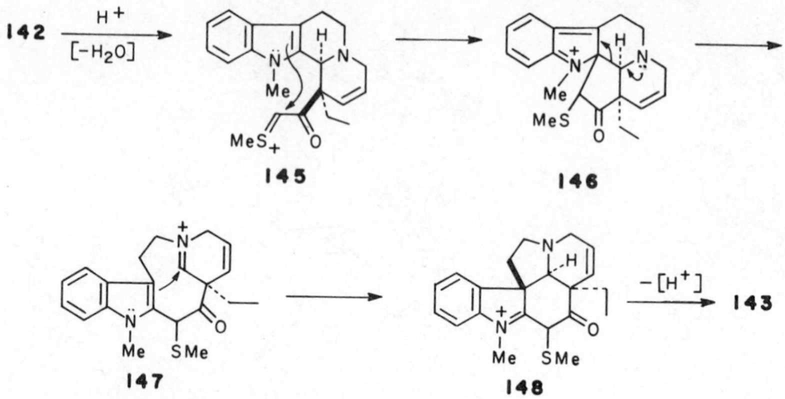

145

146

147

148

the dianion of ethyl acetoacetate to imidazolinium salt **149** provided imidazolidine derivative **150** in 85% yield. Imidazolidine **150** was readily converted to
enaminoketone **153** in the presence of either acid or base. However, when **150**
was allowed to react with tryptamine in refluxing acetic acid, the desired
(*Z*)-enaminoketone **151** was obtained in 56% yield. Conversion to the tetracyclic
ring system was accomplished by the method of Büchi [4], which involves
N_b-acylation followed by BF_3-catalyzed cyclization, and provided tetracyclic
intermediate **152** in 89% yield. The stereochemistry of intermediate **152** was
established by ^{1}H NMR nuclear Overhauser difference spectra, and interestingly
the C(19) hydrogen was found to have the less stable β-configuration. This
stereochemistry is opposite to that found by Büchi [4] in related intermediates,
and epimeric with the natural alkaloids. Pandit suggests that the β-orientation at
C(19) may arise by a kinetically controlled intramolecular Diels–Alder cyclo-

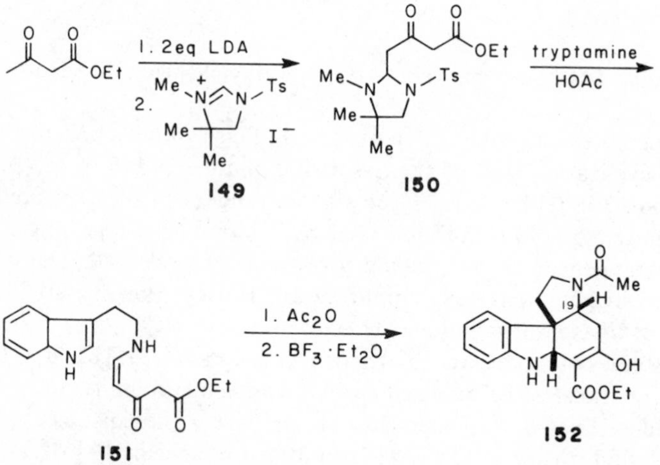

149

150

151

152

Scheme 19.   Pandit's synthesis of a 19-epi-*dl*-vindorosine intermediate.

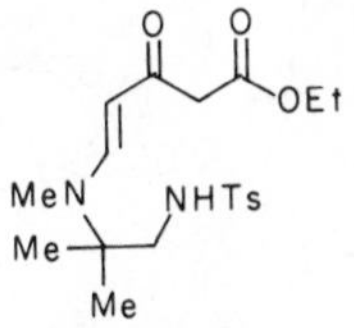

153

addition, although proof of this postulate awaits further studies. Epimerization of the less favored C(19) β-isomer to the thermodynamically preferred α-configuration has been demonstrated by Takano [52] with closely related systems, during an earlier formal total synthesis of *dl*-vindorosine. The synthesis of tetracyclic intermediate **152** was accomplished in four steps and 41% overall yield from ethyl acetoacetate.

4.2. Other Partial Total Syntheses

Martin and co-workers have developed the intramolecular Diels–Alder reactions of enamides with unactivated dienes as an entry to several azacyclic systems [53, 54]. The use of this approach to prepare hydrolilolidine **161**, an intermediate in the original Stork [3] synthesis of *dl*-aspidospermine, is outlined in Scheme 20 [53].

Reaction of 3,5-hexadienoic acid with sulfur dioxide and conversion of this product to the acyl chloride provided the masked diene **154** in 73% yield [54].

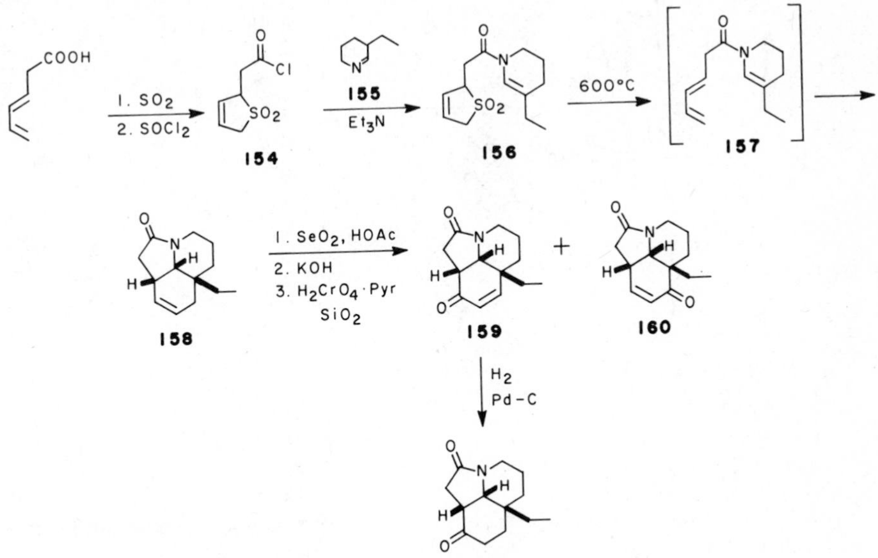

Scheme 20.   Martin's formal synthesis of *dl*-aspidospermine.

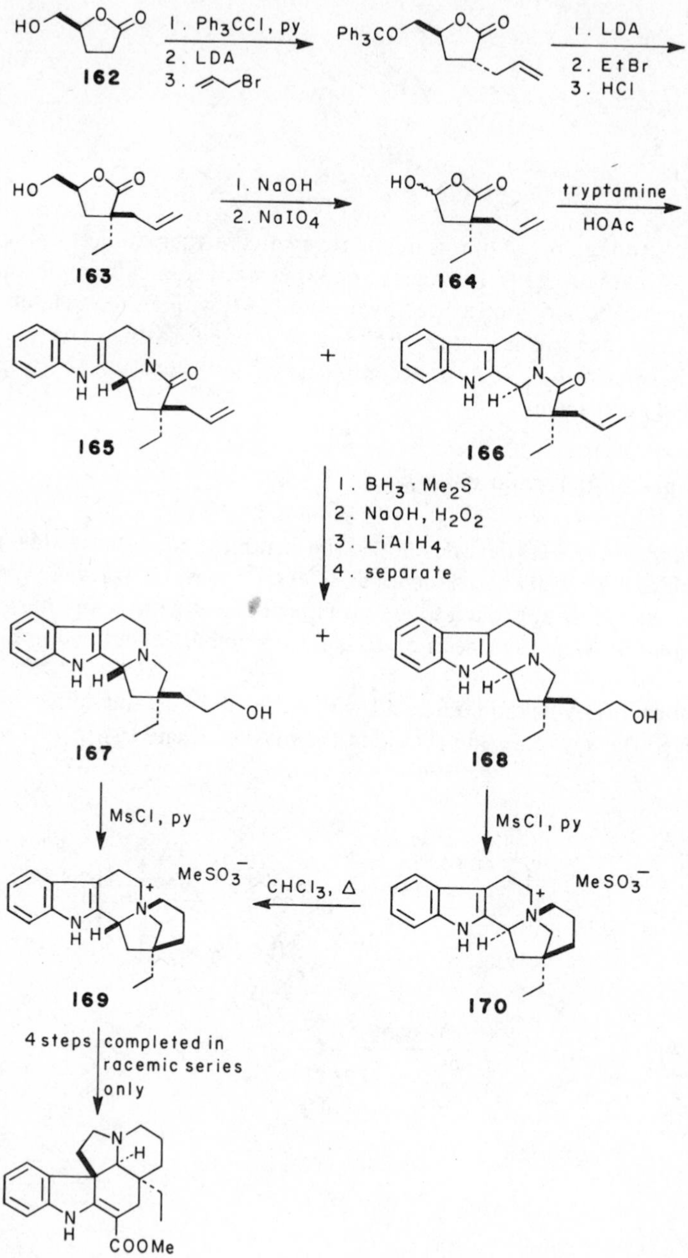

Scheme 21. Takano's synthesis of an enantiomerically pure (−)-vincadifformine intermediate.

Acylation of 3-ethyl-3,4,5,6-tetrahydropyridine (155) [55] with 154 provided enamide 156, which was passed through a vertical packed column at 600° C to give hydrolilolidine 158 in 58% yield [53]. This conversion involves [53] cheletropic extrusion of sulfur dioxide to give intermediate 157, which undergoes intramolecular Diels–Alder closure to 158. Introduction of the required carbonyl group was not straightforward. This transformation was accomplished by nonregioselective allylic oxidation of 158, hydrolysis of the derived allylic acetates, oxidation of the resulting allylic alcohols to the enones, and chromatographic separation. The desired enone lactam 159 was obtained in 26% yield, together with 20% of regioisomer 160. The formal total synthesis of dl-aspidospermine and dl-aspidospermidine (96) was completed by catalytic hydrogenation of 159 to provide the known [3] keto amide 161. The synthesis of hydrolilolidine 161 was accomplished in eight steps and 11% overall yield from 3,5-hexadienoic acid.

With all the attention that pentacyclic Aspidosperma alkaloids have received in recent years, it is surprising that no enantioselective total synthesis has been formally reported. However, Takano [56] has described (Scheme 21) the enantioselective synthesis of the pentacyclic salt 169, which was an intermediate in his recent synthesis of (+)-quebrachamine. Since the racemic mesylate salt 169 has been previously converted by Kutney [57] to dl-vincadifformine (15), repetition of this sequence with enantiomerically pure 169 could, barring an unforeseen racemization step, constitute a formal total synthesis of (−)-vincadifformine.

The chiral starting material for Takano's synthesis of 169 was L-glutamic acid, which had previously been converted to lactone 162 [58]. Tritylation of 162, followed by stepwise alkylation of the lactone enolate with allyl bromide and ethyl bromide, and finally removal of the trityl group provided 163 in 53% overall yield. In each alkylation step the bulky trityl group effectively shields the β-face of the lactone enolate. Periodate cleavage provided 164, which was condensed with tryptamine to give, in 48% yield, a 1:1 mixture of tetracyclic intermediates 165 and 166. Hydroboration of this mixture, followed by reduction with LiAlH₄ and chromatographic separation provided alcohols 167 and 168 in 24 and 22% yields, respectively. Reaction with methanesulfonyl chloride gave the quarternary mesylate salts 169 and 170, and interestingly 170 could be isomerized to the more stable β-isomer 169 by simply heating in refluxing chloroform. Intermediate 169 [59], which was available in 12 steps and ~12% overall yield from 162, was enantiomerically pure since optically pure (+)-quebrachamine was obtained from 169 [58].

REFERENCES

1. (a) G. A. Cordell, in The Alkaloids, Vol. 17, R. H. F. Manske and R. Rodrigo, Eds., Academic, New York, 1979, Chap. 3. See also earlier reivews in this series. (b) J. E. Saxton Indoles—The Monoterpenoid Indole Alkaloids, J. E. Saxton, Ed., Part 4, Intersciences, New York, 1983, Chap. 8.

2.   Cf. N. Neuss, in *Indole and Biogenetically Related Alkaloids,* J. D. Phillipson and M. H. Zenk, Eds., Academic, New York, 1980, Chap. 17.

3.   G. Stork and J. E. Dolfini, *J. Am. Chem. Soc.* **85,** 2872 (1963).

4.   (a) M. Ando, G. Büchi, and T. Ohnuma, *J. Am. Chem. Soc.* **97,** 6880 (1975); (b) G. Büchi, K. E. Matsumoto, and H. Nishimura, *J. Am. Chem. Soc.* **93,** 3299 (1971).

5.   (a) J. LeMen, W. I. Taylor, *Experientia* **21,** 508 (1965); (b) With the exception of the syntheses outlined in Schemes 16 and 21, all synthetic intermediates and products were racemic. Structures are written in the absolute configurations of the natural products.

6.   E. Wenkert, *J. Am. Chem. Soc.* **84,** 98 (1962); A. I. Scott, *Acc. Chem. Res.* **3,** 151 (1970); A. I. Scott *Biorg. Chem.* **3,** 398 (1974); J. P. Kutney, *Heterocycles* **7,** 593 (1977).

7.   M. E. Kuehne, D. M. Roland, and R. Hafter, *J. Org. Chem.* **43,** 3705 (1978).

8.   M. E. Kuehne and R. Hafter, *J. Org. Chem.* **43,** 3702 (1978).

9.   G. Stork, A. Brizzolara, H. K. Landesman, J. Szmuszkovicz, and R. Terrell, *J. Am. Chem. Soc.* **85,** 207 (1963).

10.   M. E. Kuehne, T. H. Matsko, J. C. Bohnert, and C. L. Kirkemo, *J. Org. Chem.* **44,** 1063 (1979).

11.   M. E. Kuehne, J. A. Huebner, and T. H. Matsko, *J. Org. Chem.* **44,** 2477 (1979).

12.   M. E. Kuehne, C. L. Kirkemo, T. H. Matsko, and J. C. Bohnert, *J. Org. Chem.* **45,** 3259 (1980).

13.   E. L. McCaffery and S. W. Shalaby, *J. Organomet. Chem.* **8,** 17 (1967).

14.   G. Stork, J. F. Cohen, *J. Am. Chem. Soc.* **96,** 5270 (1974).

15.   M. E. Kuehne, F. J. Okuniewicz, C. L. Kirkemo, and J. C. Bohnert, *J. Org. Chem.* **47,** 1335 (1982).

16.   M. E. Kuehne, T. H. Matsko, J. C. Bohnert, L. Motyka, and D. O. Smith, *J. Org. Chem.* **46,** 2002 (1981).

17.   J. P. Kutney, R. A. Badger, J. F. Beck, H. Bosshardt, F. S. Matough, V. E. Ridaura-Sanz, Y. H. So, R. S. Sood, and B. R. Worth, *Can. J. Chem.* **57,** 289 (1979); S. Raucher, J. E. MacDonald, and R. F. Lawrence *J. Am. Chem. Soc.* **103,** 2419 (1981).

18.   J. Hajicek and J. Trojanek, *Collect. Czech. Chem. Commun.* **47,** 2448 (1982).

19.   M. E. Kuehne and J. C. Bohnert, *J. Org. Chem.* **46,** 3443 (1981).

20.   P. A. Wehili and N. Chu, *Org. Synth.* **58,** 79 (1978).

21.   J. P. Kutney, Y. Karton, N. Kawamura, and B. R. Worth, *Can. J. Chem.* **60,** 1269 (1982).

22.   M. Nakazaki, *Bull. Chem. Soc. Jpn.* **32,** 588 (1959); M. W. Bullock and S. W. Fox, *J. Am. Chem. Soc.* **73,** 5155 (1951).

23.   A background description of this chemistry can be found in ref. 24 and L. E. Overman, L. Mendelson, and E. J. Jacobsen, *J. Am. Chem. Soc.* **105,** 6629 (1983).

24.   L. E. Overman, M. Sworin, L. S. Bass, and J. Clardy, *Tetrahedron* **37,** 4041 (1981).

25.   L. E. Overman, M. Sworin, and R. M. Burk, *J. Org. Chem.* **48,** 2685 (1983).

26.   T. Masamune, M. Takasugi, A. Murai, and K. Kobayashi, *J. Am. Chem. Soc.* **89,** 4521 (1967).

27.   B. V. Samant, *Chem. Ber.* **75,** 1008 (1942); H. R. Frank, P. E. Fanta, and D. S. Tarbell, *J. Am. Chem. Soc.* **70,** 2314 (1948).

28.   Y. Ban, H. Kinoshita, S. Murakami, and T. Oishi, *Tetrahedron Lett.,* 3687 (1971).

29.   Y. Ban, K. Yoshida, J. Goto, and T. Oishi, *J. Am. Chem. Soc.* **103** 6990 (1981).

30.   T. Gallagher and P. Magnus, *Tetrahedron* **37,** 3889 (1981).

31.   C. Exon, T. Gallagher, and P. Magnus, *J. Am. Chem. Soc.* **105,** 4739 (1983).

32.   T. Gallagher, P. Magnus, and J. C. Huffman, *J. Am Chem. Soc.* **105,** 4750 (1983).

33.   T. Gallagher and P. Magnus, *J. Am. Chem. Soc.* **105,** 2086 (1983).

34.   A. J. Pearson, *Acct. Chem. Res.* **13,** 463 (1980).

35.   A. J. Pearson and D. C. Rees, *J. Chem. Soc., Perkin* **1,** 2467 (1982).

References

36. A. J. Pearson, P. Ham, and D. C. Rees, *J. Chem. Soc., Perkin* **1,** 489 (1982).

37. G. Lawton, J. E. Saxton, and A. J. Smith, *Tetrahedron* **33,** 1641 (1977).

38. A. J. Pearson, D. C. Rees, and C. W. Thornber, *J. Chem. Soc., Perkin* **1,** 619 (1983).

39. J.-Y. Laronze, J. Laronze-Fontaine, J. Levy, and J. LeMen, *Tetrahedron Lett.,* 491 (1974).

40. J. Levy, J.-Y. Laronze, J. Laronze, and J. LeMen, *Tetrahedron Lett.,* 1579 (1978).

41. S. Takano, K. Shishido, J. Matsuzaka, M. Sato, and K. Ogasawara, *Heterocycles* **13,** 307 (1979); S. Takano, K. Shishido, M. Sato, K. Yuta, and K. Ogasawara, *J. Chem. Soc., Chem. Commun.,* 943 (1978).

42. Y. Ban, Y. Sekine, and T. Oishi, *Tetrahedron Lett.,* 151 (1978).

43. W. N. Speckamp, S. J. Veenstra, J. Dijkink, and R. Fortgens, *J. Am. Chem. Soc.* **103,** 4643 (1981).

44. S. J. Veenstra and W. N. Speckamp, *J. Am. Chem. Soc.* **103,** 4645 (1981).

45. J. B. P. A. Wijnberg and W. N. Speckamp, *Tetrahedron* **34,** 2399 (1978).

46. W. N. Speckamp, *Recueil* **100,** 345 (1981).

47. J. B. P. A. Wijnberg, H. E. Schoemaker, and W. N. Speckamp, *Tetrahedron* **34,** 179 (1978).

48. R. Z. Andriamialisoa, N. Langlois, and Y. Langlois, *J. Chem. Soc., Chem. Commun.,* 1118 (1982).

49. Y. Oikawa and O. Yonemitsu, *J. Org. Chem.* **41,** 1118 (1976).

50. J. E. D. Barton and J. Harley-Mason, *Chem. Commun.,* 298 (1965).

51. H. C. Hiemstra, H. Bieraugel, and U. K. Pandit, *Tetrahedron Lett.* **23,** 3301 (1982).

52. S. Takano, K. Shishido, M. Sato, and K. Ogasawara, *Heterocycles* **6,** 1699 (1977).

53. S. F. Martin, S. R. Desai, G. W. Phillips, and A. C. Miller, *J. Am. Chem. Soc.* **102,** 3294 (1980).

54. S. F. Martin, T. Chou, and C. Tu, *Tetrahedron Lett.,* 3823 (1979).

55. H. Zondler and W. Pfleiderer, *Justus Liebigs Ann. Chem.* **759,** 84 (1972).

56. S. Takano, K. Chiba, M. Yonaga, and K. Ogasawara, *J. Chem. Soc., Chem. Commun.,* 616 (1980).

57. J. P. Kutney, K. K. Chan, A. Failli, J. M. Fromson, C. Gletsos, A. Leutwiler, V. R. Nelson, and J. P. de Souza, *Helv. Chim. Acta* **58,** 183 (1975).

58. M. Taniguchi, K. Koga, and S. Yamada, *Tetrahedron* **30,** 3547 (1974).

59. S. Takano, S. Hatakeyama, and K. Ogasawara, *J. Am. Chem. Soc.* **101,** 6414 (1979).

60. M. E. Kuehne and W. G. Earley, *Tetrahedron* **39,** 3707, 3715 (1983).

Subject Index

Organism Index

321